Science and the Founding Fathers

Also by I. Bernard Cohen

Benjamin Franklin's Experiments (1941)

Roemer and the First Determination of the Velocity of Light (1942)

Science, Servant of Man (1948)

Some Early Tools of American Science (1950, 1967)

Ethan Allen Hitchcock of Vermont: Soldier, Humanitarian, and Scholar (1951)

General Education in Science (with Fletcher G. Watson) (1952)

Benjamin Franklin: His Contribution to the American Tradition (1953)

Isaac Newton's Papers & Letters on Natural Philosophy (1958, 1978)

A Treasury of Scientific Prose (with Howard Mumford Jones) (1963, 1977)

Introduction to Newton's *Principia* (1971)

Isaac Newton's *Principia,* with Variant Readings (with Alexandre Koyré & Anne Whitman) (1972)

Isaac Newton's Theory of the Moon's Motion (1975)

Benjamin Franklin: Scientist and Statesman (1975)

The Newtonian Revolution (1980)

An Album of Science: From Leonardo to Lavoisier, 1450–1800 (1980)

The Birth of a New Physics (Rev. & Exp. 1985)

Revolution in Science (1985)

Benjamin Franklin's Science (1990)

The Natural Sciences and the Social Sciences: Some Critical and Historical Perspectives (1994)

Interactions: Some Contacts between the Natural Sciences and the Social Sciences (1994)

Newton: Norton Critical Edition (edited with Richard S. Westfall) (1995)

Science and the Founding Fathers

Science in the Political Thought of Jefferson, Franklin, Adams, and Madison

.....

I. BERNARD COHEN

W·W· NORTON & COMPANY

NEW YORK LONDON

First published as a Norton paperback 1997
All rights reserved
Printed in the United States of America

The text of this book is composed in New Baskerville
with the display set in Caslon Open
Composition and manufacturing by
The Maple-Vail Book Manufacturing Group.
Book design by Chris Welch

Library of Congress Cataloging-in-Publication Data

Cohen, I. Bernard, 1914–
Science and the founding fathers : science in the political
thought of Jefferson, Franklin, Adams & Madison / I. Bernard Cohen.
        p.   cm.
Includes index.
ISBN 0-393-03501-8
1. Franklin, Benjamin, 1706–1790—Knowledge—Science.   2. Franklin, Benjamin, 1706–1790—Political and social views.   3. Jefferson,
Thomas, 1743–1826—Knowledge—Science.   4. Jefferson, Thomas,
1743–1826—Political and social views.   5. Madison, James, 1751–
1836—Knowledge—Science.   6. Madison, James, 1751–1836—Political and social views.   7. Adams, John Quincy, 1767–1848—Knowledge—Science.   8. Adams, John Quincy, 1767–1848—Political and
social views.   9. Political science—United States—History—18th century.   10. Science—United States—History—18th century.   I. Title.
        II. Title: Science and the founding fathers.
E302.5.C62   1995
973.3′092—dc20   94-26731
                                                                          CIP

ISBN 0-393-03501-8
ISBN 0-393-31510-X pbk.
W.W. Norton & Company, Inc., 500 Fifth Avenue, New York, N.Y. 10110
W.W. Norton & Company Ltd., 10 Coptic Street, London WC1A 1PU
        1 2 3 4 5 6 7 8 9 0

Contents

· · · · ·

3   Benjamin Franklin: A Scientist in the World of Public Affairs 135

4   Science and Politics: Some Aspects of the Thought and Career of John Adams 196

Preface

. . . . .

The guiding principle of this book is the conviction that science is an important part of history, that the history of science can illuminate some of the central aspects of American history. As is often the case, this book is the product of a line of research that originally was centered on a seemingly unrelated topic, namely the different kinds of interactions that have occurred between the physical and biological sciences and the social sciences. One of the principal fruits of that research has been a volume edited by me on *The Natural Sciences and the Social Sciences: Some Critical and Historical Perspectives* (Dordrecht/Boston/London: Kluwer Academic Publishers, 1994); my own chapters have been printed separately under the title *Interactions: Some Contacts between the Natural Sciences and the Social Sciences* (Cambridge/London: MIT Press, 1994), and in the Italian as *Scienze della Natura e Scienze Sociali* (Rome/Bari: Editori Laterza, 1993).

The social sciences treated in the above works were primarily economics and sociology, with some consideration given to theories of the state. Again and again in my research, however, I came upon examples of the ways in which political thought and even the course of political action either were determined by a scientific vector or drew on science as a source of models or of metaphors and analogies. In particular, I found that in the eighteenth century, the Age of Reason, when science was esteemed as the highest expression of human reason, the sciences served as a font of analogies and metaphors as well as a means of transferring to the realms of political discourse some reflections of the value system of the sciences. Thus James Harrington, who a century later was to become an influential figure in the political thought of the American Founding Fathers, drew on the works of the pioneer physiologist William Harvey in proposing his system of government.

For many years I have been fascinated by what I detected to be Newtonian echoes in the Declaration of Independence, a topic I have discussed again and again with the students in my graduate seminar. My researches led me to an appreciation of Jefferson's command of the technical aspects of Newtonian science and mathematics, a feature of Jefferson's thought that goes far deeper than the customary levels of discussion, which are limited to his allegiance to some form of "Newtonianism" or his admiration for Newton. In the end I came to recognize that science was a feature of consequence in his political thought and action.

I have studied the scientific thought of Benjamin Franklin for many decades and have long been aware of the way in which Franklin's theory of demography influenced his political ideas. It was thus an easy step to consider science in relation to his political career. Ever since graduate days, I have been troubled by the alleged Newtonianism of the Constitution and have sought in vain for evidence that any part of the Constitution was a conscious or unconscious imitation of Newtonian science. In the end I discovered the source of this allegation in the writings of Woodrow Wilson. As can be seen in Chapter 5, in large measure Wilson's presentation—like those of the scholars who have repeated his argument—is based on a fundamental misunderstanding of what the science of the *Principia* is. My study of science in relation to the Constitution led me by a direct path to the thought of James Madison. The subject of John Adams's use of the sciences in a political context was an unexpected bonus. I was also delighted to find how the sciences were used by others of that age, such as James Wilson and Thomas Pownall.

Again and again, in the course of writing this book, I stood in awe as I faced the massive literature concerning the Declaration of Independence and the Constitution, and as I took cognizance of the great body of scholarly writings and textual editions relating to the careers and thought of Franklin and Jefferson, to say nothing of the growing scholarship relating to works of Adams and Madison. I had to ask myself the question recorded by Merrill Peterson in his biography of Jefferson. "In the course of the work," he wrote, "I have sometimes been asked, in tones ranging from curiosity to dismay, if anything important remained to be said about Jefferson." In my own case, I had to ask myself an additional question: whether an outsider who was not a specialist in American political thought, American intellectual history, or American diplomatic or social history could add anything of consequence to the monument of established scholarship and interpretation. This very question, however, defined my mission.

A historian of science, trained in a discipline quite distinct from that of most American historians, cannot help but bring to the subjects of this book a point of view that differs from those that are current in the historical literature. One example will suffice. Countless scholars, graduate students, and ordinary readers have gone through Jefferson's book, *Notes on the State of Virginia* (1787), but no one to my knowledge has noticed that Jefferson made use of an exact statement from Newton's *Principia*, a seemingly unlikely source for a book whose subject is primarily natural history. This experience emboldened me to believe that only someone trained in Newtonian science would recognize echoes of Newton's *Principia* in the Declaration of Independence.

One of the points made in the opening chapter, and repeated again both explicitly and by implication, is that concentration on issues related to science introduces new viewpoints into our understanding of our history. For example, the themes of Washington's Farewell Address at the end of his presidency are well known. All readers will be aware that the main point made by Washington was to warn against entangling foreign alliances. Some will know that another major theme is the importance of preserving the Union. But few indeed will be aware that Washington also strongly advised his fellow Americans to support "the diffusion of knowledge," using a phrase that may be familiar in the statement of purpose of our first national scientific institution, ". . . for the increase and diffusion of knowledge."

One of the positions taken in the following pages is that scientific issues were related to the political thought and also the political action of our Founding Fathers, primarily Jefferson, Franklin, Adams, and Madison. I argue that Franklin's enormous scientific reputation was a factor in his later diplomatic career. But did his fellow Americans consider his scientific eminence as an important part of his credentials for diplomacy? When Franklin was sent by the Congress, along with Samuel Chase and Charles Carroll, to attempt to win over the Canadians to the American cause, his name was put before the others in a reversal of alphabetical order. Chase was designated "Samuel Chase Esquire, one of the delegates of the colony of Maryland," while Carroll was referred to as "Charles Carroll of Carrollton in the said colony of Maryland," but Franklin was named "Benjamin Franklin, Member of the Royal Academy of Sciences at Paris, F R S, &c. &c. one of the delegates of the province of Pennsylvania." His honorary degrees and title of "Doctor" were placed in a secondary position as part of the "et cetera," but stress was put on his having been elected a member of the two most prestigious scientific societies in the world, the Paris Académie Royale des Sciences and the Royal Society of London. He was, in fact, the only person in the New World who could claim membership in both societies. No other could claim that honor until a century had passed, when Louis Agassiz became a member of these two societies.

My thinking about the various aspects of colonial American history, the history of ideas, and the development of political thought, which are discussed in this book, has been influenced by the work of many scholars. Primary among them are Bernard Bailyn, Edmund Sears Morgan, Merrill Peterson, and J. G. A. Pocock. Among some notable others are Douglass Adair, Carl Becker, Adrienne Koch, Pauline Maier, Richard B. Morris, and Gordon S. Wood.

I am happy to acknowledge with deep gratitude the support of the research for this book by the Richard Lounsbery Foundation. Its founding director and president, Mr. Alan McHenry, was a warm and sympathetic supporter of my work. An unusual man of many talents, Alan McHenry became a true friend of all the scholars and scientists who were sponsored by the Lounsbery Foundation. One of the joys of my years of support by the Lounsbery Foundation was to become a friend of Alan McHenry.

In the preparation of this work I have profited by the research assistance of many students, foremost among them Katharine Downes and Amory Downes. Meredith St. Sauveur produced typescripts from almost undecipherable longhand versions and from tapes. Above all, I have profited from the research assistance of Julia Budenz, who has worked closely with me on every phase of the research and writing and whose critical comments and suggestions have been of the greatest value. I have also greatly profited from the research assistance and critical readings of drafts of chapters by Professor Elaine Storella of Framingham State College (Massachusetts). I have been constantly encouraged and stimulated by the concern of my publisher, Ed Barber.

* * * * *

For the paperback edition I have corrected a few minor errors and I have altered the discussion of Adams's use of Newton's second law and his dismissal of the atomic-molecular hypothesis.

Science
and the
Founding
Fathers

1

Science and American History

. . . . .

Study of the Relation between Science and the Founding Fathers

T he scholarly literature on the Founding Fathers is vast and is ever increasing at a rapid rate. Almost every aspect of the life and thought of such figures as Thomas Jefferson, Benjamin Franklin, John Adams, and James Madison has been the subject of a book, a scholarly article, a doctoral dissertation, or a master's thesis. And yet one may search in vain for any full-length study devoted to the role of science in the political thought of Jefferson, Franklin, Adams, or Madison. Although a number of authors have alleged that the Constitution was in some degree influenced by the science of Isaac Newton, their arguments tend to be unconvincing because they either give no specific examples or allege incorrectly that Newtonian science was centered on concepts of static balance or equilibrium. There are no works that seek to determine the degree to which scientific considerations were in any real sense a guide to the political actions of the Founding Fathers.

The reader should not conclude, however, that the topic of science in relation to the careers of the Founding Fathers is wholly absent from the scholarly literature on American history. To a certain degree this topic enters many studies of colonial and early federal

American history. This is hardly surprising since the American nation was conceived in a historical period that is generally known as the Enlightenment, or the great Age of Reason, and science was then esteemed as the highest expression of human rationality. Scholars generally agree that the science of Isaac Newton represented the supreme achievement of the human mind. It is simply inconceivable that thinking men and women of the eighteenth century would be uninfluenced by the ideals, the concepts, the principles, and even the laws of the science that Newton created or by other achievements in the physical and life sciences and medicine.

It is easy to demonstrate that science in general, and the science of Newton in particular, was part of the mind-set of the Founding Fathers. Benjamin Franklin was a scientist who admired Newton and tried to meet him when he was in London as a young man. Thomas Jefferson had a real passion for science all during his life; Newton's *Principia* was one of his favorite books, and a portrait of Newton adorned his gallery of the great men of history. John Adams studied the Newtonian natural philosophy as an undergraduate at Harvard College and took pride in being able to cite Newton's laws of motion in an argument about the form of the state and national legislatures. One of his accomplishments was his participation in the founding of the American Academy of Arts and Sciences in rivalry to Benjamin Franklin's American Philosophical Society, an aspect of his career that is not fully stressed in the two scholarly biographies of this notable Founding Father. James Madison while an undergraduate at Princeton studied the science of Newton and its consequences, and he wrote an essay on parallelism between the world of nature and the world of human affairs. Hence it cannot be doubted that the mind-set, including the political thought, of these four men was influenced by the science of their day. The primary assignment, therefore, is to identify the actual occasions on which science entered their political thought and to discern the nature of that influence and to evaluate its relative importance.

One point must be made at the very outset. If science was indeed a significant element in the political thought of the Founding Fathers, how could this topic have been missed by the many generations of American historians? The answer to this question has a number of different facets. First, many American historians have indeed recognized that science provided analogies and metaphors for political discussion; after all, the documents themselves declare this role of science. Furthermore, a number of scholars—among them Carl Becker, Morton White, and Garry Wills—have included science or scientific philosophy in their discussions of the ideas of the Declaration of Independence. The subject of science and the Founding Fathers is, therefore, not invented in this book but is rather explored in a new way, in full scope, in its many dimensions, and to a degree not previously attempted.

Even more important may be the fact that the questions studied in this book represent an approach to the problems of American history from a point of view that differs from that of traditional scholars, a point of view that comes from a different background and training. That is, my preparation is not primarily in American political thought, American literature or philosophy, American diplomatic and political history, American economic history, or American cultural history. Rather, I come to this subject from a background in science, more particularly the history of science. This training makes possible the identification of scientific sources in political documents that might otherwise remain unrecognized. Two examples will serve as illustrations. James Harrington, the seventeenth-century political theorist whose *Oceana* has been called a "constitutional blueprint," described his method as a "political anatomy." He was a great admirer of William Harvey, discoverer of the circulation of the blood, and made use of a number of analogies based on Harvey's work. Thus, he argued for a bicameral legislature on the analogy with Harvey's discovery of the functions of the two ventricles of the human heart, even making use in his argument of the difference in size and strength between the ventricles. This political analogy included Harvey's discussion of the difference between the blood pumped out by each ventricle, concluding that the two houses of the legislature would similarly have different functions. Only a historian of science or someone really familiar with the nature of Harvey's discoveries would fully appreciate this analogy. Furthermore, only a historian of science who knew Harvey's work on animal generation would recognize the Harveyan origins of other political analogies used by Harrington, including one introduced by the bold statement, Everything comes from an egg, taken directly from the motto "Ex ovo omnia" in the frontispiece to Harvey's Latin treatise *De Generatione Animalium.* Similarly, only a historian of science or someone familiar with the text of Newton's *Principia* would recognize the Newtonian source of Jefferson's paraphrase in his *Notes on the State of Virginia.*

I have mentioned that a good number of historians have had the insight to recognize that Newtonian science must be of some importance in the context of American thought in the age of the Founding Fathers. It is a fact, however, that most students of American history do not have a profound knowledge of either the physical or the life sciences and that fewer still are to any real degree versed in the history of science or trained to deal with the questions of science of bygone days. The result is that our library shelves groan under the burden of scholarly and amateur investigations of almost every conceivable facet of the personal lives, thought, and achievements of the founders of our country except for science. Although a few historians of science, not necessarily primarily concerned with American history, have produced a body of important scholarly literature on Benjamin Franklin, no attempt has been made to link his scientific and political concerns. There are four books on aspects of Jefferson and science, but no one has adequately investigated the scientific component of his political thought and action. I know of no serious attempts to explore the significance of science in the careers of John Adams and James Madison.

The case of Jefferson is somewhat different from the others because his biographers have recognized that he had a consuming passion for science and that he devoted much of his time and energy to scientific pursuits. Thus Dumas Malone, Merrill Peterson, and others have included science among the topics which they explore in their critical and narrative presentations. But, despite four books and innumerable articles and scholarly monographs, no one—to my knowledge—has thus far identified elements of the science of Newton's *Principia* either in Jefferson's *Notes on the State of Virginia*, his only book, or in his most famous composition, the Declaration of Independence.

Furthermore, the scientific component of Jefferson's life and career is not today given importance consonant with the significance of science in his intellectual life. One example will suffice to show the gap between the reality of Jefferson as an ardent devotee of science and the image of him projected today. When Jefferson went to Philadelphia in 1797 to be inaugurated as vice-president of the United States, he brought with him a collection of fossil bones to illustrate a lecture on paleontology that he was to give to the American Philosophical Society, America's oldest scientific society, of which he had been elected president in succession to the astronomer David Rittenhouse. It has even been argued that he took more pride in the presidency of the American Philosophical Society than in the vice-presidency of the United States.[1] One of Jefferson's three qualifications for writing the Declaration of Independence, according to the later recollections of John Adams, was his knowledge of science. Yet, in 1993, when commemorating the 250th anniversary of Jefferson's birth, that same society to which Jefferson lectured on fossil bones and of which he was president mounted a program in which the only aspect of Jefferson's thought and career to be celebrated was his interest in classics, not his science. In that same year, the National Endowment for the Humanities sponsored a set of lectures to be given as a series in various cities in celebration of this anniversary. The subjects included such topics as Jefferson and religious freedom and Jefferson and architecture, but not Jefferson and science.

There are, I believe, a number of reasons why Americans do not in general consider science to be an important component of American history and why scholars have not more fully explored the role of science in the founding of the American republic. The first of these is the obvious fact that, in the historical development of America, science has not been a path leading to the kind of public recognition and fame achieved by military heroes, political leaders, social innovators, business tycoons, inventors, and even some writers and artists. Those who have won the American presidency have included military officers, schoolteachers, university professors, engineers, and lawyers. Teddy Roosevelt was sufficiently versed in biology to give a scientific lecture at Oxford, but thus far the only other president of the United States with any serious claim to scientific expertise has been Thomas Jefferson.

In the days of the Founding Fathers, only one political leader, Benjamin Franklin, had a real scientific reputation; he was recognized the world over as one of the foremost scientists of his day. Many of our historians, however, have not had sufficient knowledge and understanding of science to evaluate his scientific achievement at its true worth. As a result, historians have generally either belittled his contributions to basic science and his scientific stature or confused the advance of knowledge with its applications, treating his scientific research as if it were on the same level as the invention of gadgets such as a rocking chair, bifocal glasses, or his "armonica."

Another reason for the relative neglect of science as a factor in American history is the fact that the history of science as an academic discipline is a very young member of the scholarly community. Even today many colleges and universities do not have departments in this field. The subject of history of science is also largely absent from curricula in American history or American studies. Not only do these programs have no courses on such topics as science and the American republic, but the textbooks and syllabi tend not to include science as a significant factor in American history. So young is the history of science as a formally recognized discipline that the first doctorate to be awarded in this new area was granted only in 1943. Small wonder, then, that we are only just beginning to take full measure of science as part of the rise of the American nation or the development of American culture.

Science in Relation to Political Thought and Action

The study of science and the Founding Fathers is part of a larger exploration of the ways in which science can affect our political thought and our political action. In relation to today's world the study of this topic would lead us directly into the technological fruits of science: the proliferation of nuclear weapons and the problems of arms control, the technological destruction of the environment and the conservation of natural resources, industrial diseases and disposal of hazardous wastes, and the new ways of preserving health and conquering disease. There are many examples that enable us to see the political consequences of technology based in science. A hundred and fifty years ago, representatives of nations were ministers plenipotentiary. They were out of immediate contact with their home bases, perhaps months away from direct communication with their capitals, and so had to be endowed with full power to act in their country's name. Before electromagnetic science spawned the great technological revolution in communications, an ambassador needed to make decisions without immediate instructions from home. So slow was the contact between Europe and America that a signal American victory in the War of 1812 was gained through a battle fought in New Orleans two weeks after the signing of a treaty of peace that supposedly ended hostilities. Today, by contrast, an ambassador will pick up the telephone and call home in order to find out from his foreign minister how he should vote on a question raised at the United Nations or even how he should respond to a proposed toast at an international banquet. Here is a notable change in our patterns of political action produced by science through technology.

But science and biomedicine have affected our thought in ways other than through the technology that they have engendered. There are many signs of the effects of science on our social and political thought of which we may not even be aware. For example, the powerful influence of science and biomedicine may be discerned in many of the metaphors that permeate our daily social and political discourse, of which a large number come from the life sciences. Even the common reference to a "head" of state is a political metaphor based on the primitive knowledge of animal life. Getting to the "heart" of the matter is another of this sort. We borrow the language of classical physics when we talk about social, political, or economic "forces." Medicine provided the concept of an economy or a society as "healthy" along with the "ills" of society and social "pathology." Our reference to "normal" (as opposed to pathological) conditions is another contribution made by medicine and the mathematical science of statistics to our daily discourse. Statistical science also gave us the important concept of "average." A metaphor from chemistry now enters discussions of the qualifications of judges: a "litmus test" to gauge their opinions on the abortion question. Borrowing a term from twentieth-century physics, we talk of a "quantum leap," erroneously supposing that a "leap" in physical quantum theory must be enormously large.

The Founding Fathers were very much aware that they were invoking metaphors and analogies from medicine and the physical and biological sciences in their political discourse. John Adams, for one, knew that the chief authority for the concept of political balance was James Harrington, and, as we shall see, he was also aware that Harrington's ideas were generated in the context of physiology. In affirming Harrington's greatness, Adams did not compare him to Newton the physicist but rather likened his political discoveries to William Harvey's discovery of the circulation of the blood, the foundation of modern physiology.

While studying the role of science in the political thought of the Founding Fathers, I shall almost exclusively be concerned with scientific ideas rather than with any dependent technology. One instance, however, the design of lightning rods, will reveal that there was a major political component to the choice of the form that this application of science might take. In this debate, there was no question of whether the rods efficiently save public and private buildings from destruction by lightning. The debate, as we shall see, was rather about the shape of the upper end of the rods. In this case politics rather than science produced the final decision that in England the rods should end in knobs rather than in points.

Scientific ideas considered apart from technological applications may also have important political repercussions. The political, economic, or social implications of a scientific theory may be of sufficient importance to produce a program of research that will challenge and eventually correct or disprove some basic scientific idea. For example, it was politically important for the future development of America that an end be put to a pernicious belief that the animals and human beings native to North America were physiologically and (for humans) intellectually inferior to their European counterparts and that a similar "degeneration" would be experienced by human beings as well as their livestock transported from the Old World to the New. This allegedly scientific theory could be exterminated only by hard scientific evidence to the contrary; only good science—in a reversal of Gresham's law—can drive out bad science. This was a mission taken on by Thomas Jefferson, whose efforts were embodied in his only published book.

Put an end to "degeneration"

T. J.

Scientific ideas may also be developed and formulated in conjunction with the development and formulation of political theory. An example can be seen in the political thought of Benjamin Franklin. The science in question was the new science of demography. Franklin developed his population theory in the context of the political problem of the relation of America to Britain. He was led to predict that there would be a new population center of the British world and that it would be located in North America. This eventually determined a line of policy including the position that the British should annex Canada rather than Guadeloupe as the spoils of victory in the war against the French.

Sometimes the use of scientific ideas in political theory or discussion is not so much to engender a political position as to provide an argument for such a position once adopted. We shall see an example of this kind of interaction in the debate between John Adams and Benjamin Franklin over whether the legislatures of the American states and the Congress of the United States should be bicameral or unicameral. In marshaling arguments against his opponent, Adams cited the highest scientific authority he could possibly muster: the

Principia of Isaac Newton, considered in Adams's day to be the greatest scientific book ever written. Here Adams was drawing on his knowledge of Newtonian physics as a source of rhetoric, using a scientific analogy in a political context in order to buttress an argument, while also calling on the highest scientific authority in support of a political position. The subject of rhetoric leads us naturally to inquire into the uses of analogies and metaphors in political exposition and debate.

Analogy in Political Thought

Among the ways in which scientific ideas are used in political thought and discourse, one of the most important is analogy. In employing what seems to be analogical thinking some social scientists have gone a step further and claimed an identity between a social or political system and some part of the physical or life sciences. Thus a number of nineteenth-century sociologists believed that society "is" an organism and looked for social counterparts of such biological entities as the cell.[2] Far more common, however, is the use of the current physical or life sciences as sources of analogues and metaphors.

It is difficult and not usually necessary to draw an exact line of demarcation between analogies and metaphors. Both involve the comparison of two entities, associating some property or properties of one entity with those of another which is otherwise different from the first. In the sciences, an analogy is usually based on the assumption that there is a similarity in function or in some other feature between two subjects or branches of science so that concepts, principles, equations, and theories can be transferred from one to the other.[3] A famous example is James Clerk Maxwell's analogy between the theory of gravitation and the theory of heat conduction. Even though the laws of "conduction of heat in uniform media," he wrote, "appear at first sight among the most different . . . from those relating to" attraction, he found that by analogy "the solution of a problem in" attraction could be "transformed into that of a problem in heat." In this example we may see how analogy becomes a useful instrument of thought for the advance of science. In fact, one of the traditional uses of analogies in the physical and biological sciences, and in the earth sciences, has been to serve as a tool of discovery.

Jeremy Bentham declared that analogy is a major tool for scientific discovery. In the *Origin of Species,* Charles Darwin not only made extensive use of analogies but also indicated that although he was aware that "analogy may be a deceitful guide," he was able to conclude "from analogy that probably all the organic beings which have ever lived on this earth have descended from some one primordial form, into which life was first breathed."[4]

This kind of use of analogy illustrates what Alfred North Whitehead has called the "logic of discovery," the path to increasing knowledge, which Maxwell called the process of "exciting appropriate mathematical ideas." We may take note that in the advancement of all the sciences, in the way that one scientist makes use of the ideas of another, there is very apt to be a creative transformation, the creation of an important new concept in the form of what seems to be a partial or imperfect replication of the original or even a dramatic recasting of the original. We can see this kind of transformation in the production of some of the most fundamental concepts of the physical and life sciences: Harvey's doctrine of the circulation of the blood, Newton's concept of inertia, Franklin's invention of a particulate electric "fluid," and Darwin's theory of natural selection. In this sense we may fully understand why the use of analogies lies at the heart of the logic of discovery, of the production of new knowledge by a radical and creative transformation rather than a cloning reproduction.

In introducing the concept of "logic of discovery," Whitehead contrasted it with what he called the "logic of the discovered." In this case, an analogy is used once a discovery has been made. There are three primary uses of analogy under such circumstances. One is to help explain a difficult concept or principle. For example, a mode of teaching the principles of electrical circuits has long been the creation of analogies from the more familiar science of mechanics and even the construction of mechanical models that simulate some principal aspects of the action of the electrical circuits. Thus, analogy can be used to make something understandable or understood.

A somewhat similar use of analogies is to validate a novel concept, to make it seem plausible. When Ernest Rutherford proposed that an atom is not a continuous piece of matter but rather consists of a central nucleus with one or more electrons circling around it, he argued for the plausibility of his new model by observing that it was a miniature analogue of the solar system, not something of a sort never known in nature. Sigmund Freud wrote about how he withheld the full presentation of one of his concepts of how the human memory functions; the reason was that he knew that his colleagues would find his idea difficult to accept. Later on, when he found a mechanical analogue, in a device called a "Mystic Writing-Pad," he made his concept public, declaring that the analogue showed that the kinds of functions he had envisaged were not absurd. Thus, analogy can be used to support something, to make it seem reasonable, to make it acceptable and accepted.[5]

Closely related to this use of analogy is one in which the likeness is presented as carrying with it a system of values. Because this utilization of analogy moves it further into the realm of rhetoric and because in this function the comparison may involve a single word or image, that is, a brief figure of speech, as well as an extended representation of similarities, analogy employed in this third way may perhaps be designated as metaphor, even though a clear and strict distinction between analogy and metaphor is neither necessary nor useful. What is especially significant is the transfer of value systems.

For example, in the mid-nineteenth century, William Stanley Jevons used an analogy to defend the introduction of the differential calculus into economics. This was a legitimate procedure, he argued, since the calculus had been used for decades in other disciplines, notably in the science of rational mechanics. Jevons even invoked specific analogies between the form of problems in mechanics and in economics, showing that equations of the same form arose in economics and in the analysis of the lever by means of the principle of virtual work. The critical reader will discern in this argument a dignifying, honorific component, the implication that a mathematical economics is like rational mechanics, the subject founded by Isaac Newton, the subject then esteemed as the highest form of scientific knowledge. In retrospect, this component, the transfer to economics of the values and legitimacy of rational mechanics, may seem to have been the most significant feature of Jevons's argument.[6]

Using the example of Jevons to illustrate the transfer of values from one science to another by way of metaphor implies that economics is a mathematical rather than a social science. Examples from within the fields of the strictly mathematical and physical and natural sciences may come to mind less readily because during the Scientific Revolution of the seventeenth century rhetoric fell into disfavor among thinkers and writers whose prose style was influenced by the "new science" or the "new philosophy," as it was sometimes called. The advocates and practitioners of the "new philosophy" held that rhetoric had no place in scientific discourse. Science was to be presented in simple language, with clear descriptions of the evidence of experiment and observation, followed by strict induction or deduction. Each step was to be set forth in language that was unadorned and clearly understood—without any rhetorical flourishes to distract the reader from the evidence and the logic. This was one of the reasons for the great esteem given to mathematics, which seemed to be the most rhetoric-free form of discourse. Today, however, a number of historians and philosophers of science have come to recognize that the sciences have a rhetoric of their own, a set of conventions and assumptions and even rules of discourse which are not provable by experiment or observation or guaranteed by mathematics. Concomitantly, a concept or process from one science may, when it is adopted into another, be regarded as a metaphor or analogy which suggests that the values or achievements of the first science are operative also in the second.

All of the various features that characterize the use of analogy also occur in the application of concepts, principles, or laws from the physical and life sciences to the realms of political and social thought and action. In some cases we may be able to make a distinction between the "logic of discovery" and the "logic of the discovered," although more often we are unable to do so. For example, when Auguste Comte adapted the medical concepts of the "normal" and the "pathological" to the analysis of society, he was very consciously invoking an analogy between the human organism and society. He wrote that he had been making specific use of the ideas of the great doctor and pathologist François Joseph Victor Broussais. But we have no evidence that would enable us to determine whether Harrington used the analogy of the two-part heart as an aid to formulating the concept of a two-part legislature or merely invoked the Harveyan analysis of the heart as an analogue to make his ideas about the legislature seem reasonable.

In the age of the American Revolution and establishment of the American nation, some outstanding examples of the use of analogy in a sociopolitical context may be found in the writings of George Berkeley, David Hume, and Adam Smith. All three of these thinkers had some relation to American political, social, or intellectual history; all three show in their writings how Newtonian ideas, forged in the context of physics and astronomy, were transferred by analogy to wholly different conceptual realms.

George Berkeley, later Bishop of Cloyne, known for his extreme "idealist" philosophical position, may command our attention especially since he spent much of his early career in America. Berkeley was an astute student of Newton's mathematics and physics and was the author of a very important critique of the foundations of the Newtonian calculus.[7] In an essay written in 1713, the year in which Newton published the second edition of his *Principia*, Berkeley set forth the principles of a social or political system which he constructed on the basis of an analogy with the physical world and the Newtonian law of gravity. He began by stating the principles of Newtonian celestial dynamics correctly. This was no mean feat since, as we shall see, some social and political thinkers of the eighteenth century, as well as a number of historians of our own time, have held an incorrect view of Newtonian celestial physics, believing wrongly that in Newtonian celestial dynamics the planets and other orbiting bodies are in a state of equilibrium because of a supposed balance between a centripetal and a centrifugal force.[8]

In his essay Berkeley asserted[9] that society is an analogue (the term he used was a "parallel case") of the Newtonian material universe and that there is accordingly a "principle of attraction" in the "Spirits or Minds of men."[10] This social force of gravitation tends to draw people together into social and political organizations: "communities, clubs, families, friendships, and all the various species of society." Furthermore, just as in physical bodies of equal mass, "the attraction is strongest between those [bodies] which are placed nearest to each other," so with respect to "the minds of men"—*ceteris paribus*—the "attraction is strongest . . . between those which are most nearly related." He drew from his analogy a number of conclusions about individuals and society, ranging from the love of parents for their children to a concern of one nation for the affairs of another, and of each generation for future ones. Although Berkeley introduced the notion of social attraction and regarded the "minds of men" and the closeness of their relation as having social roles similar to those of mass and distance, he did not attempt to develop an exact homology of concept, nor did he quantify his law of moral force.[11] He did not, in other words, attempt to find an exact mathematical equivalent of the Newtonian law. What he did do was to create and exploit an accurate and powerful analogy on a general level.

David Hume's *Treatise of Human Nature* (1738) provides an example, similar to Berkeley's, in which there is a general analogue of the Newtonian law of universal gravity without any proposal of an equivalent mathematical law. His thought is of particular interest here because he was an associate of Benjamin Franklin and because some of his ideas seem to have been of considerable influence among American political thinkers of the age of the American Revolution.[12] Hume's goal was to produce a new science of individual human moral behavior that would be equivalent to Newton's natural philosophy.[13] He stated that he had discovered in the psychological principle of "association" a "kind of ATTRACTION, which in the mental world will be found to have as extraordinary effects as in the natural, and to shew itself in as many and as various forms."[14] In short, he believed that there was an analogy between psychological phenomena and physical phenomena so that both would exhibit aspects of mutual attraction. But he did not propose a law of mental gravity as a direct counterpart to Newton's law, nor did he propose concepts equivalent to Newtonian "mass" or any other fundamental concepts of Newton's *Principia*.[15] His law for human beings was not meant to achieve the form of Newton's law of universal gravitation but rather to serve a science of human actions in much the same general way as Newton's law had done for material objects. His law was an analogue of Newton's because it provided the same kind of fundamental organizing principle.

The two foregoing examples show how analogies have led thinkers to apply aspects of Newtonian science to other domains. In each case there was an attempt to create an analogue of Newtonian physical science by introducing concepts or laws intended to be the counterpart of those used by Newton in his rational mechanics. Both Berkeley and Hume aimed at creating a system of thought that would be an analogue of Newton's system of natural philosophy to the extent of organizing the phenomena in a way similar to Newton's organization of the science of the physical and cosmic realms. This same analogical mode of using Newtonian science may be seen in a different kind of example, drawn from the early nineteenth century, the creation of Fourierism, a social doctrine of special interest to anyone concerned with American history because the Fourier system formed the basis for the establishment of Fourierist colonies in early nineteenth-century America.[16] Charles Fourier believed he had created a political and social system of thought that was a direct analogue of

Newtonian rational mechanics. He claimed to have discovered an equivalent of the gravitational law, one that applied to human nature and social behavior. He even boasted that his own "calculus of attraction" was part of his discovery of "the laws of universal motion missed by Newton."[17] He went a step further. In announcing his discovery in 1803 of a "calculus of harmony," he declared that his "mathematical theory" was superior to Newton's, since Newton and other scientists and philosophers had found only "the laws of physical motion," whereas he had discovered "the laws of social motion." At the same time Fourier's Newtonianism had the advantage of being based on a very general Newtonian analogy and containing no exact or formal equivalents of concepts or laws from Newtonian physics.

A somewhat different kind of example occurs in Adam Smith's *Wealth of Nations* (1776), in the discussion of his celebrated concept of "natural price." Smith wrote that the "natural price" is "the central price, to which the prices of all commodities are continually gravitating."[18] The words "all" and "continually gravitating" invoke Newtonian science and universal gravity. Newton's law of universal gravitation as applied to the solar system or the system of Jupiter and its satellites could actually be paraphrased in Smith's own language: There is a central body to which all of the circulating bodies are continually gravitating. This passage, in fact, has been cited by some historians of economics as an instance of Smith's use of Newtonian science in formulating his system of economics, and the Newtonian interpretation of Smithian economics finds some support in Smith's having written a history of astronomy in which he displayed some understanding of Newton's scientific achievements.[19]

Smith's use of gravitation in relation to the natural price is based on a postulated analogy between the physical universe and the economic world. We may note, however, that Smith's principle differs in one important feature from Newtonian or physical gravitation. A basic axiom of Newton's physics is his third law of motion, that action and reaction are always equal, the principle that we shall see was cited by John Adams in relation to the form of legislature best suited for the American states. A consequence of this third law is that "all" bodies are not only "gravitating" toward some central body but also mutually "gravitating" toward one another. That is, the central body is always gravitating toward all the others. In our solar system, all planets are continually gravitating toward the Sun, but the Sun is equally gravitating toward all the planets; similarly, each of Jupiter's satellites is continually gravitating toward Jupiter, while Jupiter is equally gravitating toward each of its satellites. This is another way of saying that Newtonian gravity is a mutual force, not merely a force of one body acting on others. In every possible example, the central body—according to Newtonian physics—must be "gravitating" toward all other bodies in the system. Hence, for Smith's economics to be a complete and accurate replication of Newton's physical theory of gravity, and for his law to be an exact equivalent of Newton's, all prices would have to "gravitate" toward one another and the "natural price" would analogically have to "gravitate" toward the "prices of all commodities." Such a conclusion would create havoc in Smith's theory of economics.

Accordingly, we may all the more admire Smith for having only partially replicated the Newtonian physical concept, for having adapted or transformed the Newtonian physical concept in a way that was of important creative use in economics. Only a brash display of historical Whiggism would fault Smith on the grounds of imperfect replication. The fact is that he was creating a concept for the science of economics, not working on a problem in celestial physics.

Smith's use of a gravitating economic force may serve as a reminder that economics and all social or political thought is never an exact clone of physics or of biology and that the concepts used in those realms need not be exact replicas of the ones found in the sciences. This principle has been stated in a most incisive manner by Claude Ménard, who makes the important point that the successful use of analogies is not simply a straightforward "transposition of concepts and methods" but always "highlights a difference." He concludes that in every "transfer" from one domain to another "these concepts take on a life of their own."[20]

There are also analogies that appear in works on politics or the state which may be regarded as failing even if they lead to a discovery or are used to explain or to support a political idea. A reason for the negative evaluation may be that the comparison is based on a very superficial or even on an incorrect notion of the science being applied. An example that is particularly noteworthy occurs in Montesquieu's celebrated *Spirit of the Laws*, a work that greatly influenced the architects of the Constitution, who especially took note of Montesquieu's presentation of the concept of the separation of powers. *The Spirit of the Laws* contains a seemingly Newtonian discussion of the "principle of monarchy," a passage which has been cited to prove both that Montesquieu was a Newtonian philosopher and that the Constitution is a Newtonian document because it was influenced by Montesquieu.

In this paragraph, Montesquieu introduces an analogy between monarchical government and "the system of the universe." In monarchical government, he writes, "there is a power that constantly repels all bodies from the center, and a power of gravitation that attracts them to it."[21] This notion of a "power of gravitation" that "attracts" all bodies to a center is strictly Newtonian, taken ultimately from Newton's *Principia*. But, as a careful reading of the *Principia* reveals, Newton's explanation of the "system of the universe" expressly denies any balance of centripetal and centrifugal forces. Montesquieu's pairing of an attracting and a repelling force is an incorrect analogy. It is not—like Adam Smith's force—a creative transformation of a Newtonian concept or even a partial version of Newton's explanation; it is, in fact, just wrong. With a balance of forces there can be no acceleration, and therefore no gravity and no Newtonian system of the world.[22] Montesquieu simply did not understand either the Newtonian concept of universal gravitation or the basic principles of Newtonian rational mechanics. There is abundant evidence that Montesquieu never fully grasped the principles of the new Newtonian natural philosophy.[23] Furthermore, this single isolated reference to "a power of gravitation" is the only portion of the whole treatise that has any kind of Newtonian overtone. Under these circumstances, Montesquieu's treatise hardly warrants being considered a Newtonian work.

We do not know whether or not Montesquieu was drawing on a non-Newtonian analogy—the balance of a centrally directed force and a centripetal force—in discovering a new principle. If so, he erred only in giving the centripetal force the designation of a "gravitation that attracts," which would make it seem Newtonian. In this case, the invocation of a Newtonian concept would be purely rhetorical, an attempt to make his point with an up-to-date scientific reference which he did not fully understand. This example may serve to remind us that the use of the physical and biological sciences in a political context must be understood in terms of the rhetoric of political discourse.

Similarly, the use of a scientific metaphor (that is, a metaphor based on one of the physical or biological sciences or mathematics) may serve in political decisions only or chiefly as an index of the respect accorded to science by a speaker or writer and the audience. Thus we shall see that Benjamin Franklin's associate James Wilson introduced the Newtonian concept of "vis inertiae" into his lectures on law without any comment, evidently supposing that his hearers or readers would recognize and appreciate and be impressed by the Newtonian metaphor.

Another purpose of a scientific metaphor in political discourse may be to legitimate a political concept or principle by making it seem "scientific." Such a purpose is served by introducing scientific terminology, cloaking the presentation in a scientistic garb in an attempt to gain some of the "kudos" of "science" for an oration or literary composition. Examples abound in eighteenth-century political discussions in which the use of a scientifically based metaphor implies that some of the value system of the sciences is being shared by political or social science. This transfer of value systems is, in fact, a very important aspect of the use of metaphors in political discourse.

Analogies and Metaphors in a Sociopolitical Context: James Wilson and Thomas Pownall

The writings of Franklin's colleague James Wilson and of his one-time associate Thomas Pownall enable us to see how scientific analogies and metaphors are used in the realm of politics in order to attract and hold the reader's or viewer's attention, to help make a political point, and to transfer the values of science to the presentation of political or social ideas. In a very real sense, these functions have always been part of political rhetoric, a way of supporting, justifying, explaining, or dramatizing a particular conviction or conclusion. We may see these several roles clearly in Wilson's "Lectures on Law," given in 1790 and 1791. Wilson was a friend and political associate of Benjamin Franklin's, a delegate to the Constitutional Convention, and later a member of the Supreme Court of Pennsylvania.

"There is a *vis inertiae* in publick bodies," Wilson wrote, "as well as in matter."[24] Here he was using Definition Three of Newton's *Principia*, in which Newton introduced the concept of a "vis Inertiae," or "force of inertia," an "inherent" force that exists in every variety of matter and that causes a body to resist any change in its state of rest or motion. In traditional or pre-Newtonian physics, it was generally believed that a moving body requires a mover, an applied force to keep the body going. But in Newtonian physics an externally applied and unbalanced force only changes the state of motion. As Newton explained in Definition Four, "a body perseveres in any . . . state [of motion or of rest] solely by its force of inertia." This principle was stated in full in the first law of motion and is generally known as the law or principle of inertia. We may take note that Newton here somewhat confusingly refers to this property of the inertness of matter not simply as "inertia" but as the "force of inertia." As a result, he uses "force" in two very different senses. One kind of force—displayed in centripetal or gravitational forces, in forces of impact (or percussion), and in forces of pressure—acts to change a body's "state" of being at rest or being in motion in a straight line with constant speed; the other kind of force is the "force of inertia," tending to keep a body in its state of being at rest or of moving in a straight line with constant speed. This "inertia" or "force of inertia" cannot be balanced by an accelerating force such as gravity or centripetal force, or percussion or pressure, since these two varieties of "force" are not compatible.[25]

Making an analogy with Newton's physical principles, Wilson said that there is a "*vis inertiae* in publick bodies." By analogy, then, just as a physical body of and by itself will, according to Newton's first law of motion, "persevere" in its state of rest or of uniform motion, so—according to Wilson—"publick bodies" that are "left to their natural propensities" will continue in whatever state they are in and "will not be moved without a proportioned propelling cause." In this analogy, furthermore, Wilson has used the second law of motion as well as the first law, even introducing the idea of proportion. It is the second law which declares that any "change in motion is proportional to the motive force impressed."

In using this analogy, Wilson was attempting to explain the behavior of "publick bodies" by transferring to them a generally accepted principle of Newtonian physics. The form of the analogy shows that he had a sound grasp of the axioms of rational mechanics, of the Newtonian laws of motion. He must have expected that his listeners would also understand his analogy; otherwise, he would have been indulging in useless rhetoric.

While alluding to Newton's laws of motion, Wilson was simultaneously asserting that his concept was reasonable, since he implied that the actions of public bodies mirror the actions of natural bodies and accord with Newton's laws of motion. And, of course, by introducing a central concept of the *Principia,* Wilson was invoking the highest possible authority, the physics of Isaac Newton. He was transferring some of the value system of accepted science to a wholly different area, the behavior of individuals or of societies. The metaphor carried the bold implication that Wilson's presentation shared features with the paradigmatic science of Isaac Newton.

It is, I believe, significant that Wilson did not find a need for an explicit reference to Newton or for a mention of the *Principia* by name. He apparently assumed that his audience would be sufficiently schooled in the Newtonian natural philosophy to recognize the source of his analogy. Similarly, he was able to use biological circulation in an analogy without having to mention either Harvey or the blood:

> I hope I have evinced, from authority and from reason, from precedent and from principle, that *consent* is the sole obligatory principle of human government and human laws. To trace the varying but powerful energy of this animating principle through the formation and administration of every part of our beautiful system of government and law, will be a pleasing task in the course of these lectures. Can any task be more delightful than to pursue the circulation of liberty through every limb and member of the political body? This kind of anatomy has a peculiar advantage—it traces, without destroying, the principle of life.[26]

In other places, however, Wilson achieves an effect of the same kind through explicit invocation of the scientific paradigm. For example, in speaking of the common law as "the law of experience," he admits that it is far

> from being without its general principles; but these general principles are formed strictly upon the plan of the *regulae philosophandi,* which, in another science, Sir Isaac Newton prescribed and observed with such glorious success—they are formed from the coincidence, or the analogy, or the opposition of numberless experiments, the accurate history of which is contained in records and reports of judicial determinations.[27]

Whether or not he names the great scientist, Wilson can use Newton's science metaphorically to enhance the presentation of an idea about law or government.

We may see features of scientific analogy and metaphor displayed in the writings of another associate of Benjamin Franklin's, Thomas Pownall, who was exposed to Newtonian science when he was an undergraduate at Trinity College, Cambridge, from 1740 to 1743. Pownall is a figure of real interest for anyone concerned with early American history. As a young man he came to America, where he served as lieutenant governor of New Jersey and then as governor of Massachusetts and finally as governor of South Carolina. He was especially interested in topography and geography and was a close acquaintance of Benjamin Franklin, who even wrote some footnotes included in Pownall's *Hydraulic and Nautical Observations on the Currents in the Atlantic Ocean,* published in 1787. Author of many books, Pownall argued unsuccessfully for the granting of more rights to the American colonies. After the outbreak of hostilities, while a Member of Parliament, he tried unsuccessfully to conclude the war by royal negotiations.[28]

The Newtonian theme is first presented in the *Principles of Polity,* published in 1752. Here Pownall states that all "the Matter of this Universe is under one common Law of Attraction and Motion." Yet our own system, that is,the solar system, has "a Principle of Individuality, compleat and perfect within itself." If there were no "one general Law as a Foundation," however, "this System would become a mere Castle in the Air, and a thing that could not exist." Similarly, he argued, in "the moral System" there is no reason why "a Number of Individuals, peculiarly placed and circumstanced, cannot form a distinct Community upon this common Communion as a Foundation, and in like Manner acquire a Principle of Individuality."[29]

Pownall's most important book was *The Administration of the Colonies,* published in 1764. Here he drew heavily on the principles of astronomy and Newtonian physics in a political context. For example, he presented the government of the colonies as a kind of astronomical system in which each colony was in "its proper sphere," receiving its "political motion" from "the first mover (the government of Great Britain)." Thus

Great Britain, as the center of this system, must be the center of attraction, to which these colonies, in the administration of every power of their government, in the exercise of their judicial powers, and the execution of their laws, and in every operation of their trade, must tend.[30]

The fifth edition of Pownall's book on the administration of the colonies, published in 1774, announced in its subtitle that the "rights and constitution" of the colonies "are discussed and stated." Among other metaphors drawn from the sciences to strengthen the political rhetoric, one made use of a specific Newtonian analogy. If there should be "an encrease of the quantity of matter in the planets," he wrote, the result would be that "the center of gravity in the solar system, now near the surface of the sun, would . . . be drawn out beyond that surface." In politics, Pownall argued, "the same laws of nature, analogous in all cases," must apply. Hence, "as the magnitude of the power and interest of the Colonies increases," the center of power will no longer remain fixed in England, but will "be drawn out from the island." As "true philosophers," knowing "the laws of nature," we should accordingly "follow, where that system leads, to form one general system of dominion by *an union of Great Britain and her Colonies; fixing, while it may be so fixed, the common center in Great Britain.*" The alternative, he prophesied, would be to "labour to keep the seat of government" in Britain "by force, against encreasing powers, which will, finally, by an overbalance from without, heave that center itself out of its place."[31]

In *A Memorial, Most Humbly Addressed to the Sovereigns of Europe, on the Present State of Affairs between the Old and New World*, first published in 1780,[32] Pownall once again introduced the authority of Newtonian physics. Explaining that there were common principles or laws in the natural and political worlds, Pownall expressed his regret that "the Government," being "wise in its own conceit," had "rejected Nature and would have none of her ways." There had, he argued, been a failure to recognize that in both the political realm and the world of physics there must be "the same spirit of *attraction*." Believing that the "spirit of attraction which Nature actuates" is but a "vision," the "Ministers" in the government turned against "THAT STATE OF UNION, which the hand of God held forth." They "said to Repulsion, Thou shalt guide *our Spirit;* to Distraction, Thou shalt be our wisdom." The result of repulsion and distraction, he concluded, has been dissolution. The consequence which he envisioned was that "the external parts of the Empire are one after another falling off," so that the nation "will be once more reduced to its insular existence."[33]

This led him to yet another metaphor drawn from science. Once again, we may see his Newtonian imagery used to convey a political message. This time he made use of the specifically Newtonian concept of planetary perturbations, each planet gravitationally affecting the motion of every other member of the solar system. According to Pownall, "North-America is become a new *primary planet* in the system of the world." While this planet naturally follows its own orbit, it "must have effect on the orbit of every other planet" and must necessarily "shift the common center of gravity of the whole system of the European world."[34]

It is easy, therefore, to feel the resonance of the sentence which opens the "Advertisement" to *A Memorial Addressed to the Sovereigns of America,* published three years later:

> The following Paper states and explains the System of the New World in America; the natural Liberty of the Individual settled there; the Frame into which the Communities of individuals (prior to all consideration of Political Society) naturally form Themselves.[35]

Nor is it surprising that a decade later, in a work called *Intellectual Physicks,* Pownall uses analogies from Newtonian science to present his own notions of human psychology and ethics.[36]

POPULAR SCIENCE

Shifts in Scientific Interest and in the Source of Scientific Metaphors

When historians consider "science" in relation to eighteenth-century intellectual history, political thought, or philosophy, they tend to concentrate almost exclusively on the science of Isaac Newton. Most often, in such endeavors, attention is not focused on Newton's science in general but rather on the laws of motion and the science of rational mechanics developed by Newton in the *Principia* and on the system of the world elaborated by Newton in the *Principia* from the principle and law of universal gravity. This limited perception has a number of major faults. The first is its failure to recognize that Newton's "science" was of two very different sorts.[37] One had itself two aspects, of which the first was the development of mathematical laws and principles and their application to problems of rational mechanics, a name given by Newton himself to the subject of forces and the motions which they produce.[38] This is the science developed by Newton in the first two of the three "books" that compose the

Principia, of which the full title was accordingly *Mathematical Principles of Natural Philosophy,* where the words "natural philosophy" embraced a number of the subjects that we today would call physics. The second division of this first, mathematically based, Newtonian science is the subject that we today would call, following Laplace, "celestial mechanics," the elaboration of the system of the world according to the principles of rational mechanics and the law of universal gravity. This is the subject of the third book of the *Principia.* There was also a second Newtonian science which was experimental and wholly non-mathematical. This empirical science was presented in Newton's other masterpiece, the *Opticks.* Whereas the *Principia* proceeded by the use of mathematical techniques—the methods of ratios and proportions, algebra, geometry, trigonometry, infinite series, and fluxions (or the calculus)—the *Opticks* used hardly any mathematical tools and rather offered, in proposition after proposition, what was boldly declared to be a "Proof by Experiments."

These two works, the *Principia* and the *Opticks,* set forth the basis of two very different scientific traditions. One was the mathematical development of physics and the mathematico-physical principles of astronomy; the other was an enlargement of our understanding of nature by experiment, by a special sort of direct questioning of nature. The first of these proceeded by deduction and mathematical demonstrations, the second by induction and the evidence of experiments. Those who adhered to the latter tradition included many important pioneers of the eighteenth century, among them Stephen Hales, the founder of the science of plant physiology; Benjamin Franklin, creator of the first satisfactory theory of electricity; Joseph Black, inventor of modern concepts of heat; and Antoine-Laurent Lavoisier, who was primarily responsible for the modern science of chemistry.

Popular interest in Newtonian science in the eighteenth century also had two parts, corresponding to the two main strands of Newton's own scientific work. Popular lectures on Newtonian dynamics or rational mechanics and the physical principles of astronomy were designed so that men and women could understand the basic mathematical principles even without mathematics. Cleverly designed experimental demonstrations were used to explain and illustrate the principles. Among the most celebrated of these were those given by William Whiston, Newton's successor as Lucasian Professor at Cambridge, and J. T. Desaguliers, an able experimenter and author of a poem on the application of Newtonian principles to the ideal form of government.[39] Desaguliers was the author of a two-volume introduction to Newtonian physics which boasted that it presented the subject without mathematics.[40]

Popular presentations of Newtonian science without mathematics stressed Newton's own nonmathematical science as presented in the *Opticks.* This work of Newton's was just as inviting to a general reading public as the *Principia* was forbidding. Whereas the *Principia* was composed in austere Latin, developing its theorems and their proofs by the use of mathematics, the *Opticks* was composed in graceful English prose without mathematics. While Newton had, in fact, made the *Principia* difficult to read, so that his doctrine would not become the butt of attacks by smatterers who had not fully grasped the principles, he wrote the *Opticks* in a manner that invited and held the attention of nonscientific as well as scientific audiences. The subject matter of the *Opticks* was especially attractive to the general reader, since it explored the nature of light and color, elucidating such phenomena as the formation of a multi-hued spectrum by a prism, the nature of the rainbow, and various beautiful color productions and shapes such as the famous "Newton's rings." This book explored the reasons why we see colored objects as we do. It concluded with a set of "Queries," enlarged in number and scope from edition to edition, in which Newton set forth his ideas on all sorts of subjects, such as the nature of light, the physiology of vision, the analysis of radiant heat, the cause of gravity, the nature of chemical reactions, the actions of the nerves in transmitting information to the brain and the action of the brain in sending out commands of the "will" to the muscles. There were even discussions of the atomic structure of matter and the way in which God had in the beginning created matter in the form of tiny atoms. Here were detailed explanations of how to reason correctly and profitably in science and a final brief treatment of morals. We can well understand why this book had so great an influence, as Marjorie Hope Nicolson was able to demonstrate in her pioneering studies on science and the English literary imagination, and even was a source of Edmund Burke's essay on "the sublime and the beautiful."

It must have been a real pleasure to attend a demonstration illustrating Newton's principles of light and color. One of the most entertaining and popular books on Newtonian science was Algarotti's *Il Newtonianismo per le dame* or *Newtonianism for the Ladies,* a work that deals exclusively with the *Opticks* and does not even mention the prin-

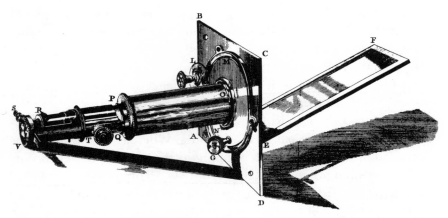

Fig. 1. Solar Microscope (Mid-1700s). This device, designed and manufactured by B. Martin, was mounted with the square plate (ABCD) fixed in a window shutter. The mirror could be adjusted so that the sun's light was reflected directly through the microscope. The image could be studied by direct examination or the image could be projected onto a wall.

ciples of dynamics. Voltaire's popular *Éléments de la philosophie de New- ton mis à la portée de tout le monde,* or *Elements of Newton's Philosophy Made Accessible to Everyone,* in many ways still the finest introduction to Newtonian science ever written, devotes the first half to a presenta- tion of Newton's *Opticks,* reserving the subjects of rational mechanics and gravitational cosmology for the second half, presumably to be read only by the hardiest readers. Other general presentations of Newtonian science, such as Henry Pemberton's *A View of Sir Isaac Newton's Philosophy* (London, 1728), devoted about half of the text to the *Opticks.*

A complete presentation of the eighteenth-century scientific sources of political analogues and metaphors must take note that by mid-century there was an enormous interest in the life sciences. This took several different forms. On the one hand, there was a real pas- sion to learn about the world of nature revealed by the newly improved microscope (see Figs. 1 and 2). In an essay published in *Poor Richard* for 1751, Benjamin Franklin described how that "admi- rable Instrument the Microscope has opened to us . . . a World utterly unknown to the Ancients." He proceeded to list in detail how the microscope shows "something curious and unexpected" in almost every manifestation of nature. One of the primary subjects of scien- tific curiosity was the actual nature and function of the spermatozoa,

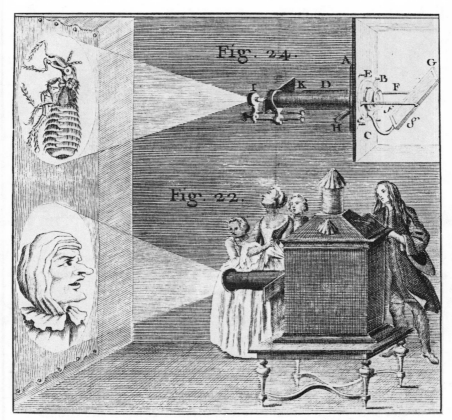

Fig. 2. Projection Microscope and Magic Lantern (Mid-1700s). The group of young women are staring in amazement at an image of a horrible face, painted on a small sheet of glass, and projected onto the wall of a darkened room. At the top is shown how a solar microscope is fastened into a window shutter and used to project the image of a louse onto the wall. This pair of demonstrations was part of a course in optics.

discovered by Antoni van Leeuwenhoek some hundred years earlier. There was enormous concern about the mechanism of generation of animals and human beings. Both scientists and amateurs were also enthralled by the revelation that plants reproduce sexually and that the stamens and pistils found in flowers are sexual organs of reproduction, comparable to those found in animals and in human beings.

In mid-century, the discoveries concerning the sexual reproduction of plants were used by Linnaeus in a wholly unexpected and extraordinary way to form the basis of a new system for the classifica-

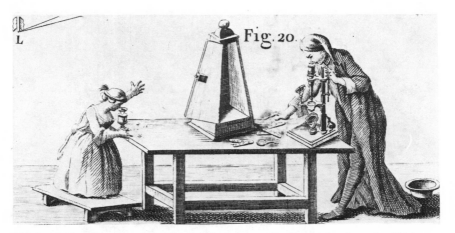

Fig. 3. Mother and Daughter Using Microscopes. An illustration from the Abbé Jean-Antoine Nollet's popular multi-volume Lessons in Experimental Philosophy *(1748) shows the great vogue of interest in microscopy and the life sciences in the mid-eighteenth century. The large pyramidal box was used for storing the compound microscope being used by the mother.*

<u>tion of plants.</u> Although today the Linnaean system seems artificial, it had the virtue of being based on the most fundamental process that made a plant what it is—the stages of generation that replicate the characteristics of the parents in the offspring. More important was the fact that the Linnaean system enabled anyone who encountered a flowering plant to find out exactly what that plant was. Men and women everywhere began to go out on botanizing expeditions, collecting flowers to be identified from one of the handbooks that made Linnaeus's work available (see Fig. 3).

Linnaeus divided all plants and animals into *classes*. Within each class, the plant or animal would be assigned to an *order*. Within the order it would have two further groupings, the *genus* and the *species*. The system worked especially well for the classification and identification of plants, although, despite the fact that many of the species and genus names used by Linnaeus have survived, his system has been extensively modified.

In identifying a particular plant, the Linnaean system required an observer to go through a formalized series of steps, similar to finding answers in the game of "twenty questions." A careful examination of a flower would reveal the number and arrangement of the plant's organs of reproduction. A plant, for example, might turn out to have one, two, three, or more free stamens, or male organs of reproduction, and would be assigned to the class called Monandria, Diandria, Triandria, and so on. Then the order would be determined according to the number of styles, that is, the number of elongations of the pistil, or female organ of reproduction. Thus the class of Monandria would have orders Monandria Monogynia (one stamen and one style), Monandria Digynia (one stamen and two styles), Monandria Trigynia (one stamen and three styles), and so on (for details see Fig. 4). As an example consider *Euphorbia apios,* also known as spurge. It is in the class Dodecandria (having a system of about twelve stamens) and order Trigynia (with a system of three styles). The *Genera Plantarum* in its fifth edition, that of 1754, describes *Euphorbia* as one of the two genera listed as Dodecandria Trigynia,[41] and the *Species Plantarum* of 1753 gives fifty-six species of *Euphorbia,*[42] with *apios* described as "no. 33."[43]

The would-be botanist, with a flower in hand, would be able to turn to one of the handbooks of the Linnaean system and find at once the complete technical description of all known plants in the relevant category. An inspection of the flower would almost at once lead the observer to a group of plants with detailed descriptions of color, size, appearance, habitat, and other characteristics that would readily determine whether the plant in question was already known and recorded or whether it was a new species or variety. Thus, people could go off on botanizing expeditions with their collecting boxes in one hand and a handbook in the other. They began to produce albums of pressed flowers, carefully identifying each one according to the system of Linnaeus by simply counting the number of stamens and the number of styles and then noting whether they were arranged in the same or in separate efflorescences. In this way the great new vogue of botanization came into being.

It should be noted that discoverers of new specimens sent samples on to Linnaeus at Uppsala to be entered into the main register and to be introduced into the next edition of his catalog. New plants from America were named after American botanists and other scientists, producing our wisteria or wistaria (named for Caspar Wistar of Philadelphia), gardenia (for Alexander Garden of Charleston), Franklinia, Jeffersonia, and others. Franklin's friend and scientific associate John Bartram gained fame not only as a collector of plants, but also as an experimental botanist, recording phenomena of plant hybridization. Linnaeus is said to have called him "the greatest living natural bota-

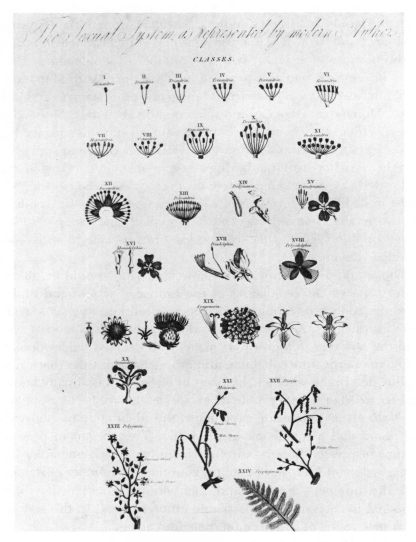

Fig. 4. The Linnaean System Delineated. The diagram illustrates the 24 primary botanical classes, of which 23 are flowering plants. The 24th is shown by a fern. The first 11 figures show flowers with 1, 2, 3, . . . , 11 stamens. Number 12 represents a flower with from 12 to 19 stamens, while number 13 has 20 or more.

nist." We should also take note that two Americans, Cotton Mather and James Logan, were among the earliest scientists to record hybridization in plants in evidence for the sexual reproduction of plants, the basis of the Linnaean system of classification. Jane Colden, the daughter of Benjamin Franklin's friend Cadwallader Colden, the lieutenant governor of New York, is said to have been the first female Linnaean botanist in America.

In the 1740s, the general interest in the life sciences found a new source of enthusiasm in the discovery of the polyp, or the hydra. This tiny organism was discovered in a water-filled ditch by Abraham Trembley,[44] a young Genevese biologist who was serving as a tutor to the sons of a ducal family in Holland.[45] At first Trembley could not decide whether the organism which he had found was a plant or an animal. It adhered to the side of a glass vessel and seemed to reproduce by budding, much in the manner of a plant, but it turned out to have the power of moving toward a source of light, a property that seemed to indicate that it was an animal. This tiny creature had an opening that seemed like a mouth, and there was an abundance of tiny hairlike appendages that gave it a strange beauty (Fig. 5). Trembley found that he could actually turn it inside out and that when he did so it would continue all its life functions (Fig. 6). Most remarkable of all was the polyp's astonishing ability to regenerate.

[margin note:] Hydra — Plant or animal?

This organism aroused the excitement of the intellectual world. One reason was that it seemed to be the key to a number of problems. For example, it seemed to be the "missing link" between animal and vegetable forms of life, partaking somewhat of both. Although it seemed to be an insect, the polyp normally reproduced by putting forth buds, much like an aquatic plant. Trembley described this process in a letter to Charles Bonnet. "The young," he wrote, "come from the body of the parent the way the branches come out of a trunk." That is, "at first only a little excrescence can be seen which grows each day; then the legs appear and at the end of some time, when the animal is complete, it detaches itself from the body of the mother."[46] The world of biology had never seen anything like this before.

Even more important, the polyp thus seemed to offer a new insight into the problem of generation. It is difficult for us today to recapture the currents of thought concerning reproduction in the mid-eighteenth century. Some biologists were committed to epigenesis, holding that the embryo developed in successive stages from a kind of

[margin note:] → Correct

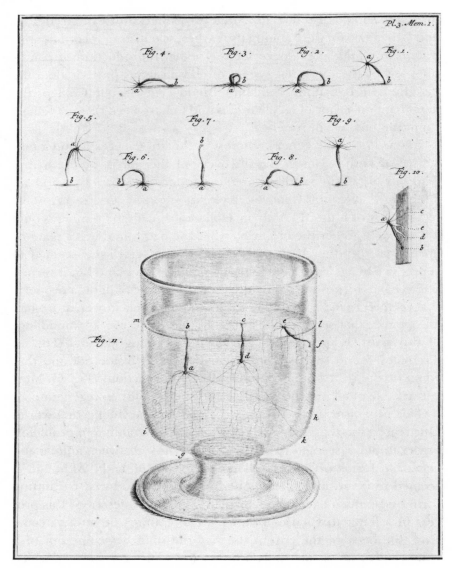

Fig. 5. The Polyp. A plate, from Trembley's Mémoires, *shows the different modes of locomotion observed in polyps and the way in which polyps anchor themselves to the sides of a glass powder jar. In the uppermost row (Figs. 1–4), the polyp has a worm-like forward motion, while the second row (Figs. 5–9) shows a "head-over-heels" form of motion. The large figure (center bottom, Fig. 11) delineates two polyps suspended from the surface of water, while a third (at the extreme right) is detaching itself from the side of the glass vessel in the process of attaining a mode of suspension from the surface of the water.*

Fig. 6. Trembley and His Pupils Studying the Polyp. A vignette, from Trembley's Mémoires *(1749), shows Trembley demonstrating to his two pupils, the sons of Count Bentinck, the way in which he can turn a polyp inside out, described as "the most difficult procedure discussed in this Memoir." Sunlight, entering through the window, provides illumination for the experimenter and for the child studying a polyp with the aid of a microscope.*

"seed," a fertilized egg. The problem was that until well into the nineteenth century, no one had ever succeeded in finding a mammalian ovum. In the seventeenth century, William Harvey devoted the greater part of his research life to an unsuccessful search for an ovum or egg in mammals such as deer. The rival theory was based on the notion of preformation. This theory involved the concept of little preformed individuals who are brought into living existence by fertilization. These preformed individuals, or "homunculi," as they were sometimes called, were thought to have within them tiny individuals which in turn contained the preformed individuals for the next generation, and so on. Sometimes this concept was referred to as "emboîtement," or being boxed in, a reference to the analogy of one tiny box within another, like a Chinese puzzle. There was some division of thought among the preformationists as to whether this succession of preformed individuals occurred in the ova or in the spermatozoa.

Two discoveries in the mid-eighteenth century stirred the debates over generation and aroused a veritable storm of popular interest.

One was the discovery made by the Genevese Charles Bonnet, following up on the work of René-Antoine Réaumur, that certain aphids reproduce parthenogenetically, that is, without the need for contact with a male. The other, even more spectacular, was Trembley's discovery of the odd properties of the polyp.

Regeneration was the most astonishing of the many unusual properties of the polyp. Today, many of us are familiar with this phenomenon in crustaceans, as when a lobster loses a claw and regenerates a new one. We also know that when a starfish is torn in two each half will regenerate the missing parts and become a whole new organism. In the eighteenth century, however, this phenomenon was not at all well known. Therefore, great astonishment was produced by the fact that if a polyp is cut into many pieces each piece will regenerate the missing parts and become a whole new creature.

We may gather some index of the excitement caused by this discovery from the report of Trembley's observations published in the history of the usually staid Paris Academy of Sciences. "The story of the Phoenix who is reborn from its ashes," it began, "fabulous as it is, offers nothing more marvelous than the discovery of which we are going to speak." Referring to such "chimerical ideas" as that of a "serpent cut in half," which, if its parts are joined together, "gives but one & the same serpent,"[47] the report asserted that "here is Nature which goes farther than our chimeras." Take "one piece of the same animal," it continued, cut it "in 2, 3, 4, 10, 20, 30, 40 pieces," and there will be "reborn as many complete animals similar to the first."[48] Perhaps it is not strange that Voltaire could not bring himself to accept these phenomena as true.

One reason why the discovery of regeneration caused such an intellectual stir was that it seemed to raise a very important question about the "soul." Charles Bonnet, following up Trembley's work on the polyp and unable to find any freshwater polyps in his region around Geneva, made similar studies of worms. He found that if certain worms were cut into parts, the separate parts would continue to live and move independently, eventually regenerating the missing parts. Actually, a portion of a worm without a head would continue to move just as if the head were still there. This raised the question of where the "principle of life" resides in such worms. Not in the head, since when the head is cut off, "they still demonstrate the same movements." Are "these worms only simple machines?" he asked, or "are they Composites in which the soul makes their springs move?"

Indeed, "if they have in them such a principle, how can this principle find itself in each portion?" Must we "admit that there are as many souls in these Worms as there are portions of these same Worms, which can themselves become complete Worms?"[49] This line of thought, inspired originally by the discovery of regeneration in the polyp, led to the notion that the animal soul is not centered in the head but may have a kind of material being, spread out through all the living matter of an organism.

The news about the polyp and its extraordinary properties spread rapidly. Summaries of Trembley's discoveries were published in popular books on the microscope, such as George Adams's *Micrographia Illustrata, or the Knowledge of the Microscope Explain'd* (London, 1746). Newspapers and magazines broadcast news about this discovery, enlarging on the accounts produced for scientific journals and books. Benjamin Franklin was so greatly impressed by the curious properties of the polyp that he wrote a summary of its properties for publication in his almanac, *Poor Richard,* for 1751, as part of his presentation of the wonders of the microscope.

Franklin, as we shall see in Chapter 3, also used the properties of the polyp in two very different political analogies. Others in the eighteenth century introduced to political discourse a variety of metaphors drawn from the life sciences. One of these centered on a discussion of the colonies in terms of the metaphor of a growing tree. In the fifth edition of his *Administration of the British Colonies* (1774), Thomas Pownall explained that he had "conceived an idea of our colonies as shoots which the old tree, in the vigour of its health had put forth." Thus the colonies could be likened to "spreading branches of the same organized plant, advancing in its natural vegetation." He found, however, that this "system" was "a mere vision." What had happened was that the "parent tree" began "to view these shoots *as a separate plant*" and that "with its over-topping branches" it cast over them "a shade . . . of jealousy and mistrust" in place of "its old affections." As an inescapable result, "the young shoot in its exuberancy *feels itself as a separate plant.*" It then finds "the old connecting layer" to be "curbing and cramping" and no longer to be "supporting its increasing vegetation."[50] Here was a biologically based plea for a friendly reform of the government for the colonies, even to the extent of introducing "free government."

In the mid-eighteenth century exciting new discoveries about electricity caused a displacement of the polyp as a primary focus of popular interest in science. This shift of interest is described in a letter written by Benjamin Franklin's London patron, Peter Collinson, to Cadwallader Colden in New York, giving a summary account of the changes going on in science. It seems that Collinson sent a similar letter to Franklin. According to Collinson, they were living in what "may . . . very justly be stiled an age of wonders." He would "just hint them." The "surpriseing phenomena of the polypus," he wrote, "entertained the curious for a year or two past." Now, however, "the vertuosi of Europe are taken up in electrical experiments."[51] Collinson's letter was actually a close paraphrase of an article by Albrecht von Haller, published in English translation in the *Gentleman's Magazine* in 1745. This article was reprinted in the *American Magazine* in that same year. According to Haller, some truly "astonishing discoveries have been made within these [last] four years." First "the polypus," as "incredible as a prodigy." And now "the electric fire," as "surprising as a miracle."[52]

Public Demos.

The electrical experiments mentioned by Collinson and Haller became part of the public demonstrations for the edification and amusement of people everywhere. Among the spectacular productions of these demonstrators was the electrification of a boy suspended from a ceiling by silk cords or standing on an insulating stool, the production of sparks in a variety of different circumstances, igniting alcohol by means of a spark, and much else (see Figs. 7 and 8). Franklin saw some electrical experiments performed in Boston by an itinerant lecturer, Archibald Spencer, in 1743. Having arranged for Spencer to come to Philadelphia and repeat his demonstrations, he records that he eventually bought Spencer's experiments and used them to get started on his own career of research in electricity.[53] Eventually Franklin's friend and fellow researcher Ebenezer Kinnersley traveled from one American city to another, giving lecture-demonstrations concerning electricity.[54] Franklin also records that a certain Samuel Domien, "a native of Transylvania, of Tartar descent, but a Priest of the Greek church," had learned about electrical experiment from Franklin in Philadelphia and had then written "from Charles-Town, that he had lived eight hundred miles upon Electricity, it had been meat, drink, and cloathing to him." Later Father Domien had written to Franklin from Jamaica, planning further travels and "proposing to support himself chiefly by Electricity."[55]

By the time of the American Revolution, recent advances in medicine and even discoveries in chemistry were of particular importance.

Fig. 7. Electrification of a Human Subject in the Mid-1740s. The Abbé Nollet is shown performing an experiment, first done in England, in which a boy is suspended by silk (insulating) threads. The experimenter holds an electrified glass rod above the boy's head, while a lady draws a spark from his nose. Both men and women are in attendance at this demonstration.

Fig. 8. Electrical Entertainment in England in the 1740s. An illustration from a book by Franklin's contemporary William Watson shows, in the lower half, two experimenters (on the right-hand side) producing a charge; one is rotating a glass globe in a frame, while the other holds his hand in contact with the globe to produce rubbing. The charge is carried along an insulated metal rod and conveyed to the experimenter on the left, who is on an insulating stand. This experimenter holds a sword that points toward a dish of warm alcohol held by the lady at the left. A spark is produced which ignites the alcohol.

These in turn attracted popular attention and became the source of new metaphors in political discussion. We shall see in Chapter 5 that the science-based metaphors used in the debates at the Constitutional Convention and in the *Federalist* reflect the many aspects of science in which the delegates were interested. It will be seen that, despite the claim made by some historians and political scientists that the Constitution is a Newtonian document, the metaphors based on science in the context of the Constitution are in no sense primarily drawn from Newtonian science.

Deductive versus Inductive Thought

American political thought in the age of the Declaration of Independence and the Constitution exhibits two seemingly different qualities. One is deductive, the other inductive. The deductive feature emulates mathematics and logic as the expression of human reason, finding a model in the geometry of Euclid and in Newton's *Principia.* An example would be some of the *Federalist* papers, in which certain general principles are held to be the axioms of political thought from which some features of the Constitution are derived or defended. So extreme did this procedure seem to critics that one of them pilloried Madison for the form of arguments in the *Federalist.* He would, the correspondent wrote, "next have recourse to CONIC SECTIONS, by which he will be enabled with greater facility, to discover the *many windings* of his favorite system."[56] This deductive quality is a feature of the Declaration of Independence, where certain general principles are set forth in the preamble, followed by particular grievances seen in consequence.

The inductive feature is more closely related to experience than to logic or mathematics. A reliance on the lessons of experience rather than on the simple powers of reason characterizes this approach. John Adams was expressing the viewpoint of many of his fellow Americans when he wrote that "the two sources of true government are reason and experience." This notion of a test by experience was a feature of Lincoln's famed Gettysburg Address, when he spoke of a test whether a "nation, conceived in Liberty, and dedicated to the proposition that all men are created equal" can "long endure."

In the days of the Declaration and the Constitution, the two chief prophets of the deductive and the inductive points of view were Isaac Newton and Francis Bacon, respectively. Jefferson's great trio of immortals, the highest incarnation of the human mind, were Bacon and Newton plus John Locke. He commissioned their portraits and hung them in a place of prominence in his office in the State Department, where they were duly admired by Alexander Hamilton, and he later displayed them prominently in his house at Monticello.

The inductive approach, with its grounding on experience, had certain features which made it seem particularly attractive to Americans. The constant regard for the lessons of experience had to be significant to citizens of the New World in a way that was not the case for Europeans, simply because in the New World there was a consciousness of a frontier, even for those who lived in urban centers or on farms and plantations far removed from the boundaries of the wilderness and the domains of the Indians. Woe to anyone who was so wedded to theory or abstractions as to neglect the hard facts of brute experience. "Experience," as Franklin had Poor Richard say, "keeps a dear school."

Experiment The inductive approach, based ultimately on experience, had a special appeal in the age of the Enlightenment. Basically, it implied an experiential test of knowledge or of system, the same kind of criterion of truth that in the sciences had become Newton's "Proof by Experiments," or a reliance on critical observations. The method of induction was premised on the principle that experiment and observation—the two facets of experience—constitute the only sound basis of knowledge. This belief in experience meant that nature was the highest authority, overruling and vetoing any human authority. This point of view was encapsulated in the motto of the Royal Society, the premier scientific society in the world: NULLIUS IN VERBA—On the Word of No One! In the words of the Declaration of Independence, the supreme authority is held to be "the laws of nature and of nature's GOD."

Locke = Induction The role of experience was a principal feature of the philosophy of the third of Jefferson's trio of immortals, John Locke. Locke, it will be recalled, held that infants are born with an empty mind, a "tabula rasa," with no "innate" ideas. Ideas come to be formed, he argued, on the basis of the impinging of the sense-data or impressions of experience. In this context it must not be forgotten that Newton's two masterpieces of physical science, the *Principia* and the *Opticks,* were both based ultimately on experience and on induction. Newton's third and fourth "Rules for Natural Philosophy" in the *Principia* are concerned with the method of induction. In the final "Query" of the *Opticks* Newton gave the reader a guide to the way of "arguing from Experiments and Observations by Induction," the method of proceeding "from Effects to their Causes, and from particular Causes to more general ones, till the Argument end in the most general." Newton even expressed the hope that "if natural Philosophy in all its Parts, by pursuing this Method, shall at length be perfected, the Bounds of Moral Philosophy will be also enlarged."

In the age of the Constitution the veneration of Newton did not preclude placing a high value on the new and post-Newtonian sciences based on direct interrogation of nature by experiment and observation. The most famous American at the Convention was Benjamin Franklin, a scientist renowned for his experiments and for a theory devised to explain the features of experiential knowledge. His celebrated treatise was, in fact, called *Experiments and Observations on Electricity*. One of the scientific heroes of the Enlightenment was Linnaeus, founder of the great system of classification of plants that bore his name. Linnaeus had proceeded by observation and not by mathematics or deduction.

Political creeds are always ultimately based on religious beliefs or political or social philosophies, a set of general beliefs or axioms from which particular conclusions are derived. In considering the political thought of the men who signed the Declaration of Independence and who framed the Constitution, therefore, we cannot neglect the impress on their mind-set of the intellectual background of the Enlightenment, the Age of Reason. Even George Washington, not usually noted for a deep reading of the philosophers, was aware of the importance of Enlightenment thinkers in making possible the political new world of America being created after independence. In a circular letter to the states, announcing his retirement as commander-in-chief of the army and declaring his return to private life, George Washington took note that the establishment of the new nation did not occur "in the gloomy age of Ignorance and Superstition," but rather "at an Epocha when the rights of mankind were better understood and more clearly defined, than at any former period." Washington attributed this change to the successful "researches of the human mind, after social happiness," to the "collected wisdom" acquired by the "labours of Philosophers," and of "Sages, and Legislatures."[57]

Conclusion

As we examine the role of science in the political thought of the Founding Fathers and the possible scientific context of the Declaration of Independence and the Constitution, we must keep in mind that the American nation formed its identity in the period known as the Enlightenment or the Age of Reason. Two great intellectual heroes of that age were the philosopher John Locke and the scientist Isaac Newton, sometimes called the "twin luminaries" of the Augustan Age. Voltaire, in his book of *Philosophical Letters concerning the English*, esteemed Locke as the modern Plato and lauded Newton as the example of human reason in its highest form. John Locke was not only the author of the influential *Two Treatises of Government;* he was also the first philosopher (in our modern sense of the word) to become a Newtonian.[58]

In an age in which reason was venerated, science was esteemed as the intellectual manifestation of human reason in action. Not only was there a veneration of Newton and his achievement, but there was also a recognition that science had been advancing and was still producing the fruits of reason. There were even notable advances in the very subjects in which Newton had pioneered, including mathematical analysis, rational mechanics, celestial dynamics, and optics. In addition, new sciences were emerging of which Newton had no idea. One of these was electricity, made into a science by the creative intellectual efforts of Benjamin Franklin. Joseph Priestley, in his history of electricity, recorded that Isaac Newton would have been astonished to learn of this new science. Furthermore, the post-Newtonian science of plant physiology had been created by Stephen Hales, and there were astonishing new discoveries in geology and a reform of chemistry.

Many of the Founding Fathers were educated in college. At Harvard, at Yale, at Princeton, at William and Mary, students were required to study mathematics and the principles of Newtonian science and were introduced to the new sciences being created or advanced in their own time. Accordingly, it should come as no surprise that these men drew on their scientific knowledge for analogies and metaphors to be used in their political discussions and writings. It does not matter whether the actual occurrence of such science-based metaphors seems to us to be large in number or small. What matters is the fact that science actually does appear in a political context in the thought of Jefferson, Franklin, and Adams and that science did provide a source of metaphors for some of the discussions in relation to the Constitution. There can be no doubt that the Founding Fathers displayed a knowledge of scientific concepts and principles which establishes their credentials as citizens of the Age of Reason.

2

Science and the Political Thought of
Thomas Jefferson: The
Declaration of Independence

·····

Jefferson and Science[1]

"Nature," Thomas Jefferson wrote, in a letter to the French economist Pierre-Samuel Dupont de Nemours, "intended me for the tranquil pursuits of science, by rendering them my supreme delight."[2] Only one other president of the United States, Theodore Roosevelt, could possibly be described as a would-be scientist in politics.[3] Both Jefferson and Roosevelt pursued science as an avocation throughout their lives. Both of these men had an abiding and primary interest in natural history, in the science of life. Neither one, however, made substantive contributions to science. Thus, unlike Benjamin Franklin, they are not considered to have been members of the scientific community.

Jefferson's intellectual interests included subjects that we would designate today as agriculture, anthropology, archeology, architecture, botany, the Greek and Latin classics, higher education, history, linguistics, literature, mechanical invention, natural history, paleontology, philosophy and rhetoric, theory of government, zoology, and the promotion of learning. We may agree with the sentiment expressed by President John F. Kennedy on the occasion of a dinner to honor some recipients of the Noble Prize. Kennedy addressed his guests as "the most extraordinary collection of talent, of human knowledge, that has ever been gathered together at the White House, with the possible exception of when Thomas Jefferson dined alone."[4]

At the end of his life, Jefferson wanted to be remembered for three accomplishments and he designed his epitaph accordingly. Visitors to his home in Monticello can read on his tombstone his wish to be known as: "Author of the Declaration of American Independence," Founder of the University of Virginia, and Author of the Virginia Statute on Religious Freedom. There is, in this list, no scientific discovery or theory to make his name immortal, even though science was always an important component of his life and thought. Nor did Jefferson refer here to his services as governor of Virginia, as first secretary of state of the new nation under Washington's administration, or as president of the United States.

Scholarly testimony to the importance of science for Jefferson lies in the existence of many scholarly articles and four fairly recent books devoted wholly to the subject of Jefferson and science.[5] One of the Jefferson Lectures, established in his honor by the National Endowment for the Humanities, was given by a physicist and historian of science, Gerald Holton, and was devoted in large measure to Jefferson's science and its significance for us today.[6] Considering Jefferson's career in a novel way, Holton argues persuasively that Jefferson "saw himself first of all as a student of the sciences, philosopher, educator, planter, and scholar." He was not a true scientist in the sense of his idol, Isaac Newton, according to Holton, in that his primary concern was not to find a single "unified theoretical structure" that would "subsume the whole world of experience." He did, however, place "the center of research in an area of basic scientific ignorance," conceiving that science might find the solution to social and political problems.[7]

The subject of Jefferson and science has usually been treated in terms of a delineation, in various degrees of detail, of his scientific interests and accomplishments, often combined with his activities as agriculturalist, inventor, and architect. The authors of these works have not attempted to see whether Jefferson's training in science and his interest in scientific matters ever had any relation to his political ideas and political career. Nor have scholars traced the occurrence of Newtonian concepts in Jefferson's political writings.

When Jefferson wrote to Pierre-Samuel Dupont de Nemours that he had been intended for the pursuit of science, he was ending a public phase of his career, completing his second and final term as president of the United States. But for the "enormities of the times," he added, he would never have given up a scientific career, he would never have embarked on the "boisterous ocean of political passions." Now, at last, he was free and he felt like "a prisoner, released from his chains."[8] Some years earlier, in 1791, during his service as secretary of state under Washington, he wrote to Harry Innes that while politics was his "duty," natural history was his "passion," a sentiment he expressed again and again.[9] In 1778, during the turbulent days of the Revolution, Jefferson—serving as governor of Virginia—wrote a letter to Giovanni Fabbroni in Williamsburg, explaining that even though he was busily engaged "in the councils of America," he still was able to make time to "indulge" his "fondness" for "philosophical" studies.[10] While preparing to sail for France as minister to the court of Louis XVI, he wrote to Lafayette of his desire to visit those countries "whose improvements in science" along with "arts" he had long admired from far away.[11] In 1791, while serving as secretary of state, he wrote to Thomas Mann Randolph, Jr., about his hatred of the "detestable" drain on his time and energy and his longing for the freedom from office that would enable him to study the "weavil of Virginia" and give him time for other scientific "pursuits of this kind."[12]

Jefferson's divided loyalties to science and the affairs of state are seen in bold relief in some circumstances of 1797 to which reference has already been made. He arrived in Philadelphia, then the seat of the national government, as the recently elected vice-president of the United States and as president-elect of the American Philosophical Society, America's foremost scientific academy or institution, of which Benjamin Franklin had been a principal founder. He was president in direct succession to the recently deceased David Rittenhouse, astronomer and clock-maker, a longtime friend. In his carriage Jefferson brought with him a valuable treasure, huge bones of a giant fossil animal which had been dug up by some miners in Virginia. Jefferson would present these bones to the American Philosophical Society, delivering a lecture on the giant animal they came from, a super-lion he believed, which he named the "Megalonyx."[13]

In a letter to his daughter Martha, written at this time, Jefferson expressed his deep feelings that "Politics are such a torment that I would advise everyone I love not to mix with them."[14] He had changed his "circle," he added, "according to my wish, abandoning the rich and declining their dinners and parties" so that he could associate "entirely with the class of science, of whom there is a valuable society here."[15]

Even during the crucial month of February 1801, in the new capital at Washington, when the Congress was debating whether Aaron Burr or Thomas Jefferson would become the third president of the United States, Jefferson kept up with his scientific pursuits and engaged in correspondence with fellow scientists. In a letter to Caspar Wistar of Philadelphia, Jefferson discussed some new fossils discovered in New York, and the possibility of purchasing them for the collections of the American Philosophical Society.[16] He had received information about them from Robert R. Livingston in New York and he hoped that Wistar might be able to select some specimen parts of the fossil animal such as tusks, vertebrae, a fragment of jawbone, plus the sternum, scapula, tibia, and part of the head containing the socket of the tusks.

Even while serving as president, Jefferson never forsook his concerns for science. The wife of Samuel Harrison Smith, founding editor of the *National Intelligencer,* described a botanizing expedition with the president of the United States. Not "a plant from the lowliest weed to the loftiest tree escaped his notice," she wrote. Dismounting from his horse, "he would climb rocks, or wade through swamps to obtain any plant he discovered or desired and seldom returned from these excursions without a variety of specimens."[17] Edwin Martin has written a vivid summary of the activities of the scientist-president. "In his Washington residence," Martin wrote, "we see him not only with his beloved flowers, plants, books, and pet mocking-bird, but also with carpenter's tools, garden implements, maps, globes, charts, a drafting board, and scientific instruments." He "filled the unfinished East Room of his residence, which Abigail Adams had used for hanging out wash, with a huge fossil collection he had gotten from the Big Bone Lick." On his "lawn passers-by might, at one time, have seen young grizzly bears brought by Meriwether Lewis from the far West." His "political enemies jeered at Jefferson's 'bear-garden.' "[18]

Dr. Samuel L. Mitchill wrote his wife that President Jefferson discussed cowpox with him "with the intelligence of a physician."[19] Jefferson, in fact, was one of two bold innovators—the other was Dr. Benjamin Waterhouse, professor of materia medica at Harvard's medical school—who were responsible for the introduction into America of the new practice of vaccination for the smallpox, discovered in England by Edward Jenner. Mitchill told his wife that Jefferson was "more deeply versed in human nature and human learning than almost the whole tribe of his opponents and revilers."

In 1804, John Quincy Adams, then a senator and later to become the sixth president, recorded how President Jefferson had shown his guests "a Natural History of Parrots, in French, with colored plates very beautifully executed."[20] We have another portrayal of an evening with Jefferson by John Quincy Adams, this one written in November 1807, after what he described as "one of the *agreeable* dinners I have had at Mr. Jefferson's." The guests that evening were "chiefly . . . members of Congress." During the course of the evening, Adams recorded, Samuel L. Mitchill "mentioned Mr. Fulton's steamboat as an invention of great importance." To this, Jefferson assented, adding, "and I think his torpedoes a valuable invention too." Adams noted also that Jefferson "then enlarged upon the certainty of their effect, and adverted to some of the obvious objections against them, which he contended were not conclusive." The subjects of the conversation ranged from chemistry, geography, and natural philosophy to "oils, grasses, beasts, birds, petrifactions, and incrustations."[21] Here we get a glimpse of Jefferson in the role in which he described himself, a "zealous amateur."

One of Jefferson's close associates and friends, the painter and naturalist Charles Willson Peale, recorded an evening in the White House with the great naturalist and explorer Alexander von Humboldt. During "a very elegant dinner at the President's," he noted, "not a single toast was given or called for, or politics touched on." Instead, the conversation was entirely on "subjects of natural history, and improvements of the conveniences of life."[22]

Jefferson, like Franklin, was an inveterate gadgeteer and inventor. He was also a skilled and gifted architect, as his home in Monticello visibly testifies. In his writings, we "find him discussing things like plows, farm machinery and conveyances, cisterns, the orrery, the polygraph, the pedometer, the odometer, a 'geometrical wheelbarrow,' . . . a 'hydrostatic waistcoat,' . . . air pumps, compasses, the use of wooden and ivory diagrams in geometrical demonstrations, canal locks, balloons, the great future possibilities—particularly for America—in the application of steam power to machinery."[23] He filled his house with books on almost every aspect of science and on many branches of technology. Artworks adorned the walls in an almost continuous tapestry and sculptures were mingled with specimens of natural history, archeological finds, fossils, minerals, and

Inventions

natural curiosities. His inventions were to be seen all over the house, from a complicated clock in the front hallway, with weights which descended in special passageways from the roof to the cellar, to a kind of dumbwaiter designed to expedite the serving of meals, and even a specially designed bed. When the Marquis de Chastellux visited Jefferson in Monticello, he wrote that "no object had escaped Mr. Jefferson." It "seemed as if from his youth he had placed his mind, as he had done his house, on an elevated situation, from which he might contemplate the universe."[24]

As America's first secretary of state, during the administration of George Washington, one of Jefferson's obligations was to supervise the awarding of patents of invention, according to the provisions of the Constitution. Jefferson must have found this an especially congenial task since he was himself an inventor and gadgeteer of the highest order. One of his most important inventions was a new form of moldboard for a plow, which will be discussed below in the context of Newtonian mathematics. Among the many mechanical devices and gadgets to be found in his home in Monticello was a revolving bookstand which could hold five large tomes, each one opened to designated pages. There was also a portable desk, with special compartments for filing notes, for holding pens and ink, and for storing versions of a document in progress. He had a special gadget of his own invention, the polygraph, for writing documents in duplicate, and a press for making copies. He installed in his home a special kind of ventilating and cooling system of his own devising. There were so many gadgets and contrivances in Monticello that it was an inventor's paradise.[25]

Polygraph

The branches of science to which Jefferson was most highly committed included scientific agriculture (for him "a science of the very first order"), mathematics, physics and astronomy, natural history, paleontology and archeology, and certain aspects of anthropology. Although by no means a vertebrate paleontologist in the literal sense of that term,[26] he did make some significant contributions to this subject, which was a constant source of interest to him. (An aspect of his concern for paleontology is presented in Supplement 5.) So notable and advanced were his views on the method of archeology that a current handbook, written by one of the world's most distinguished archeologists, Sir Mortimer Wheeler, calls him "the first scientific digger."[27] He finds that Jefferson's *Notes on the State of Virginia* contains the first recorded instance "of the observation of archaeological stratification." Jefferson's "clear and concise report," according to Wheeler, "describes the situation of the mound [under study] in relation to natural features and evidences of human occupation." Jefferson "detects components of geological interest in its materials and . . . their sources." He "indicates the stratigraphical stages in the construction of the mound" and "records certain significant features of the skeletal remains." And, finally, "He relates his evidence objectively to current theories." We may agree with Wheeler's conclusion, "No mean achievement for a busy statesman in 1784!"

Anthropologists find Jefferson's writings on Native Americans, the Indians, to be especially noteworthy. He was particularly intrigued by the question of their origin. An accomplished student of Indian languages, he was a true pioneer in applying linguistic criteria to anthropological questions. His method of studying native languages was essentially modern: to begin by constructing vocabularies and only later to wrestle with problems of grammar and syntax. In his *Notes on the State of Virginia,* he described the nobility of American Indians and compared their great skill in oratory with that of the eminent Demosthenes and Cicero.

Jefferson's science intersected with his political career on a number of occasions. One of the most significant was his formulation of instructions for the Lewis and Clark expedition, authorized by Congress in 1803. Lewis and Clark were being sent out to explore the territory to the west, spending two and a half years of travel and study. Of course, a major interest was the possibility for expansion of the territorial boundaries of the United States and the commercial potentialities of these western lands. Ahead lay the negotiations for the famed Louisiana Purchase, which was to double the continental territory of the United States. Being a man of science, Jefferson was aware that here was a fabulous opportunity for scientific exploration, and he wanted detailed and precise information about topography, the geography and geology of this unexplored region, the occurrence of useful minerals, the different tribes of Indians and their customs, the varieties of animals and plants encountered, and even the occurrence of fossils. In preparation for the expedition, Jefferson arranged for the chosen leader, Meriwether Lewis, to go to Philadelphia, in order to be instructed in science. His teachers included the botanist Benjamin Smith Barton, the anatomist and zoologist Casper Wistar, the mathematician Robert Patterson (who was to instruct Lewis in astronomy), and the medical doctor Benjamin Rush.[28] He was also to learn the principles of surveying and mapping from Major Andrew Ellicott in nearby Lancaster. Jefferson also asked Barton to instruct Lewis in those aspects of botany, zoology, and Indian history "which you think most worthy of enquiry and observation."[29] All of these teachers were Jefferson's fellow members of the American Philosophical Society.

Jefferson's actual instructions are a scientific model of their kind. They display Jefferson's mastery of the principles of geographical exploration, of anthropology and archeology, of the problems of the fauna and flora of an unexplored region, and of mineralogy. There was a clearly expressed goal "to extend the boundaries of science." No other document in Jefferson's career so fully attests to his command of the sciences at large and the needs in scientific exploration.[30]

The preparation for the scientific component of the Lewis and Clark expedition was but one of many occasions on which Jefferson's knowledge of science proved to be significant in a strictly political context. Another was the solving of the mathematical problem of designing an equitable mode of apportioning representatives in the Congress. Yet another was the application of Newtonian principles in the design of a standard of length for the new nation. Of a wholly different kind was the refutation of a current biological theory that all forms of life in the New World exhibited a degree of "degeneration." Demolishing this theory had obvious important implications for the future development of America. Finally, we shall see that Jefferson's most renowned political statement, the Declaration of Independence, exhibits signs of his commitment to the Newtonian philosophy.

Jefferson's Scientific Education

Jefferson's formal higher education began when he entered the College of William and Mary in Williamsburg in 1760 at the age of seventeen (see Fig. 9). The two subjects he then wanted to pursue were the classics and mathematics. He was fortunate that one of his teachers, the professor of natural philosophy, was William Small. Although Small was a well-trained scientist, his job at William and Mary was to teach not just science (or "natural philosophy"), but also a variety of other subjects, including metaphysics and mathematics (see Figs. 10 and 11). Later in life, Jefferson said that "Small probably fixed the destinies" of his life.[31]

Fig. 9. Teaching Natural Philosophy (Science) in 1759. A print, published a year before Jefferson entered the College of William and Mary as a freshman, shows the master demonstrating to his students the features of the orrery, an eighteenth-century planetarium. On the shelves are various scientific instruments, including a compound microscope; preserved specimens of natural history hang from the ceiling.

Fig. 10. A Mid-Eighteenth-Century Demonstration in Experimental Philosophy in France. The Abbé Nollet, from whose multi-volume set of Lessons in Experimental Philosophy *this illustration is taken, is shown performing an experiment for a group of young ladies and gentlemen. On the shelves there are a vacuum pump, various types of bell jars, and other bits of apparatus. The course was intended for women as well as for men.*

Appointed a teacher at William and Mary in 1758, Small stayed on at Williamsburg for only six years, after which he returned to England where he was listed as "William Small, M.D." Small settled in Birmingham, where he became an associate of Josiah Wedgwood the potter, James Watt, Erasmus Darwin (grandfather of the founder of the theory of evolution), and other scientists and amateurs of science who organized themselves into an organization known as the Lunar Society. They took their name from the fact that they met

Fig. 11. A Lecture in Experimental Philosophy. This illustration from the Universal Magazine *for 1748, a dozen years before Jefferson went to college, shows some young men being instructed in experimental physical science. The subject of the lecture-demonstration (the first in the course) is chemistry. On a shelf on the back wall may be seen a telescope and other instruments for later demonstrations.*

regularly at the time of the full moon, when there was sufficient light for them to travel the highways to and from meetings. Later on, Joseph Priestley became the society's most famous scientific member.[32]

Dumas Malone, Jefferson's distinguished biographer, has pieced together some bits of information concerning Small's teaching and influence. It was Small (by birth a Scot) who introduced Jefferson and others to the thought of the Enlightenment and especially the ideas of the Scottish school of philosophy. He taught Jefferson not only science and mathematics, but also ethics, rhetoric, and even belles lettres. One of his Virginia students, John Page, called Small "illustrious," referring to him as his "ever to be beloved professor." It was no doubt Small who showed Jefferson that mathematics and natural philosophy—the science of Isaac Newton—were "peculiarly engaging and delightful." Jefferson later wrote that mathematics had always been his favorite subject. "We have no theories there," he wrote, "no uncertainties remain on the mind," but "all is demonstration and satisfaction." I have "forgotten much," he added, "and recover it with more difficulty than when in the vigor of my mind I originally acquired it." He was grateful for "the good foundation laid at college by my old master and friend Small," which enabled him to do mathematics "with a delight and success beyond my expectation."[33] Small would have stressed the rigors of Euclid as the foundation of mathematics and all rigorous thought. Jefferson would have learned and applied Euclid's system of definitions, postulates, and axioms—the "self-evident" foundations of all sound knowledge and reasoning.

At the beginning of the Revolution, when—in Dumas Malone's words—"public dissension threatened to divide him from his friend as the ocean had already done," Jefferson sent Small a token of his continued admiration and gratitude, three dozen bottles of good Madeira which he had aged for eight years in his own wine cellar. We may agree that even if Small was only "a minor torchbearer of the Enlightenment," he was "by any reckoning . . . one of those rare men who point the way, who show new paths, who open doors before the mind."[34]

It was through Small that Jefferson became acquainted with George Wythe of Williamsburg, with whom Jefferson later studied law. Wythe himself was very much interested in the sciences. He collected natural curiosities and he owned a good astronomical telescope (which can still be seen in his house at Williamsburg). Small, Wythe, and Francis Fauquier (the governor of Virginia) became Jefferson's first triad of great influences. Fauquier was a true son of the Enlightenment, interested in the sciences and a constant observer of curious natural phenomena. He was the son of a Huguenot physician who had fled from France to England, gaining employment at the mint under Isaac Newton.[35]

A Naturalist in the Service of His Country: Confutation of the Theory of "Degeneration" in the New World

Jefferson's scientific interests were marshaled in the service of his country in his only book, *Notes on the State of Virginia*. This work has been studied by many scholars, who have taken note of its beautiful portrayals of nature and of such elements of the American landscape as Virginia's Natural Bridge. Jefferson's *Notes* has also attracted attention because of its meticulous description of American minerals, plants, and animals. Here we are not concerned so much with establishing Jefferson's credentials as a naturalist, or with his expressions of a native response to the American landscape. Rather, our purpose is to explore Jefferson's systematic rebuttal of a widely held "scientific" theory that plants and animals, and even human beings, of the New World were inferior to those of the Old. The theory also asserted that there would be a degeneration in plants and animals transplanted from the Old World to the New. To demonstrate the falsity of this theory was a matter of real political consequence.

Jefferson's *Notes on the State of Virginia* was composed in response to a questionnaire about conditions in Virginia, sent out in 1780 by François de Barbé-Marbois, secretary of the French legation in Philadelphia. Jefferson records[36] that he began to assemble replies to Marbois in early June 1781, when he retired as governor of Virginia, using his newly found leisure to organize the many notes and bits of statistical data that he had been collecting over many years. As he put his text together, Jefferson found it necessary to enlarge the number of Marbois's queries from twenty-two to twenty-three. The full text was ready for Marbois in December.[37]

Jefferson circulated his first manuscript among friends, taking note of their comments and revising and enlarging his text until it had grown to book length. Jefferson contemplated having it published in Philadelphia, but the costs were prohibitive. When he went to Europe in 1784 as a commissioner in the negotiations of treaties with friendly European powers, remaining as the American minister to the French government, he took the manuscript with him.[38] The text was printed in France in 1785 in a small edition for private distribution. A French translation was published in 1787 and a proper English version was published in London in that same year.

Jefferson's *Notes* has been described as "probably the most important scientific and political book written by an American before 1785," the work on which "much of Jefferson's contemporary fame as a philosopher was based."[39] The *Notes* is a repository of all sorts of statistical data, of information concerning mines and minerals, of details on climate, of information concerning plants and animals, and of the habits and customs of Native Americans. Here also the reader would find Jefferson's "ideas concerning religious freedom or the separation of church and state, his analysis of the ideals of representative government versus dictatorship, his theories of art and education, his attitude concerning slavery and the Negro, his interest in science."[40]

From the point of view of politics and science, the most important chapter of Jefferson's *Notes* is the sixth, bearing the innocuous title "Productions Mineral, Vegetable and Animal."[41] The introductory part of Chapter 6—a rather dry account of minerals, of "medicinal springs" and "syphon fountains," followed by long lists of plants—hardly prepares the reader for the force of the final section on animals (including human beings). Here Jefferson changes his style rather abruptly. First, he gives detailed descriptions concerning American fauna, much like the previous presentation of flora, and then rather consciously sets out to demolish the theory of "degeneration" in the New World. Espoused by Buffon, at that time one of the world's foremost naturalists, this theory was embodied in a number of widely read works about Americans, of which the most important were Cornélius de Pauw's *Recherches Philosophiques sur les Américains (Scientific Researches on the Americans)* and the Abbé Raynal's *Histoire Philosophique et Politique des Établissements et du Commerce des Européens dans les deux Indes* (translated as *Philosophical and Political History of the Settlements and Trade of the Europeans in the East and West Indies*). Abbé Raynal's *History* was a "history" somewhat in the sense of a "natural history" and not merely a chronology of events or circumstances.

Degeneration The theory of degeneration had a number of primary propositions, among them the following:

1. There exist fewer species of animals in the New World than in the Old.
2. Whenever the same or a similar species exists in the New World and the Old, the animals of the Old World are larger than those of the New.
3. Domesticated animals transplanted from the Old World to the New are smaller than their progenitors.
4. Animals which exist only in the New World tend to be smaller than their Old World counterparts.
5. All forms of life (animal and human) in the New World tend to be "degenerate."[42]

Two examples of supposed American degeneration will indicate the sense of this term. The nightingale, according to Oliver Goldsmith, exists in both Europe and America, but the American variety does not sing, it is mute.[43] Buffon considered the tapir to be the "elephant" of America, observing that it is only the size of a small cow.[44]

Every American would be doubly disturbed by Buffon's theory. First of all, as Jefferson demonstrated, and as every citizen of America knew, Buffon's "facts" were wrong. A simple assembly of the true facts concerning animals of the New World showed plainly that the theory must be false. Second, a widespread belief in the theory of degeneration, especially in its application to men and women, would daunt many prospective immigrants to the Americas. Jefferson, therefore, in his rebuttal, was acting in his dual role as scientist and statesman.

One of the reasons advanced by Buffon for the occurrence of degeneration in the New World was what he held to be the high degree of humidity as compared with conditions in Europe. Jefferson argued in rebuttal that America is not more humid than the other continents, although he did admit that there were no data sufficient to decide for or against this supposition. But even should Buffon be right, he explained, then the consequence should be the opposite of Buffon's conclusion, since experience shows that vegetable growth is made lush by heat and moisture. It is for this reason, he noted, that in humid regions we find that animals have greatly "multiplied in their numbers" and have "improved in their bulk." He concluded, with evident satisfaction, that Buffon himself had said that the world's largest cattle were to be found in two humid regions, Denmark and the Ukraine.

In order to combat Buffon's notions of comparative size, Jefferson drew up "three different tables," showing "A Comparative View of the Quadrupeds" in Europe and America. He berated Buffon for accepting reports of travelers who certainly did not "measure or weigh the animals they speak of" and who may not even have been "acquainted with the animals of their own country, with which they undertake to compare them." Jefferson's evidence led him to this conclusion:

> The result of this view then is, that of twenty-six quadrupeds common to both countries, seven are said to be larger in America,

seven of equal size, and twelve not sufficiently examined. So that the first table impeaches the first member of the assertion, that of the animals common to both countries, the American are smallest, "et cela sans aucune exception." It shews it [is] not just, in all the latitude in which its author has advanced it, and probably not to such a degree as to found a distinction between the two countries.

Jefferson's lengthy tables (a list of birds occupies five pages) appear convincing. He shows that if the tapir is to be considered the "elephant" of the New World, then the wild boar must be the elephant of Europe. The latter "is little more than half" the size of its American counterpart. Jefferson does admit that some domesticated animals in America may have "become less than their original stock,"[45] but the explanation is not to be found in the climate, but in the fact that these particular animals were poorly fed. It is the same in Europe, he observed, wherever "the poverty of the soil, or poverty of the owner, reduces them to the same scanty subsistance." It would be a mistake, Jefferson concluded, a case of erring "against that rule of philosophy, which teaches us to ascribe like effects to like causes," if we should "impute this diminution of size in America to any imbecility of want of uniformity in the operations of nature."

Since Jefferson scholars are not familiar with the text of Newton's *Principia,* it has not been recognized that in Jefferson's invocation of a "rule of philosophy," he is quoting directly from that work. His statement about effects and causes is taken from the second of the set of "Regulae Philosophandi," or "Rules for Natural Philosophy," with which Book Three of the *Principia* begins. The strength of Jefferson's argument may be seen in the fact that Newton's rules are presented in the *Principia* as guides for the proper conduct of natural philosophy. Jefferson's "rule of philosophy," to "ascribe like effects to like causes," uses the actual words of Newton's "rule of philosophy." Newton's text declares that "effectuum naturalium ejusdem generis eaedem assignandae sunt causae," which in Newton's day had been rendered by Andrew Motte as an injunction that "to the same natural effects we must . . . assign the same causes."[46] The cognoscenti of Jefferson's day would at once identify the source of his methodological principle, especially since he calls it a "rule of philosophy," the very name used by Newton in the *Principia.* What is perhaps the most striking aspect of this direct citation of Newton's *Principia* is that Jefferson did not find a need to mention either the name of Newton or the title of his book. He assumed that Newton's rules were so well known to his readers that to mention either Newton's name or the title of his treatise would be supererogatory, a breech of good taste in rhetoric.

Jefferson was aware that simple numbers, however effective, would not be sufficient to demolish Buffon's views concerning the character, physiology, and behavior of the Indians as an example of the degeneration supposed to be characteristic of life in the New World. The "savage of the New World," Buffon wrote, "is about the same height as man in our world," but he is not in any significant way "an exception to the general fact that all living nature has become smaller on that continent." The savage, according to Buffon, "is feeble." He "has small organs of generation." He "has neither hair nor beard" and he has "no ardor whatever for his female." Buffon described the American Indian as "timid and cowardly," without "vivacity" or "activity of mind." Any "activity of his body is less an exercise, a voluntary motion, than a necessary action caused by want." Relieve him "of hunger and thirst, and you deprive him of the active principle of all his movements; he will rest stupidly upon his legs or lying down entire days." Buffon explained that the "original defect" in Native Americans appears in every aspect of their relations to others; they "are indifferent because they have little sexual capacity." It is "this indifference to the other sex," he insisted, "which weakens their nature, prevents its development, and—destroying the very germs of life—uproots society at the same time." Buffon concluded in summary: "Man is here [i.e., in America] no exception to the general rule. Nature, by refusing him the power of love, has treated him worse and lowered him deeper than any animal."[47]

In reply to this farrago of error and misinformation, Jefferson displays the character of a true scientist. He confines his discussion to North American Indians, of whom he says he can speak "from my own knowledge" and from reporters "on whose truth and judgment I can rely." He will say nothing about the Indians of South America, because he neither knows them at firsthand nor has reliable information concerning them. He at once characterizes Buffon's account as an "afflicting picture," but one which, "for the honor of human nature, I am glad to believe has no original." The Indian, Jefferson says, is not "more defective in ardor, nor more impotent with his female, than the white reduced to the same diet and exercise." The Indian is brave. He "is affectionate to his children, careful of them, and indulgent in the extreme." His "friendships are strong and faithful to the uttermost extremity." Jefferson admits that Indian "women are submitted to unjust drudgery," but this is true for "every barbarous people," which leads him to observe, "It is civilization alone which replaces women in the enjoyment of their natural equality." Indian women do have fewer children than whites, but the reason is to be found in the social conditions of their life. Indian women married to white traders, "who feed them and their children plentifully and regularly, who exempt them from excessive drudgery, who keep them stationary and unexposed to accident, produce and raise as many children as the white women." Jefferson replies to Buffon's charge that "Indians have less hair than the whites, except on the head," by pointing out that "with them it is disgraceful to be hairy on the body." Accordingly they "pluck the hair as fast as it appears."

Jefferson found a deep flaw in Buffon's logic. If "cold and moisture be the agents of nature for diminishing the races of animals," Jefferson asked, how can it be that these agents of nature act in two contrary ways at the same time? Since, as Buffon admits, the humans of the New World are of "about the same size" as the humans of the Old World, it would seem that the agents of nature "suspend their operation" of degeneration with respect to human physiology. And yet, at the same time, these agents (according to Buffon) act strongly on the "moral faculties." Jefferson could not see how any "combination of the elements and other physical causes," which supposedly prevents "the enlargement of animal nature in this new world," has "been arrested and suspended" so as "to permit the human body to acquire its just dimensions." By "what inconceivable process," he wondered, could the action of these causes of diminution and degeneration have "been directed" to act on the human mind but not on the human body?

Jefferson was aware that data were not available on which "to form a just estimate" of the "genius and mental powers" of Native Americans. He was fully convinced that, once all the facts were known, and due allowance was made for the particular economic and social circumstances of the life of the Indians, "we shall probably find that they are formed in mind as well as in body, on the same module with the 'Homo sapiens Europaeus.'" As proof of the high level of Indian accomplishments, Jefferson cited the example of a "speech of Logan, a Mingo chief, to Lord Dunmore, when governor of this state." Jefferson dared to "challenge the whole orations of Demosthenes and

Cicero, and of any more eminent orator, if Europe has furnished [any] more eminent, to produce a single passage" that would be "superior to the speech of Logan."[48]

Jefferson's *Notes on the State of Virginia* has been praised for two centuries as a splendid example of the naturalist's presentation of almost every aspect of a region and especially for its high level of scientific accuracy and the masterful prose in which it is written. Today's critical reader, however, cannot help but be struck by the extreme contrast between Jefferson's treatment of Native Americans and of African-Americans. The extremity of this contrast becomes evident in Jefferson's praise of Indian oratory, his opinion that the Indian chief Logan was to be compared to a Cicero. Even more astonishing is the bald fact that while Jefferson berated Buffon for repeating prejudiced hearsay concerning Native Americans, he himself gave expression to generalities about African-Americans which were equally unfounded and merely repeated the kind of prejudices current among southern plantation owners, an espousal of racialist opinions which he later came to regret. (This topic is explored further in Supplement 8.)

Jefferson concluded his rebuttal by observing that Buffon's "new theory of the tendency of nature to belittle her productions on this side of the Atlantic" has been applied by the Abbé Raynal "to the race of whites, transplanted from Europe." He quoted from the Abbé Raynal's history to the effect that one "must be astonished (he says) that America has not yet produced one good poet, one able mathematician, one man of genius in a single art or a single science."[49] In his response to this charge, Jefferson allowed his patriotic enthusiasm to outrun his scientific judgment. In the realm of science, he could point to Benjamin Franklin, recognized the world over for his contributions to knowledge. As for a "man of genius in a single art," he noted that in "war we have produced a Washington." In future ages, he wrote, Washington will be reckoned "among the most celebrated worthies of the world, when that wretched philosophy shall be forgotten which would have arranged him among the degeneracies of nature." And then Jefferson turned to his third candidate for excellence, David Rittenhouse, who was—Jefferson declared—"second to no astronomer living." In "genius," furthermore, Rittenhouse "must be the first, because he is self-taught."

Rittenhouse was a most remarkable figure. A clock-maker by profession, he taught himself mathematics and science and became a first-rate astronomer. His contemporaries set a high value on his observations of eclipses and of the transit of Venus in 1769.[50] We shall see below that Jefferson turned to Rittenhouse's knowledge of Newtonian dynamics when exploring the possible use of a pendulum for establishing a standard of length. One of Rittenhouse's most celebrated achievements was the design and construction of a complex mechanical device, run by clockwork, his own improvement of what was known as an orrery, a kind of planetarium. Such devices were a common feature of collections of scientific instruments in the eighteenth century. Each of the American colleges sought to obtain such an orrery, which could be used to demonstrate the principal movements of the planets and their satellites. The orreries were used in public lectures on science and in college courses as part of the demonstration of Newtonian principles (see Fig. 12). Harvard was proud of its orreries, two of which were made by the famous London instrument-maker B. Martin (see Fig. 13). It was in reference to the orreries which Rittenhouse designed and manufactured, rather than to his clocks, that Jefferson described him in the *Notes* as "an artist" (that is, an artisan), who "has exhibited as great a proof of mechanical genius as the world has ever produced." In designing and constructing the orrery, or model of the solar system, Jefferson concluded, Rittenhouse "has not indeed made a world; but he has by imitation approached nearer its Maker than any man who has lived from the creation to this day."[51] This is high praise indeed.

Rittenhouse's orrery, the object of Jefferson's praise, was certainly remarkable. Most orreries, such as those made by B. Martin (Fig. 13) and Joseph Pope (Figs. 14 and 15), were designed to exhibit some of the principal features of the solar system in an approximate manner, but Rittenhouse's machine simulated the details of the motions of the heavenly bodies with great accuracy. His expressed goal was to have an orrery in which the motions would "correspond exactly with the celestial motions; and not to differ several degrees from the truth, in a few revolutions, as is common in Orreries."[52] Rittenhouse actually constructed two such orreries, both of which survive, one at Princeton University, the other at the University of Pennsylvania (see Figs. 16–18). In addition to simulating the motions of the solar system, Rittenhouse's device was a kind of computer. By means of "an Easy Motion of the hand," Rittenhouse wrote, the orrery "will in the space of a few Minutes, point out the times of all remarkable phaenomena of the Heavenly Bodies for years to come."[53]

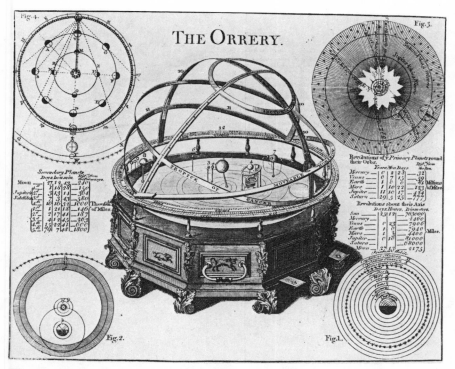

Fig. 12. The Orrery: A Mechanical Model of the Solar System. The orrery was named after the Fourth Earl of Orrery by Richard Steele, of Addison and Steele fame. The first orreries, constructed in London around 1713, showed only the motions of Earth and Moon. By mid-century this popular device, pictured in the Universal Magazine (1749), displayed the orbital motions of the planets and showed their satellites. It was used for teaching principles of astronomy and illustrating some features of the Newtonian system.

The Pennsylvania orrery is contained in a large rectangular case with a glass front, mounted on legs. There are three independent parts, of which the largest was designed to represent "all the primary Planets . . . in their true proportional Distances from each other, and from the Sun."[54] Each planet "moves in its own Plane, different from that of any other" and "the several Angles of their Inclination are truly adjusted and preserved." Furthermore, each planet "varies in its Distance from the Sun in its Revolutions, according to the Quantity of its Excentricity," while "the Velocities likewise vary with the Distances, so as to describe *equal Areas in equal Times,* agreeable to the true Laws of the Newtonian System." The axes of Jupiter, Mars,

Fig. 13. B. Martin Orrery. Purchased by Harvard College in the 1760s, this instru-
ment was used by generations of undergraduates in their study of astronomy. Entirely
made of brass, it is geared so that tiny "planets" attached to concentric rings will com-
plete their revolutions about the central "Sun" in times that are proportional to the
actual periods of revolution of the planets. There are tiny moons for Jupiter and Sa-
turn and a ring for Saturn, but these are not geared. A gearing system gives some
sense of the motion of the Moon about the Earth.

Fig. 14. The Pope Orrery. Constructed by a Boston clock-maker, Joseph Pope, this large and complex machine was purchased for Harvard College by the proceeds of a public lottery. Far more complex than the Martin orrery (Fig. 13), this machine showed the motions of planetary satellites as well as the motions of the primary planets.

Earth, and Venus are properly oriented and turn so as to show the revolutions of the planets; the poles of Earth's axis "revolve about the Poles of the Ecliptic," just as "in the world of Nature." At that time, no rotational motion had been discovered in Mercury and Saturn. The orrery would show accurately the occurrence of transits of Mercury and Venus and the occurrence of conjunction. A calendar consisting of a set of three dials enabled the orrery to be set for any year, month, and day. Turning a crank would set the world system to correspond to any time between 4000 B.C. and A.D. 6000.

Wholly separate mechanisms were devoted to showing respectively the motions of Saturn and Jupiter and the motions of the Moon (see Fig. 17). One device, known as a "lunarium," was designed to repre-

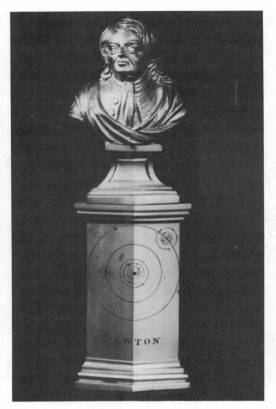

Fig. 15. Portrayal of Newton on the Pope Orrery. A notable feature of the orrery was a set of twelve decorative figures, placed at the corners of the twelve-sided mahogany and glass case. Each of the twelve sides corresponds to one of the twelve signs of the zodiac. There are four castings in bronze of three subjects: Isaac Newton, Benjamin Franklin, and James Bowdoin (governor of Massachusetts and an amateur scientist). These castings were apparently made by Paul Revere. For Newton, there is a bust shown in relief, above a diagram of the solar system.

sent most of the irregularities of "the Moon's motions, according to Sir Isaac Newton's Theory." So accurately was the lunarium made that it was said to be "capable of shewing all *Eclipses of the Moon,* in their precise Quantity," not only on "the Day, but the Hour of the Day, on which they will" occur. The lunarium also showed the occurrence of solar eclipses. In a "Minute or two, without any Calculation," the lunarium could be set "to the Latitude and Longitude of any particular Place upon Earth, and will then truly represent Eclipses of the Sun as they will be seen at that Place."[55] It was claimed that "if the

Fig. 16. Rittenhouse's Orrery at the University of Pennsylvania. The relative size of the planetary orbits are very close to those found in the solar system, but the relative sizes of the Sun and planets and the dimensions of the orbits of planetary satellites could not be those of nature. The inclinations of the planetary orbits are maintained as well as the orientation of the planetary axes of rotation and the planetary rotations (so far as they were known). Planetary configurations are shown for any time between 4000 B.C. and A.D. 6000. The right-hand panel contains the "lunarium" or model of the motion of the Moon.

present Order of Nature subsists," Rittenhouse's orrery "will not vary a Degree from the Truth in less than *Six Thousand Years.*"[56]

Rittenhouse's orrery must elicit high praise for its intricate and accurate mechanism and for its display of knowledge concerning the principal features of the solar system. We must take note, however, that Rittenhouse never made any significant astronomical discoveries. Although he became president of the American Philosophical Society (he was Jefferson's predecessor in that office), Rittenhouse was never elected a Fellow of the Royal Society. Since he did not hold an academic post, as did his contemporary, Professor John Winthrop at Harvard, Rittenhouse did not produce a series of disciples. However much we may agree that Jefferson's admiration of Rittenhouse's

Fig. 17. The "Lunarium" of the Rittenhouse Orrery at the University of Pennsylvania. A separate and independent mechanism activated the model of the Moon's motion in relation to the Earth and the Sun. Rittenhouse alleged that his device took account of each of the principal irregularities in the Moon's motion that had been set forth in the second edition of the Principia *(1713).*

abilities was fully justified, we must also admit that the fulsomeness of his praise was an example of allowing his patriotism to overcome his normal critical faculties.

Part of Jefferson's campaign to demolish the theory of degeneration was to provide real evidence concerning the existence of large animals in the New World. At his own expense he had an expedition mounted in order to obtain the skin and skeleton of a moose from New Hampshire to be forwarded to Buffon. Over time he sent other specimens to prove his point.[57] James Madison contributed evidence to support Jefferson by sending accurate measurements of the organs of a female weasel, together with similar measurements of a

Fig. 18. Gears of the Rittenhouse Orrery at Princeton University. Some of the many toothed wheels that activate the Princeton orrery were displayed during the restoration in 1952. These give some idea of the kind of precision needed in constructing the orrery and are a reminder that Rittenhouse was a skilled clock-maker as well as a gifted student of astronomy.

mole, as numerical proof that American varieties are larger than their European counterparts.[58] Later on, Buffon retracted, but his books continued to express his prejudiced and unfounded theory. He seems to have promised to introduce a correction in the next volume of his large work on natural history, but he died before being able to do so.[59] The Abbé Raynal, in later editions of his work, removed the passage about America having produced no single man of genius or real talent.

In 1787, William Carmichael sent Jefferson a report on a dinner party hosted by Benjamin Franklin. One of the guests asked Franklin what he thought of the theory that animals and humans in the New World were not the equal of those from the Old World. Franklin noted that the American guests were huge and very muscular compared to their French counterparts. "In fact," according to Carmichael's account, "there was not one American present who could not have tost out of the Windows any one or perhaps two of the rest of the Company, if this Effort depended merely on muscular force." Franklin asked his interrogator to look around him at the dinner guests and then "Judge whether the human race had degenerated by being transplanted to another section of the Globe."[60] Franklin's demonstration may possibly have been as effective an argument against "degeneration" as the scientific evidence amassed by Jefferson in his *Notes*.

Franklin's Dinner Party!
→ Funny

Mathematics and Politics: Apportionment of Representatives in the Congress

Jefferson's intellectual world and his daily life were regulated by numbers to a degree that seems astonishing to a reader in the twentieth century. He was skilled in mathematics and delighted in numbers and in calculation. Almost every aspect of his life was reduced to numerical observations and calculations. It is one of the real merits of Garry Wills's book on Jefferson and the Declaration of Independence to have spelled out in detail Jefferson's commitment to numerical rules and quantitative observations and to trace them to their possible sources, among them Sir William Petty, one of the founders of social statistics and the first prophet of a numerically based polity. The earliest preserved writing of Jefferson's centers on quantitative considerations. In a letter to his guardian in 1760, at the age of sixteen, Jefferson requested permission to depart from the Shadwell estate to go to the College of William and Mary, computing exactly how much time this move would save. As "long as I stay at the Mountains," he wrote, "the Loss of one fourth of my Time is inevitable," principally because of "Company's coming here and detaining me from School."[61] Later in life, he wrote of himself in an amusing numerical fashion in a letter to Abigail Adams, noting that he had "ten and one-half grandchildren and two and three-fourths great-grandchildren," adding that "these fractions will ere long become units."[62] He kept detailed numerical records of his farming activities, of meteorological conditions, and of various other aspects of daily life. He even drew on an argument based on numbers to analyze Shays's Rebellion. "The late rebellion in Massachusets," he wrote, "has given more alarm than I think it should have done." He calculated that "one rebellion in 13 states in the course of 11 years" is not very great, being the same as "one for each state in a century and a half." In fact, "No country should be so long without one," and there is no "degree of power" in any government that can "prevent insurrections." Writing from Paris, he took note that in France, despite "all its despotism" and its "two or three hundred thousand men always in arms," he had been witness to "three insurrections in the three years" he had been there, "in every one of which greater numbers were engaged than in Massachusets and a great deal more blood was spilt."[63]

Jefferson's concern for numbers and his arithmetical skill proved to be of real importance for political action during his service as secretary of state under the presidency of George Washington. The numerical problem addressed by Jefferson was an apparently simple one: how to assign to each state a number of representatives in the Congress that would accord with the provisions of the Constitution. The Constitution demanded that apportionment be made on the basis of a decennial census of the population. The Constitution did not go into details concerning the way in which the apportionment was to be made, but it did set forth certain rules or conditions. The first was that the representation was not to exceed one for every thirty thousand in the population (reckoned according to a count of slave population at three-fifths); the second that each state should have at least one representative; the third that the House of Representatives not exceed a certain number. These requirements for apportionment were also to be applied to the distribution of taxation.

Most Americans assume that the composition of the House of Representatives is determined by a just and fair apportionment of representatives according to the populations of the individual states, but not one person in thirty thousand understands how that apportionment is determined in actual practice. It may come as a surprise to learn that the actual method of deciding just how many representatives are to be assigned to each state is far from simple and poses a mathematical as well as a political problem, one that has been the cause of intense argument, debate, and analysis during most of our country's history. The method of determining the apportionment of representatives eventually has enlisted the creative efforts of some of the country's most able mathematicians. Apparently simple and obvious solutions have proved to involve major political issues that have raised great passions of debate in the Congress.

The problems of dividing taxes did not present the same issues that beset apportionment. For example, if a given state has 1/13th of the total national population, its share of the total tax burden can be easily and fairly reckoned at 1/13th of the whole. In the case of apportionment, however, there is no such even division because of the problem of fractions. Let us suppose that there are to be 120 seats in the House of Representatives, and that one of the states has 1/13th of the national population. Simple arithmetic yields the result that this state should be entitled to 1/13th of the total of 120 representatives. Dividing 120 by 13 yields $9\frac{1}{13}$ or 9.231, a little more than 9 representatives. Making similar calculations for each state similarly will yield an integral number of representatives plus a fractional or decimal part.

At once there arises a serious problem since, obviously, there cannot be "fractional" representatives. A number of possible solutions at once suggest themselves. One is to ignore fractions altogether. This would be unfair to a state with, say, 1.987, which would lose almost "half" of what could be considered its just share. The loss of the fractional or decimal part would be less significant for a state with 12.124 since its loss would be about 1/10th rather than half. Another possible solution would be to give an extra seat to the state with the highest fraction, then to give an extra seat to the state with the next highest fraction, continuing until the total number of representatives adds up to 120. An alternative would be to use a similar procedure, but starting with the most populous state, rather than the state with the highest fraction.

There are many other possibilities, among them to pick some other common divisor than 120 and discard all fractions. By trial and error a divisor can then be found that will yield a total number of 120 representatives. Objections can be raised to each of these methods, perhaps for favoring either the small or the large states or the South versus the North, or for some other reason.

When the inaugural Congress first attempted to devise a system of apportionment in 1792 in an "act for the apportionment of Representatives . . . according to the first enumeration," the proposed plan gave rise to such serious objections by Thomas Jefferson that it was vetoed by George Washington. We may gather how important an event this was by the simple fact that this was the first presidential veto in the history of our country, one of only two times when Washington exercised the veto power.

This incident provides a striking example of the way in which blueprints for government may introduce technical—that is, scientific and mathematical—problems outside the range of knowledge and forethought of founders. It also displays the mathematical acumen of Thomas Jefferson, whose analysis of the proposed method of apportionment provided the grounds for Washington's veto and determined the basis for the system that was finally adopted.

In order to understand the nature of the problem and Jefferson's solution, let us look at a relatively simple hypothetical situation. Suppose that the Union is composed of only five states—Massachusetts, Connecticut, New York, Virginia, and Delaware—and that representatives are to be apportioned according to one of the simplest methods to understand.[64] In fact, this is the method adopted by the first Congress and vetoed by Washington, the method endorsed by Alexander Hamilton. In the hypothetical example to which the method will be applied, let the total population of the country be 26,000 and let there be 26 seats in the Congress. Then, as a first step, the total population of the country (26,000) is divided by the total number of seats (26) to give the common divisor, which in this case turns out to be 1,000. Suppose the population of Massachusetts is 5,259; then the number of seats assigned to Massachusetts by simple division is 5,259 ÷ 1,000 or 5.259. This number is sometimes known as "the ideal number of representatives."[65] Now, suppose the population of Connecticut to be 3,319; division by 1,000 gives 3.319 as the ideal number of representatives. In this first round, we discard all the fractional (or decimal) parts, keeping only the integral part of each number. Massachusetts would then be assigned 5 seats, while Connecticut would have 3. The results for our hypothetical country are shown in the following table: It is to be observed that the total number of seats assigned by this method is not the full 26, but only 25. On what basis shall the extra seat be assigned to bring this total to 26?

A first decision might be to assign the 26th seat to the state with the largest population, in this case, Virginia. Another would be to assign the 26th seat to the state with the largest decimal fraction in the list of the ideal number of representatives. In this case, the extra

State	Population	Ideal Number of Representatives	First-Round Assignment
Virginia	9,061	9.061	9
New York	7,179	7.179	7
Massachusetts	5,259	5.259	5
Connecticut	3,319	3.319	3
Delaware	1,182	1.182	1
Total	26,000	26	25

seat would go to Connecticut, because 319 is larger than any of the other decimal fractions. The political implications of the choice of method of apportionment can now be seen. The first decision would favor a southern state by giving Virginia the extra seat, whereas the second decision would favor a northern state by giving the extra seat to Connecticut. A third possibility would be to limit the House to only 25 members, but this decision would rob some state (in this case Virginia or Connecticut) of the extra vote it might have gained. Each of these decisions would affect the division of power within the House of Representatives in a different way. It is obvious that the actual choice of rules for assigning the number of representatives to the states has major political implications.

In the method adopted by the Congress in 1792, following the first census of the population, made in 1791, the final decision was to assign the extra seat to the state with the largest decimal fraction. In our hypothetical example, this would mean giving Connecticut and not Virginia one more seat. The results are shown below:

State	Population	Ideal Number of Representatives	First-Round Assignment	Second-Round Assignment
Virginia	9,061	9.061	9	9
New York	7,179	7.179	7	7
Massachusetts	5,259	5.259	5	5
Connecticut	3,319	3.319	3	4
Delaware	1,182	1.182	1	1
Total	26,000	26	25	26

This mode of assigning seats in the House of Representatives is sometimes known as the "method of quota." Endorsed by Alexander Hamilton, it has recently been renamed the "method of Hamilton."[66] In case the total number of seats assigned in the first round would fall short by two or more rather than one, an extra seat would be assigned to the two states with the highest fractions, and so on to three, or however many would be needed in order to bring the total to the predetermined number. In our example, if the total were short by two, then extra seats would be assigned to the two states with the highest fractions, Connecticut and Massachusetts. Plainly, in this case, the northern states would be greatly favored at the expense of the southern states.

Hamilton's method, initially approved by the Congress, may seem to be so simple and logical that the reader may wonder why Jefferson objected to it so strongly that he was able to persuade Washington to veto the bill proposing it. Jefferson's primary objection to this bill was on grounds of basic principles of good government: the bill contained no statement whatever of how the proposed assignment of seats had been proposed. There was no disclosure of the actual method that had been used. Hence, Jefferson quite correctly argued, when the next apportionment was made following the census of 1800, there was no firm guide as to how to apportion representatives.[67] Since the method had not been specified, the way was open to change the procedure at will the next time around. As Jefferson wrote, the bill "seems to have avoided establishing [the procedure] . . . into a rule, lest it might not suit on another occasion." Perhaps, he observed, "it may be found the next time more convenient to distribute them [the residuary representatives] *among the smaller states;* at another time *among the larger states;* at other times according to any other crotchet which ingenuity may invent, and the [political] combinations of the day give strength to carry."[68] Jefferson wanted the bill to contain an explicit method which "reduces the apportionment always to an arithmetical operation, about which no two men can ever possibly differ."

The failure of the bill to specify the actual rules used in getting the results must have seemed to Jefferson to be contrary to all principles and procedures of good science, where sound method is of supreme importance. As one who was well grounded in the physical and the biological sciences, Jefferson always stressed method in all subjects involving any aspect of science or its applications. The requirement of having a sound and unambiguous method was a fundamental characteristic of all good Enlightenment science, expressed simply and beautifully by Linnaeus: "Method [is] the soul of science."

Since the apportionment bill did not specify the method that had been used, Jefferson had to apply his skill in computation and his experience in solving numerical problems in order to figure out just how the numbers contained in the bill had been obtained. By carefully analyzing the apportionments assigned in the bill, he was able to decode the mathematical method, working backward from the results to disclose the procedural rules that had been used by the Congress, but never stated explicitly. His analysis was verified soon afterward when the method of apportionment used by the Congress was stated explicitly by Hamilton in a memorandum to Washington in support of the bill.

It should be noted that many southerners, including Jefferson and Madison, believed the apportionment in the bill favored the northern states, since it assigned an extra seat to five northern states (Connecticut, Massachusetts, New Hampshire, New Jersey, and Vermont) but only three to southern states (North Carolina, South Carolina, and Virginia), while an extra seat was also given to a border state (Delaware). In his memorandum to Washington, however, Jefferson admitted that the representation suggested in the bill gave a "tolerably just" division between northern and southern states and between "great and small" states. His objections were not to be considered a mere partisan reaction to the particular results, but rather to the way in which they had been obtained.[69]

Jefferson particularly objected to the fiddling required in the second and later rounds in Hamilton's method, when the extra seats are assigned on the basis of the fractional remainders. He proposed an alternative method that eliminates any dependence on fractional remainders and so avoids arbitrary decisions concerning how to allocate the extra seats. His own method requires a first step of choosing a size for the House. By trial and error, using simple arithmetical procedures, the largest possible divisor is found that will have the following property: when this divisor is divided into the population of each state, the results (all fractional or decimal parts being discarded) will add up to the chosen size of the House. In our previous hypothetical example, the divisor 906.1 will satisfy this condition. Divide this number into 9,061 (the population of Virginia); the answer is 10. Divide it into 7,179 (the population of New York); the answer is 7.923, which is rounded off to 7; and so on. In the following table, these results are listed and compared with the Hamilton allotment:

State	Population	Result of Dividing by Jefferson's Divisor of 906.1 for 26 Seats	Jefferson's Allotment	Hamilton's Allotment
Virginia	9,061	10.000	10	9
New York	7,179	7.923	7	7
Massachusetts	5,259	5.804	5	5
Connecticut	3,319	3.663	3	4
Delaware	1,182	1.304	1	1
Total	26,000		26	26

It should be noted that in this particular example, Jefferson's method gives a southern state, Virginia, one more seat than it would receive according to Hamilton's method, which would give this seat to Connecticut. This example, though hypothetical, plainly demonstrates that the choice of method is heavily laden with political consequences.

Jefferson's method of apportionment of representatives has traditionally been known in American writings on the subject as the "method of greatest divisors," while in Europe it is generally called the "method of d'Hondt," after Victor d'Hondt, a nineteenth-century Belgian lawyer who devised the method long after Jefferson had done so. The general attribution to Jefferson of the discovery of the method dates only from 1975, when this fact was turned up in the course of the historical and mathematical investigations of the subject by M. L. Balinski and H. P. Young.[70]

One reason why Jefferson objected to the apportionment bill, as he explained to Washington, was that the actual assignment of number of representatives contravened the strict injunction of the Constitution "that representatives shall be apportioned among the several states according to their *respective numbers.*" This sentence, Jefferson explained, means that "they shall be apportioned by some common ratio. For *proportion,* and ratio, are equivalent words." According to the "definition of *proportion among numbers,*" it is required that there be "a *ratio common to all,* or in other words a *common divisor.*" A little arithmetic was all that was needed in order to prove that in the proposed bill, as Jefferson said, "representatives are *not* apportioned among the several states, according to their respective numbers." An exercise in arithmetic showed that the bill made use of "*two ratios,* at least": that of 30,026 for Rhode Island, New York, Pennsylvania, Maryland, Virginia, Kentucky, and Georgia, and of 27,770 for Vermont, New Hampshire, Massachusetts, Connecticut, New Jersey, Delaware, North Carolina, and South Carolina. If "*two* ratios may be applied," Jefferson asked, why not fifteen or any other number? In this way "the distribution [has] become arbitrary, instead of being apportioned to numbers."

Finally, Jefferson argued, Hamilton's method contravened a direct prescription of the Constitution, namely, that "the number of representatives shall not exceed one for every 30,000." The bill would give eight states a number that is equivalent to one representative for every 27,770 inhabitants, exceeding the constitutional limit.[71] The position of Hamilton and the proponents of the bill was that the constitutional ratio of one to 30,000 was not intended to be applied to each individual state but rather was to be the ratio of the total number of representatives to the population of the country considered as a whole. Washington favored Jefferson's interpretation.

Jefferson's carefully reasoned critical analysis runs on for many pages. We need not follow him through every detail, it being sufficient to take note that Washington was swayed by Jefferson's mathematics and reasoning. Washington was also aware that the bill had passed in each house of Congress by only a small majority. After calling for the opinion of all the members of the Cabinet, Washington decided to veto the bill. Jefferson's alternative to Hamilton's plan was then adopted by the Congress and was used to determine apportionment for many decades.

After following the controversy between Jefferson and Hamilton, the reader may wonder what the fuss was really about: can a single vote really make a great difference? Was it ever worth the time and energy to be so concerned over what may seem to be a simple matter of arithmetic? The answer to these questions lies in history, psychology, and politics. First of all, history. Because of the American political system, the president has traditionally been elected by the electoral college, rather than by a simple majority of the total vote. Each state has a number of votes in the electoral college determined by the sum of its number of representatives and senators. A difference of one vote could decide which of two candidates would become president. In fact, this did happen once, in the election of 1876, a contest between Rutherford B. Hayes and Samuel J. Tilden. Tilden gained a majority of the national popular votes, but he lost to Hayes in the electoral college by a single vote.

Second, let us turn to psychology. Inhabitants of a state take pride in their power and influence, determined (in the House) by the number of seats. To lose a seat is a matter of hurt pride and one that is not suffered gladly since voters feel a distinct sense of loss. This is especially the case when, as a result of the loss of a seat, there follows a redistricting of the state. This brings us to the realm of politics. A consequence of the loss of a single seat may signal the end of service of a popular representative, whose seat has been abolished. Another result is that a number of voters will have a change in their district and a consequent change in their representative, perhaps from someone who has been attentive to local interests to someone who is a complete stranger. A shift in a single seat may also change an important sectional or regional balance. It is easy to understand, therefore, why the choice of a method of apportionment has been a perennial problem of great political concern.

Jefferson and Newtonian Science

The previous two episodes, the confutation of the Buffon-Raynal theory of degeneration and the design of a method of apportionment for the Congress, show Jefferson's knowledge of science and his skill in mathematics in the service of his country. A third such episode will display his command of the science of Newton's *Principia*. Jefferson was also well trained in higher mathematics, the differential and integral calculus or the "calculus of fluxions," as Newton called it. An appreciation of his familiarity with the principles of Newtonian science and the actual text of Newton's *Principia* will prepare us to detect Newtonian echoes in Jefferson's most famous composition, the Declaration of Independence.

Thomas Jefferson was surely the only president of the United States who ever read Newton's *Principia*. He esteemed Isaac Newton as one of the greatest minds the world had produced. In his gallery of immortals in Monticello, he assigned a high place to a set of three portraits: Isaac Newton, mathematician and natural philosopher; Francis Bacon, jurist and codifier of the methods of science; and John Locke, philosopher of "common sense" and author of the influential *Two Treatises of Government*. He had obtained these portraits from England, having asked the painter John Trumbull to order copies for him.[72] These three portraits were hung in his office while he was secretary of state. Every visitor would be at once aware of Jefferson's admiration for Newton.

We may see another token of Jefferson's admiration for Isaac Newton in his possession of one of the dozen or so death masks of Newton (see Fig. 19). We do not know how or when this death mask was obtained but it was among Jefferson's papers and belongings that passed into the Coolidge family through Jefferson's granddaughter Ellen, who married Joseph Coolidge in 1825. These Jefferson materials eventually ended up in the Boston Athenaeum and the Massachusetts Historical Society. The death mask of Newton was given by the Athenaeum to Babson College to join the great collection of New-

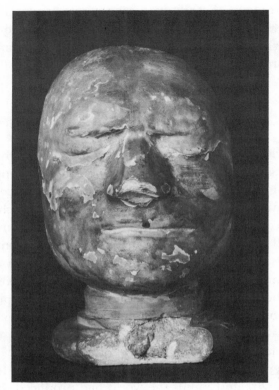

Fig. 19. Death Mask of Isaac Newton from Jefferson's Collection. At Newton's death, some dozen examples were cast of a death mask made by the Flemish sculptor John Michael Rysbrack. Many of Jefferson's manuscripts and memorabilia came into the possession of the Coolidge family when Jefferson's granddaughter Ellen married Joseph Coolidge in 1825. The death mask was given to the Boston Athenaeum by Joseph Coolidge in 1837.

ton's writings, manuscripts, and other memorabilia assembled by Roger Babson and his wife, Grace. Babson was enamored of Newton because he said that it was by applying Newton's third law of motion that he was able to track future economic trends. His most spectacular prediction was the stock market crash of 1929.

Jefferson indicated his esteem for Newton and his delight in studying the writings of that great scientist in a letter written on 21 January 1812, three years after he had left the presidency. In this letter, addressed to his former political rival, John Adams, Jefferson wrote of his great relief in having "taken final leave" of the affairs of state and the issues of politics. He expressed his great happiness in having "given up newspapers in exchange for Tacitus and Thucydides, for Newton and Euclid."[73] Here Jefferson stated clearly his devotion to ancient and modern learning. The ancients were represented by a great Roman historian and a great Greek historian, along with the Greek founder of geometry. The only modern worthy of mention in such exalted company was Isaac Newton. Will there ever be another president of the United States who would read Newton for pleasure?

Mathematics and physics are ideal subjects for retirement, challenges to keep the mental facilities honed. Jefferson was well prepared to read Newton's works because he had been trained in physics, astronomy, and mathematics as a college student. We cannot be absolutely certain of his course of study, but we may be certain that Small would have had him read some of the leading expositions of Newtonian natural philosophy and the textbooks of that science, and would have introduced him to the texts of Newton's two great works, the *Principia* and the *Opticks*. Jefferson mastered the calculus, studying a textbook of Newtonian "fluxions" written by William Emerson, a copy of which he later obtained for his personal library.[74]

All students of science in the days of Jefferson's youth would have studied Newton's *Opticks,* a delightful book to read. Written in an elegant English style, without making any mathematical demand on its readers, the *Opticks* was literally a handbook of the method of experiment, showing not only how to devise and perform experiments, but also how to draw conclusions from them. At the beginning of the *Opticks* as at the conclusion of the *Principia,* Jefferson would have found a warning about indulging in hypotheses. Jefferson would also have learned from the *Opticks* the way in which Newton constructed his scientific system on a set of "Axioms." In this work, Newton elaborated his theory of light and colors, including the analysis of sunlight and the explanation of why we see colored objects in their particular hues. A major application made by Newton of his theory of light and colors was in his explanation of the formation of the rainbow. Later in life, reporting on current debates in France, Jefferson made use of his recollection of the Newtonian explanation of the rainbow.[75] His library contained the corrected fourth edition (1730) of Newton's *Opticks.*

Jefferson, like other students of Newtonian science of the mid-eighteenth century, would have read some portions of Newton's *Principia,* either in the English translation of Andrew Motte, *Mathematical Principles of Natural Philosophy* (first published in 1729), or in the Latin edition, *Philosophiae Naturalis Principia Mathematica,* perhaps with the help of the extensive annotations in the so-called Jesuits' edition, prepared by two Minim Fathers, Thomas Le Seur and François Jacquier. His library contained a Latin edition in two volumes of this particular text and commentary (Geneva, 1760), and a later edition of Motte's English version (London, 1803) in three volumes, with a commentary by William Emerson, author of the textbook on the calculus he had used as a college student. In his report on weights and measures, prepared for the Congress while he was secretary of state, Jefferson—as we shall see—quoted extracts from the *Principia* in Latin.

From the text of the *Principia,* Jefferson would have learned how Newton had organized the subject of rational mechanics, beginning—like Euclid—with a series of definitions, followed by the famous "Axioms, or Laws of Motion," from which the succeeding propositions were to be mathematically deduced. In the final part of the *Principia,* Jefferson would have encountered Newton's "Rules for Natural Philosophy" or "Regulae Philosophandi," which introduce Book Three on the Newtonian "System of the World." In this third and final "book," he would have learned how the solar universe functions through the action of universal gravitation according to the "axioms" or laws of motion and the principles of rational mechanics which Newton had previously set forth.

As an undergraduate, Jefferson would also have studied certain major textbooks of Newtonian science, most likely those by J. T. Desaguliers, W. J. 'sGravesande (translated by J. T. Desaguliers), Pieter van Musschenbroek, and John Keill. Copies of the Desaguliers, Keill, and Musschenbroek texts were in Jefferson's later library. No doubt he would also have studied parts of the textbooks of Newtonian astronomy by David Gregory and Benjamin Martin, both of whose works he later obtained for his library. Most likely he would also have read parts of the two most important general works on Newtonian science, both of which were in his later library, Henry Pemberton's *A View of Sir Isaac Newton's Philosophy* and Colin Maclaurin's *Account of Sir Isaac Newton's Philosophical Discoveries.* Jefferson also owned books on the Newtonian form of the calculus by Colin Maclaurin, John Colson, and William Emerson.

Jefferson was genuinely interested in rational mechanics, as Newton styled the science of his *Principia,* and kept up with the latest works on the subject. For example, in a letter to Joseph Willard, president of Harvard, written from France in 1789, Jefferson called attention to J. L. Lagrange's *Méchanique Analitique,* just published (in 1788), calling it "a very remarkable work," the publication of which was one of the most significant events of recent science. Lagrange, he added, was "allowed to be the greatest mathematician now living," one whose "personal worth" is "equal to his science."[76] Jefferson did not limit himself to generalities, but showed his deep understanding of Lagrange's opus by a summary of "the object of his work." Lagrange's goal, he wrote, was "to reduce all the principles of Mechanics to the single one of the Equilibrium, and to give a simple formula applicable" to all. Jefferson recognized one of the most revolutionary features of Lagrange's approach to mechanics, his treatment of the subject in the algebraic or analytical mode "without diagrams to assist the conception." Very few, if indeed any, of the American statesmen of that era would even have heard of Lagrange, much less have been able to appreciate the radical importance of his treatise.

Jefferson's mastery of Newtonian mathematics is disclosed to us in what will seem to many readers a wholly unexpected occasion, the design of an improved form of plow. Although Jefferson's improve- Plows ment of the design of plows has been discussed by all his biographers, his specific use of Newtonian mathematics in this endeavor has not attracted much attention. One of the features of Jefferson's plow was the shape of the moldboard, the part of the plow that turns over the sod which has been cut as the coulter moves through the sod or earth. Jefferson recognized that for greatest efficiency, the moldboard needs to have a shape that offers the least resistance. At first, he had simply worked on this part of the design by trial and error, but in correspondence with Robert Patterson, professor of mathematics at the University of Pennsylvania in Philadelphia, he learned that this was a problem requiring the calculus for its solution. Patterson told Jefferson that this very problem of a shape of least resistance had been explored in William Emerson's *Doctrine of Fluxions* (originally published in London in 1747). It was then that Jefferson told Patterson that he had actually learned the calculus of fluxions as a student by reading Emerson's textbook.

The problem of finding the shape of a solid of least resistance was originally set forth in Newton's *Principia,* and the methods for solving the problem were explained by John Machin in an appendix to Andrew Motte's translation of the *Principia,* a copy of which was in Jefferson's library. Jefferson made use of Emerson's work and so the problem of designing a plow became in good part a Newtonian exercise in applied mathematics. This episode is of importance in the present context because it demonstrates Jefferson's mastery of higher mathematics, in particular, the mathematics of Isaac Newton. (Further information concerning the application of the Newtonian calculus to the design of the plow is given in Supplement 6.)

Jefferson's familiarity with Newtonian science was also exhibited in his program to establish a standard of length. One of his assignments as secretary of state was to establish standards of weights and measures to be used in the new United States. At that time, 1790, each state determined its own standards, which could introduce confusion whenever there was any commerce between states. Jefferson had two primary goals. He wanted to rationalize the whole system, to have a system of weights and measures of length and area and volume that would be based on decimal divisions, similar in form to the decimal base that had already been accepted for money—dollars divided into dimes and cents. He also wanted to have the units based on nature. Thus, in seeking for a standard of length, his aim was to base the fundamental unit on some natural phenomenon, not on some arbitrary human artifact. In those days the standards of length and other measures were not as yet uniform. Thus, a Paris "foot" was not the same as a London "foot." As a matter of fact, the unit of length was not identically the same in different regions of a single country.

Jefferson finally decided to base the standard of length on a seconds pendulum, that is, a pendulum for which the time of swing from the extreme left position to the extreme right position, or from extreme right to extreme left, is exactly one second of "mean time."[77] This proposal had an obvious appeal to a person of Jefferson's scientific and mechanical bent. The basic principles involved a theoretical understanding of Newtonian physics, while the practical elaboration of the proposal required the practical artisanship of a skilled mechanic.

There is much misunderstanding of the role played by Newton in forming Jefferson's conception of a standard pendulum. For example, it is said that this program comes "straight out of Newton's *Principia*," although a careful reading of the text of Newton's *Principia* does not reveal any such suggestion by Newton. In fact, the proposal of such a standard is pre-Newtonian, going back to Christiaan Huygens's treatise on the pendulum clock, published in 1673, fourteen years before the *Principia*. Newton's investigations of the pendulum in relation to the shape of Earth, reported in Book Three, prop. 20, of the *Principia*, would have tended to discourage rather than support the use of the pendulum in establishing a universal standard of length.

Another common error is to believe that in the *Principia*, Newton gave 39.2 inches as the length of a seconds pendulum for the latitude of London. This fault can be traced back to Jefferson himself, who wrote an early draft of the report at a time when he did not have a copy of the *Principia* available to him.[78] In the discussion of pendulums in the *Principia*,[79] all of the numerical examples are for a seconds pendulum in Paris.

The basic idea of using the length of a seconds pendulum as a standard of length, proposed by Christiaan Huygens, was abandoned, apparently because of certain practical problems. The idea was independently revived by a number of scientists, artisans, and public figures at the time when Jefferson undertook to devise a system of weights and measures.[80] In the end, Jefferson's proposal was not accepted by the Congress.[81] This episode is of interest, therefore, primarily in illustrating Jefferson's knowledge of physics and his command of the science of the *Principia*.

Jefferson's report is devoted to two aspects of the use of the pendulum—one theoretical, the other practical. First of all, there is the matter of definition of the "length" of a pendulum. The simple law of the pendulum holds exactly for an ideal or theoretical or mathematical pendulum, one in which a "point mass" hangs at the end of a weightless and inextensible cord. It is only for such a pendulum that the simple law holds, relating the period to the length. Mathematical physics enters the discussion at this stage because the dimensions of any real (or physical) pendulum must be converted into the length of an equivalent ideal or mathematical pendulum. This is done by determining what is known as a "center of oscillation." The equivalent length is then the distance from the point of suspension to the center of oscillation, which—as Rittenhouse remarked to Jefferson—can readily be found by calculation for any pendulum with a "regular figure."[82] On this score, he noted, "Perhaps no other part of Mathematick affords so many theorems, beautiful for their simplicity, as the doctrine of Oscillations."[83] Jefferson's knowledge of mathematical physics is proved by his ability to deal with the center of oscillation.

Another physical problem of pendulums is to design an apparatus so that the period of the physical pendulum will be kept constant. A mechanism much like that in clocks is needed, providing a series of impulses to keep the width of vibrations constant. Another problem is the effect of temperature. Since the period of a pendulum depends on its length, and since the length of a rod or cylinder increases or decreases with a rise or fall in temperature, the period will be affected by changes in temperature. Accordingly, the temperature must be specified at which the length of a standard seconds pendulum is determined. In the *Principia*, Jefferson found that Newton himself had addressed this problem in presenting the physics of the pendulum in Book Three, prop. 20. In his final report, therefore, Jefferson quoted a paragraph from Newton's *Principia* about the expansion of an iron rod with changes in temperature, giving Newton's text in the Latin original rather than in an English translation.[84]

Two remaining problems are directly related to the physics of Newton's *Principia*. The first is the fact, recorded by Newton, that the length of a seconds pendulum is not the same for different latitudes. A pendulum which has an accurate period of one second in the latitude of Paris will no longer have that same period if it is transported to a place with a different latitude. The second, and related, problem is the possible variation of the length of a seconds pendulum with altitude, that is, with distance from the center of the Earth. Both of these problems—latitude and altitude—introduce fundamental aspects of Newtonian physics.

The change in the period of a pendulum as an observer moves from one latitude to another was discovered in the decades just before the *Principia*. A number of observers—notably Edmond Halley and Jean Richer—encountered this phenomenon while on astronomical expeditions. Richer had gone from Paris to Cayenne in 1672 in order to observe Mars in its perigee. He found, to his dismay, that a pendulum which had accurately marked seconds in Paris did so no longer in Cayenne.[85] He had to shorten its length. Halley had a similar experience when he transported a pendulum from London to St. Helena. Newton's explanation of this phenomenon is one of the glories of the *Principia*. Newton showed that the change in the length of a seconds pendulum arose from a change in the weight of the pendulum bob as it was moved from one latitude to another. He showed theoretically why such a change in the period of a pendulum implies a concomitant change in weight. By the time of the third edition of the *Principia* (1726), the discussion of this topic in Book Three, prop.

20, came to include many additional examples of scientists who had observed changes in the period of a pendulum as a function of latitude.

The fact that weight varies with terrestrial latitude is linked to one of the greatest novelties in the conceptual framework of science, as expounded by Newton in the *Principia,* the introduction of the new concept of mass. Newton abandoned the older use of weight as a measure of the "quantity of matter" in a body. Such experiences as Richer's and Halley's indicated that weight is an "accidental" rather than a fundamental property of any given body. The weight depends on the "accident" of where the body happens to be. Newton's concept of mass is not, in this sense, "accidental." The mass of any body, or any sample of matter, is the same no matter where it happens to be. The mass of a body remains the same even if it should be transported to the Moon or to one of the planets. When a pendulum is taken from one latitude to another the period changes, the weight changes, but the mass remains constant. Similarly, a body transported to the Moon will have a "Moon-weight" much less than its ordinary Earth-weight, even though the mass will be the same. There can even be a condition when a body will be weightless, although it will still have the same mass. These statements are commonplaces today because we have seen astronauts on the Moon who have greatly diminished "weights" and we have seen many telecasts of astronauts performing experiments as well as commonplace tasks in a condition of weightlessness. But in Newton's day, there was no such evidence of experience, only the fact of the change in period in the length of a seconds pendulum. This episode may well serve to illustrate the power of the creative thought that produced the science of the *Principia.*

Newton's concept of mass was introduced in the *Principia* (Definition One) as a measure of matter, or as "quantity of matter." The mass of any body is invariable in Newtonian physics and does not change if the body is heated or chilled, compressed or expanded, twisted, bent, or transported to another latitude or to any other place in the universe. Of course, we now know that the mass of any body is not universally constant; that is, it is not the same under all possible conditions. At speeds approaching the speed of light, there must be a relativistic correction. It is mass which, in Newtonian physics, is responsible for a body's "persevering" in its state of rest or of uniform motion, according to Newton's first law of motion. Mass is also the factor that determines how greatly a body will be accelerated when an external force is applied, according to Newton's second law. Newton found by experiment that the weight of bodies at any given place is proportional to the mass. As a result, a body's mass is the factor which determines its reaction to a given gravitational field, according to Newton's law of universal gravity.[86] It is thus easy to see why this new concept of mass is considered to be one of the most basic and original ideas of the *Principia.*

In Book Three, prop. 20, of the *Principia,* Newton proves that the weight of a body is a function of the body's distance from the center of the Earth. Since the period of a pendulum depends on the weight, Jefferson had to take account of the possibility that there would be a difference in a body's weight at sea level and at a high elevation. If there should be a significant difference in weight at two altitudes, then there would be a concomitant difference in the length of a seconds pendulum. Accordingly, as a consequence of Newtonian principles, he had to make calculations of the possible effects of altitude. His calculations showed that "the highest mountain in the U.S. would not add 1/1000 part to the length of the earth's radius, nor 1/128 of an inch to the length of the pendulum." He even computed that "the highest part of the Andes . . . might add about . . . 1/25 of an inch to the pendulum."[87]

The most important Newtonian aspect of the pendulum proposal was the problem of latitude, the fact that the length of a seconds pendulum varies from one latitude to another. In practice, this means that in defining a unit of length, there must be some designated latitude. For practical purposes, no knowledge of the elaborate Newtonian theory is needed, only the rules of calculation which are results of Newton's investigations. Indeed, John Whitehurst's book (London, 1787) on the use of a pendulum in setting a standard of length does not even mention Newton's name, nor does it refer to the *Principia* or to the principles of Newtonian physics. It is a "nuts-and-bolts" manual, written by a clock-maker.

For Jefferson, however, mere results and rules would never suffice. As a student of Newtonian physics, and especially as one who had actually studied the *Principia,* Jefferson wanted to base his system on more than mere rules, to go beyond mere practice to theory and to have the rules be related to fundamental physics. In addition, he seems to have believed that for the computing rules to be accepted, they must be justified by sound physics. Newton's exposition of the physics of the pendulum is, as a matter of fact, one of the most beautiful presentations in the *Principia,* since it not only embraces the laws of pendulums but extends the subject to include the actual shape of the Earth in a manner that elicited the admiration of the cognoscenti the world over. Voltaire, for example, held that a sign of Newton's great genius was his ability to determine the shape of the Earth without having to undertake great measuring expeditions, as Maupertuis was to do later on, but merely by using his powerful intellect and pencil and paper in his own chambers. Newton's primary evidence that the Earth is not a perfect sphere, but is an oblate spheroid, that is, flattened at the poles and bulging out at the equator, came from an analysis of the gravitational pull of the Moon on the Earth,[88] producing the phenomenon of precession.

In presenting the subject of pendulums in the *Principia,* Book Three, prop. 20, Newton gives an elegant demonstration of how the weight of a body on the surface of a nonspherical Earth will be inversely proportional to the distance from the center.[89] He not only derives a rule for the variation in length of a seconds pendulum, but actually computes the length of a seconds pendulum in various latitudes, that is, for every five degrees from 0° to 90°, and for each degree from 40° to 50°. Since Newton's table gives these lengths in French measures, Jefferson needed to convert them into the English inches. As he pointed out in a long footnote in his final report, this was no easy matter since various scientists had proposed quite different conversion factors.

In his report, Jefferson computed that Newton's value for the length of a seconds pendulum at latitude 45° (36 *pouces*, 8.428 *lignes*) is the equivalent of 39.14912 inches. He also computed the value for an equivalent rod, one which would "vibrate in the same time." Jefferson was aware, of course, that most experimental determinations of the length of a seconds pendulum would not be made at exactly the latitude of 45°. Accordingly, he gave Newton's determination of the length of a seconds pendulum at latitude 30°, 35°, 40°, 41°, 42°, 43°, 44°, and 45°, so that wherever the determination might be made, the results could be corrected "for the latitude of the place, and so brought exactly to the standard of 45°."

In the course of the correspondence with Rittenhouse about the use of the seconds pendulum in determining a standard of length, Rittenhouse sent Jefferson an elegant gloss on the *Principia,* Book Three, prop. 20. Unfortunately, Rittenhouse was in a hurry and made a fundamental error. When Jefferson read Rittenhouse's gloss, he recognized the fault and bracketed the incorrect phrase, indicating that it should be deleted, writing the correct statement in between the lines. Then, keyed to his correction, he wrote out an extract (in Latin) from the *Principia,* showing that he had correctly interpreted the passage.[90] In this event, we may see yet another instance of Jefferson's sure hand with the physics of the *Principia.*

Jefferson's final proposal was so heavily laden with physics that it was difficult, if not impossible, for ordinary readers to understand. He even included the mathematics of determining the center of oscillation of a pendulum with a spherical bob and of a vibrating rod or cylinder. It is no wonder that the report was not accepted, that Jefferson's recommendations were not put into practice. What seemed at first a simple and easy proposal for a standard of length turned out to involve so many kinds of complications that it could hardly be considered a practical measure. Jefferson's proposal is, therefore, of interest primarily as an index of his command of physics and his use of the text of Newton's *Principia.* By basing his system of measures on Newtonian science rather than on the simple technologies of measurement, Jefferson made his *Report on Weights and Measures* a document for the Age of Reason rather than a blueprint for the age of machines.

Jefferson and the Declaration of Independence

The Declaration of Independence is one of the most celebrated documents in American history. It has been subjected to a minute scrutiny, including the careful collation of the various manuscript and printed texts and a word-for-word examination of the several versions. In addition to the usual galaxy of scholarly articles, there are at least six full-length historical studies,[91] and of course the Declaration is examined in every biography of Jefferson. It would hardly seem that anything further remains to be said about this document. And yet, despite the mountain of scholarly exegesis, I know of no attempt to ascertain in Jefferson's words the obvious reverberations of the Newtonian natural philosophy. There are, however, two major studies, by Carl Becker and Garry Wills, that link the Declaration with the scientific background of the age. But neither Becker nor Wills, nor any other commentator with whose work I am familiar, has detected specific echoes of the science of Newton's *Principia* in the actual language used by Jefferson.

When the Second Continental Congress met in Philadelphia in the summer of 1775, Jefferson (aged thirty-two) was almost the youngest member. John Adams later recorded a conversation between himself and Jefferson on why Jefferson was chosen to be the author of the Declaration rather than Adams. One reason was that Jefferson could "write ten times better."[92] In a separate document, Adams recorded that the delegates knew Jefferson's "reputation for literature, science, and a happy talent of composition."[93] Surely the expectation would have been that a document on liberty produced by a man with these three talents would be well written, would show influences of great literature, and would also reflect the scientific worldview of the author.

The Declaration opens with a preamble, the purpose of which was ostensibly to justify to the world the basis of the separation from Britain. As Merrill Peterson says, Jefferson "seized the occasion to advance in axiomatic terms a political ideology for the new nation." He expressed the major premise that "just governments are founded on equal rights and the consent of the governed."[94] Whenever, in Jefferson's words, "any Form of Government becomes destructive of these ends, it is the Right of the People to alter or to abolish it, and to institute new Government. . . ." The basis is Jefferson's declaration that there are "unalienable Rights" which are epitomized in the celebrated phrase, "Life, Liberty and the pursuit of Happiness."[95]

The "strategy" of Jefferson's argument, Peterson tells us, turns on the "asserted 'self-evidence' of the major premise," which was a "moral truth beyond powers of demonstration" and which "carried immediate conviction to men of reason and good will."[96] In a minor premise, Jefferson charged that George III had "repeatedly usurped the rights of the Americans." There followed a series of formal charges, constituting a kind of "bill of indictment" against the king. Jefferson's conclusion was that "these United Colonies are, and of Right ought to be Free and Independent States."[97]

Many years later, Jefferson wrote that his task had not been "to find out new principles, or new arguments, never before thought of," nor "merely to say things which had never been said before." Rather, his purpose was "to place before mankind the common sense of the subject, and to justify ourselves in the independent stand we are compelled to take." He neither aimed "at originality of principle or sentiment" nor copied "from any particular or previous writing." His intention was to produce "an expression of the American mind, and to give to that expression the proper tone and spirit called for by the occasion."[98] After he completed a draft, he showed it to his fellow committee members, primarily Adams and Franklin, who made some minor corrections.

The delegates to the Congress eventually went over Jefferson's text line by line and practically word by word for two and a half days. The introductory portions were not heavily revised and were readily approved, but many alterations were made in the main body of the document. Some of these were stylistic, while others were intended to make the statements more accurate and less polemical.[99] The Congress deleted completely what has been described as "the longest, angriest, and climactic count in Jefferson's indictment"[100]: the king's "cruel war against human nature itself,"[101] that is, the trade in human slaves.[102] Nevertheless, the preamble, the part containing Newtonian overtones, stands in the final document almost as Jefferson wrote it.

The Declaration (in the final version) begins with the following stirring words: "When in the Course of human events, it becomes necessary for one people to dissolve the political bands which have connected them with another, and to assume among the powers of the earth, the separate and equal station. . . ."[103] Then Jefferson defines the "separate and equal station" as one to which the people are entitled by "the Laws of Nature and of Nature's God."[104] We may take note that Jefferson does not invoke the principles of international law, as—for example—expounded by Hugo Grotius. Nor does he appeal to Holy Writ or the God of revelation. Rather his authority consists expressly of "Laws" attributed both to "Nature" and to "Nature's God."

This attribution was the occasion for Carl Becker's exploration of the eighteenth-century concept of nature as part of the background of the Declaration. As a sensitive scholar, well versed in the thought of the eighteenth century, Becker at once recognized that the concept of nature in Jefferson's day was strongly influenced by the scientific philosophy of Isaac Newton. Becker, however, had no real understanding of the science of Isaac Newton, nor did he appreciate the degree to which Jefferson was actually steeped in Newtonian science. Unable to pursue in depth the implications of his own brilliant insight, Becker did not take cognizance of the concepts and principles of Newtonian science and was limited largely to citing titles of books popularizing this science, without too much serious discussion of their contents.[105] Becker trivialized the science of the *Principia* to such a degree that it is difficult to believe that this characterization could have been written by one of the major historians of the twentieth century.[106]

Garry Wills's study of the Declaration, like other works by this historian and critic, displays enormous scholarly learning and is replete with important insights of all sorts. For example, no one has previously so fully understood the background and significance of Jefferson's obsession with reducing every possible aspect of life to numbers. We have seen an instance of such concern for numbers in the design of a system for apportioning representatives. One of Wills's insights is the recognition that there is a sense in which the Declaration may be read as "a scientific" document, exhibiting reflections of the science of Isaac Newton. He declared that the "Declaration's opening is Newtonian." But he did not relate the Declaration to any specific Newtonian concepts or principles. Rather, he supposed that "the word with full Newtonian *pondus* in Jefferson's opening phrase is 'necessary.'" Wills did not regard Jefferson's reference to "the Laws of Nature" as having a Newtonian context, and he relegated his discussion of it to a section of his book called "A Moral Paper."

The usual interpretation of the Declaration is that Jefferson's appeal to the "Laws of Nature" is to be read in the context of the then-current concept of "natural law." In Jefferson's day, and for centuries earlier, "natural law" meant a supreme moral law known to humankind through the exercise of reason. Natural law is not easy to define, in part because the two words that are here combined, "natural" and "law," are not themselves subject to "univocal objective definition" or even "commonly accepted usage." The concept of natural law is also difficult to define in simple terms because it has reflected a changing intellectual environment from Cicero to Aquinas to John Locke. A recent study concludes that there are at least 120 different meanings that have been given to this term.[107] Basically, however, "natural law" has always been a "higher" and more fundamental law than the laws of science or the laws of humankind or the laws of the state.

Closely associated with natural law is the concept of natural right. In a work which presaged, if it did not directly influence, the Declaration of Independence, Samuel Adams wrote in 1772 of "the right to life, liberty, and property" as "a natural right, a branch of the first law of nature."[108] During the winter of 1758–1759, while John Adams was an undergraduate at Harvard College, he recorded some thoughts on "natural law" and on "civil law." Referring to natural law in Latin as "jus naturale,"[109] Adams declared that "the Law of Nature" is "taught by Nature to all living things." The "Law of Nature," he wrote, "includes the Laws of Reason," just as much as "Self Love and Desire of Propagation and education" include "those Rules of Temperance, Prudence, Justice, and Fortitude which Reason, by the help of Experience, discovers to be productive of the Happiness and Perfection of human [Nature]."[110]

In the eighteenth century, natural law was closely associated with international law. Jefferson's library, for example, had a single category for "Law of Nature and Nations," which included the treatise by Hugo Grotius, often considered the founder of modern international law. A primary text on natural law at the time of the Declaration was the second of John Locke's *Two Treatises of Government* (1689/1690 and many later editions), a work which was present in Jefferson's library (but cataloged under "Politics"). Versions of Locke's ideas were to be found in the influential treatise on natural law by the Swiss jurist Jean Jaques Burlamaqui. Two copies of the latter's treatise on this subject were in Jefferson's library and cataloged under "Law of Nature and Nations."[111] Trained in the law, Jefferson was well acquainted with the literature on natural law and would certainly have drawn on this corpus of legal thinking in considering the problems of natural rights while composing the Declaration.

A recent study of the Declaration by Morton White devotes a long chapter to natural law. White finds that Locke's direct influence was less than is often supposed and that his ideas were disseminated in somewhat altered form by others, notably Burlamaqui. Although White discusses the concept of scientific law, he does not discuss the Declaration as a Newtonian document, nor does he explore any aspects of Newton's science in relation to the Declaration.[112]

The importance of the concept of "natural law" in the general political philosophy of the Declaration has been well documented.[113] My own reading of Jefferson's reference to "the Laws of Nature," however, differs from the commonly accepted interpretation, although it does not deny or supplant it. Rather, I propose a supplementary point of view that enhances the meaning of Jefferson's words. There can be no question that Jefferson, a lawyer and student of political philosophy, was influenced by the current ideas of natural law and the allied concept of natural right. But I shall argue that there is another possible gloss on Jefferson's phrases, one that indicates that the interpretation should not be strictly limited to considerations derived exclusively from natural law.

We shall see below that the evidence is overwhelming that the words "laws of nature" would have been familiar to Thomas Jefferson and to others of that day as a resonant echo of the science of Isaac Newton. It should be noted that in its Newtonian context, the plural form "laws" is used rather than the singular "law." Jefferson's personal library contained many books whose subject, as expressed in the title, was natural law; they were cataloged—as has been mentioned—in a separate division which he called "Law of Nature and Nations." It is notable that all but one of the books on this subject in Jefferson's library bore a title referring to the subject in the singular as "natural law" or "the law of nature"; only one—by Richard Cumberland—declared its subject to be "laws of nature."

The full title of Cumberland's book indicates the author's dual purpose: to attack the ideas of Thomas Hobbes and to explore the subject of "the laws of nature."[114] The first part of the book uses the phrase "laws of nature" in the general sense of scientific laws as found in Descartes and Hobbes, whose writings are cited and discussed. This leads, toward the end of the first chapter, to an easy transition from such laws to the supreme moral law, *the* law of nature, from which all others derive their validity. Cumberland adopts the common usage: "laws" for the laws of nature which are scientific laws and "law" for natural law or *the* law of nature.

We have seen that Jefferson referred to "laws" in the plural when writing about the laws of science as the "laws of nature" in the *Notes on the State of Virginia*. This usage may be contrasted with his use of the singular "law" in arguing a case as a young lawyer in 1770. In his statement to the court, Jefferson said that "under the law of nature, all men are born free." In addition, "every one comes into the world with a right to his own person, which includes the liberty of moving and using it at his own will."[115] Here Jefferson was asserting the kinds of rights that he would later set forth in the Declaration. In this instance, however, Jefferson was unambiguously referring to natural law, to *the* "law" of nature and not to "nature's laws."

Newtonian Echoes in the Declaration of Independence: "Laws of Nature"

The canonical interpretation of the opening sentence of the Declaration not only sees Jefferson writing in the tradition of natural law or the law of nature, but declares specifically that a major inspiration was John Locke. Here the primary source would be Locke's second *Treatise of Government.* Of course, this work need not have been a direct source of Jefferson's ideas, since Locke's principles (or principles very similar to Locke's) were available in the writings of other authors. A challenger to this standard viewpoint, Garry Wills, views Jefferson as more strongly influenced, especially in his political and philosophical thinking, by representatives of the eighteenth-century Scottish "Enlightenment," notably Adam Ferguson, Francis Hutcheson, and Thomas Reid. Although Wills makes out a fair case for a Scottish input, most students of American history have not fully accepted this conclusion.[116]

In the face of a massive scholarly literature on the Declaration and natural law, one would have to be hardy indeed to offer yet another interpretation of Jefferson's reference to the laws of nature. Yet anyone trained in the history of science, particularly anyone who has read deeply in the Newtonian literature of Jefferson's day, will have a different and complementary reading of Jefferson's words. We have no way of determining exactly what was in Jefferson's conscious or unconscious mind when he wrote the Declaration. We can, however, easily detect some of the scientific overtones of what he wrote, some unambiguous resonances of his own phrases and those he had been encountering in works we know him to have read. When Jefferson read his own words, there must have been a direct train of association arising from his reference to the laws of nature, a sequence leading by direct steps to the realms of science and to scientific laws. Indeed, the evidence is overwhelming that in books with which Jefferson was familiar, books that he had read or studied, and books that were in his personal library, the expression "laws of nature" appeared prominently in the sense in which we today would refer to laws of science or scientific laws. Also, as we shall see in a moment, the laws of nature tended to be a very specific set of laws, namely, Newton's three laws of motion, found in the opening sections of the *Principia.* Such a train of association would be unavoidable for anyone who, like Jefferson, was schooled in Newtonian science and in the science of that day.

Year after year, in my seminar on eighteenth-century science, I would read the opening sentence of the Declaration and ask the students what the words "laws of nature" suggested to them. My students were all deeply immersed in the scientific literature of Jefferson's day; they had been studying the very books that had been used by Jefferson and that were present in his library. Without a moment's hesitation, they would reply that Jefferson had in mind what we would call scientific laws. Some students would add that the laws in question were Newton's laws of motion, the axioms of the *Principia*.

Jefferson himself, in his *Notes on the State of Virginia*, used this same phrase, "laws of nature," in the course of an argument about fossils. He could not believe that there had ever been a great flood, with waters so high that shells could have been deposited on mountaintops. In order to deny any such possibility, he put forward evidence that a flood which would reach as high as the North Mountain or Kentucky would "seem out of the laws of nature."[117] By this phrase he meant that a flood of such proportions would be impossible according to the laws of science, according to the established "laws of nature." In short, Jefferson himself interpreted the words "laws of nature" as the principles of science in much the same sense in which we today would talk of scientific laws.

Newton himself did not write of "leges naturae" or "leges naturales" in the *Principia*, but he did refer to "laws of nature" in the *Opticks*. Toward the end of this work, near the conclusion of the final Query 31, Newton wrote of "Principles" which he considered "as general Laws of Nature, by which . . . Things themselves are form'd; their Truth appearing to us by Phaenomena." In all the copies of the *Principia* in Jefferson's library, both in the original Latin and in Motte's English translation, there were references to the "laws of nature" in this sense in the preface which Roger Cotes wrote for the second edition and which was reprinted in all later editions. The antepenultimate paragraph began with a discussion of the way in which "the laws of Nature" flow from what, at the end of the previous paragraph, is called "the perfectly free will of God directing and presiding over all." This is almost immediately followed by a reference to "the true principles of physics and the laws of natural things." These are, as

Cotes says earlier in the preface, the "laws on which the Great Creator actually chose to found this most beautiful Frame of the World."

Newton used the phrase "laws of nature" very dramatically, in a sense similar to our "scientific laws," in a notorious book review that appeared in the *Philosophical Transactions* of the Royal Society in 1715.[118] Published anonymously, this review purported to be an account of the report of the Royal Society on the invention of the calculus. The review was actually written by Newton, who felt the need to stress and to document further the conclusions of the report, asserting his own priority of invention and endorsing the verdict that Leibniz must have been a plagiarizer.[119] What interests us here, however, is not the chicanery of men of science in the seventeenth century but rather Newton's statement in the review concerning "the constant and universal Laws of Nature."[120] For Newton, here as in the *Opticks*, "laws of nature" meant the laws of natural philosophy, the laws of science. This is the sense in which those who were scientifically literate in the eighteenth century read this phrase, esteeming Newton's three laws of motion as the supreme example.

The phrase "laws of nature," especially as it occurs in the plural, had acquired a very special significance in the scientific literature by the time that Jefferson wrote the Declaration. These words—"leges naturae" (in Latin)—had been a bold feature of one of the most revolutionary and fundamental works of the Scientific Revolution, the *Principia* of René Descartes, a copy of which was in Jefferson's library. Descartes used this phrase to denote the basic rules (or "regulae") of natural philosophy, which he conceived to be "regulae quaedam sive leges naturae," that is, "certain rules or laws of nature." As set forth by Descartes, these were in fact laws of motion. We shall see in a moment that when Newton's *Principia* displaced Descartes's *Principia*, Newton's "leges motus," or "laws of motion," came in turn to be considered the "laws of nature."

The reader may wonder how Descartes or anyone else came to believe that laws of motion are of so fundamental a character that they can constitute "the laws of nature." In what sense did the next generation of the scientifically literate conceive the Newtonian three laws of motion to be the fundamental laws of nature? The answer is given by history. Throughout the development of modern thought, motion was considered to be the primary phenomenon of nature. Hence its laws would be fundamental in any system of understanding natural phenomena. John Keill, the professor of natural philosophy at Oxford in the early eighteenth century, addressed this problem in his textbook of Newtonian science by giving a quotation from Aristotle in Greek. In translation this reads, "To be ignorant of motion is to be ignorant of nature."[121] Expressed differently, this would mean that a knowledge of motion and its laws is the key to a knowledge of nature.

When Isaac Newton composed his own *Principia,* he transformed both the title and the foundations of Descartes's previous work. Thus he added two qualifying words to the title of Descartes's treatise, which was called *Principia Philosophiae,* or *Principles of Philosophy.* Newton's additions made the subject "Natural Philosophy" rather than just "Philosophy" and specified "Mathematical Principles" rather than "Principles" in general. Hence the full title of Newton's great work, *Philosophiae Naturalis Principia Mathematica,* or *Mathematical Principles of Natural Philosophy,* is a simple enlargement of the title of the book of Descartes which it replaced (see Fig. 20). Newton similarly transformed Descartes's "regulae," or "leges naturae," that is, "rules," or "laws of nature," into "axiomata," or "axioms," which were "leges motus," or "laws of motion." The complete phrasing makes the transformation obvious, from "regulae quaedam, sive leges naturae," to "axiomata, sive leges motus." And if the evidence of the choice of title were not clear enough, Newton's first law is almost the same as Descartes's first law, in concept and even in the actual words used to express it (see Fig. 21).

So great was the impact of this originally Cartesian phrase, "laws of nature," that it became universally adopted for the fundamental principles of science.[122] It was employed in this sense by Thomas Hobbes, who knew and used the writings of Descartes. In Richard Cumberland's book, which we have seen was the only work on natural law in Jefferson's library to have "laws" of nature in its title, the words "laws of nature" mean scientific laws and principles in the sense in which "laws of nature" was used by Descartes and Hobbes.

A very large number of books on science in the eighteenth century, and notably books on the Newtonian natural philosophy, made reference to Newton's "axioms, or laws of motion" as the "laws of nature." For example, John Harris's *Lexicon Technicum* (London, 1704), of which a copy was in Jefferson's library, states boldly and simply that the "Three Laws of Motion" given by the "Incomparable Mr. *Isaac*

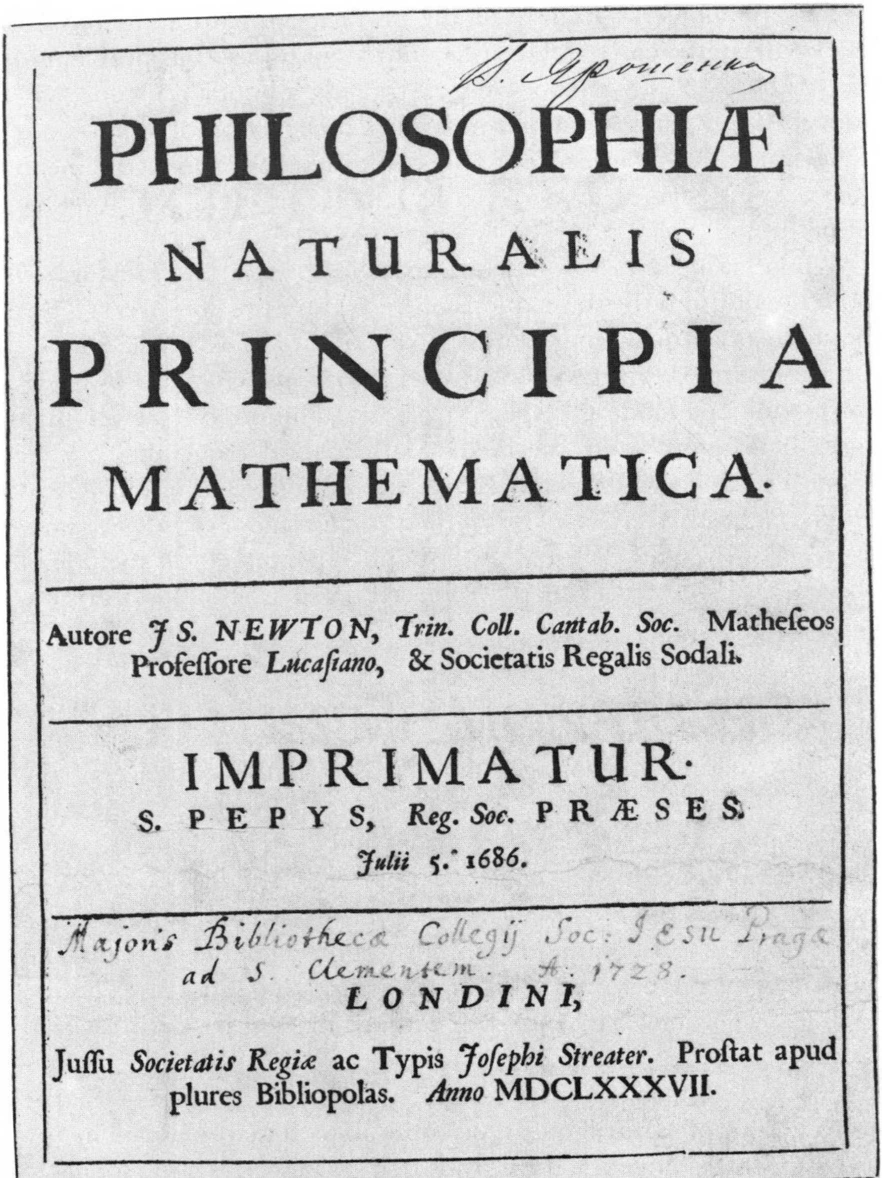

H. Громенко

PHILOSOPHIÆ
NATURALIS
PRINCIPIA
MATHEMATICA.

Autore *JS. NEWTON*, *Trin. Coll. Cantab. Soc.* Matheseos
Professore *Lucasiano*, & Societatis Regalis Sodali.

IMPRIMATUR·
S. PEPYS, *Reg. Soc.* PRÆSES.
Julii 5. 1686.

*Majoris Bibliotheca Collegij Soc: Jesu Praga
ad S. Clementem. A. 1728.*
LONDINI,

Jussu *Societatis Regiæ* ac Typis *Josephi Streater*. Prostat apud
plures Bibliopolas. *Anno* MDCLXXXVII.

Fig. 20. Newton's Principia. *The title page of the first edition of Newton's* Principia *(1687), indicating the prominence given to the words "PHILOSOPHIAE"
and "PRINCIPIA." In the final approved edition (1727), these two words were
printed in large letters in red ink.*

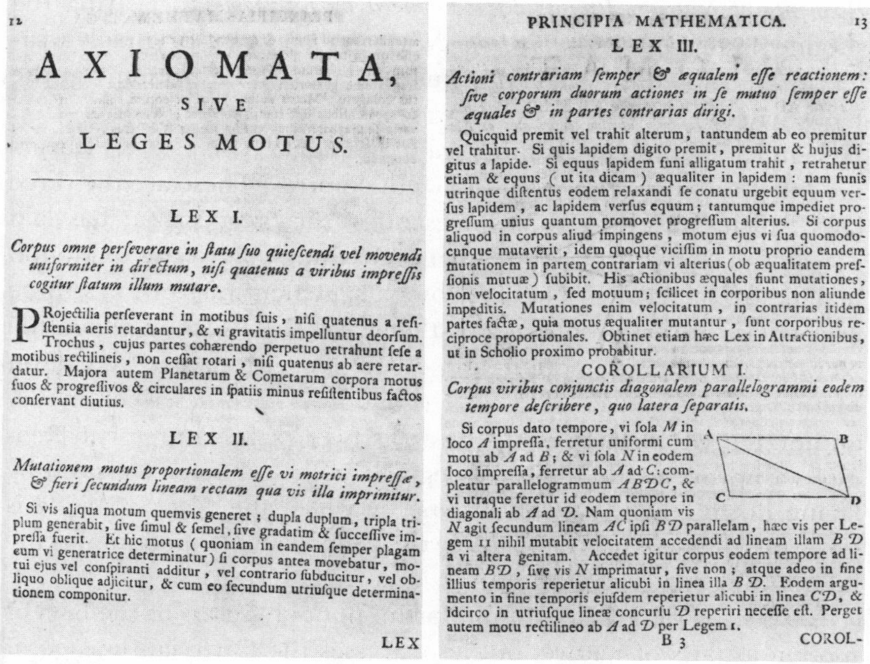

12

AXIOMATA,

SIVE

LEGES MOTUS.

LEX I.

Corpus omne perſeverare in ſtatu ſuo quieſcendi vel movendi uniformiter in directum, niſi quatenus a viribus impreſſis cogitur ſtatum illum mutare.

PRojectilia perſeverant in motibus ſuis, niſi quatenus a reſiſtentia aeris retardantur, & vi gravitatis impelluntur deorſum. Trochus, cujus partes cohærendo perpetuo retrahunt ſeſe a motibus rectilineis, non ceſſat rotari, niſi quatenus ab aere retardatur. Majora autem Planetarum & Cometarum corpora motus ſuos & progreſſivos & circulares in ſpatiis minus reſiſtentibus factos conſervant diutius.

LEX II.

Mutationem motus proportionalem eſſe vi motrici impreſſæ, & fieri ſecundum lineam rectam qua vis illa imprimitur.

Si vis aliqua motum quemvis generet; dupla duplum, tripla triplum generabit, ſive ſimul & ſemel, ſive gradatim & ſucceſſive impreſſa fuerit. Et hic motus (quoniam in eandem ſemper plagam cum vi generatrice determinatur) ſi corpus antea movebatur, motui ejus vel conſpiranti additur, vel contrario ſubducitur, vel obliquo oblique adjicitur, & cum eo ſecundum utriuſque determinationem componitur.

LEX

PRINCIPIA MATHEMATICA. 13

LEX III.

Actioni contrariam ſemper & æqualem eſſe reactionem: ſive corporum duorum actiones in ſe mutuo ſemper eſſe æquales & in partes contrarias dirigi.

Quicquid premit vel trahit alterum, tantundem ab eo premitur vel trahitur. Si quis lapidem digito premit, premitur & hujus digitus a lapide. Si equus lapidem funi alligatum trahit, retrahetur etiam & equus (ut ita dicam) æqualiter in lapidem: nam funis utrinque diſtentus eodem relaxandi ſe conatu urgebit equum verſus lapidem, ac lapidem verſus equum; tantumque impediet progreſſum unius quantum promovet progreſſum alterius. Si corpus aliquod in corpus aliud impingens, motum ejus vi ſua quomodocunque mutaverit, idem quoque viciſſim in motu proprio eandem mutationem in partem contrariam vi alterius (ob æqualitatem preſſionis mutuæ) ſubibit. His actionibus æquales fiunt mutationes, non velocitatum, ſed motuum; ſcilicet in corporibus non aliunde impeditis. Mutationes enim velocitatum, in contrarias itidem partes factæ, quia motus æqualiter mutantur, ſunt corporibus reciproce proportionales. Obtinet etiam hæc Lex in Attractionibus, ut in Scholio proximo probabitur.

COROLLARIUM I.

Corpus viribus conjunctis diagonalem parallelogrammi eodem tempore deſcribere, quo latera ſeparatis.

Si corpus dato tempore, vi ſola *M* in loco *A* impreſſa, ferretur uniformi cum motu ab *A* ad *B*; & vi ſola *N* in eodem loco impreſſa, ferretur ab *A* ad *C*: compleatur parallelogrammum *ABDC*, & vi utraque feretur id eodem tempore in diagonali ab *A* ad *D*. Nam quoniam vis *N* agit ſecundum lineam *AC* ipſi *BD* parallelam, hæc vis per Legem II nihil mutabit velocitatem accedendi ad lineam illam *BD* a vi altera genitam. Accedet igitur corpus eodem tempore ad lineam *BD*, ſive vis *N* imprimatur, ſive non; atque adeo in fine illius temporis reperietur alicubi in linea illa *BD*. Eodem argumento in fine temporis ejuſdem reperietur alicubi in linea *CD*, & idcirco in utriuſque lineæ concurſu *D* reperiri neceſſe eſt. Perget autem motu rectilineo ab *A* ad *D* per Legem I.

B 3 COROL-

Fig. 21. "Axioms or Laws of Motion." Newton's "leges motus" or laws of motion were presented in the Principia *as "axiomata" or Axioms. Note the bold printing of "AXIOMATA."*

Newton" may be "truly called *Laws of Nature.*" John Keill's *Introduction to Natural Philosophy,* one of the most popular and often reprinted textbooks on this subject in the first half of the eighteenth century, also referred to Newton's laws of motion as "laws of nature." Jefferson owned both the Latin and the English texts of this work, and there is good reason to believe that it was one of the books he studied as an undergraduate at William and Mary. Lecture XI of Keill's *Introduction* is entitled simply, "Of the Laws of Nature." The "Laws of Nature," according to Keill, are "such Laws, as it is necessary that all natural Bodies do obey." Keill says that he will "deliver these in the same order, and in the very same Words, as they are laid down by the Illustrious Sir *Isaac Newton.*" He then proceeds to quote and to discuss and illustrate each of the three Newtonian laws of motion found at the beginning of the *Principia.*

In Jefferson's day, one of the most widely used and authoritative works on natural philosophy was W. J. 'sGravesande's *Mathematical*

Elements of Natural Philosophy, confirmed by Experiments, bearing the subtitle *An Introduction to Sir Isaac Newton's Philosophy.* Originally written in Latin by the professor of mathematics and astronomy at the University of Leyden, this work was translated into English by J. T. Desaguliers and was reprinted again and again. Chapter XVI bears the title, "Of Sir *Isaac Newton's* Laws of Nature." These are the three laws which, according to 'sGravesande, "Sir *Isaac Newton* has laid down" and "by which we think that every thing that relates to Motion may be explained."[123] There follows a statement of Newton's three laws and a discussion of each of them. Another work in which the Newtonian laws of motion are called laws of nature is Cotton Mather's *The Christian Philosopher* (London, 1721), described by the author as "A Collection of the Best Discoveries in Nature, with Religious Improvements." Mather begins the first chapter ("essay") by laying down "those *Laws of Nature,* by which the *Material World* is governed." These prove to be Newton's three laws of motion plus the law of universal gravity.

Another work, by Benjamin Martin, not only refers to the laws of motion as laws of nature, but also stresses the Cartesian lineage of the Newtonian laws. Benjamin Martin was an ardent Newtonian of the mid-eighteenth century, a celebrated maker of scientific instruments and author of many popular books on Newtonian science, including *The Philosophical Grammar,* which was in Jefferson's library.[124] In this work, written in the form of a dialogue, Martin has a student ask how many laws of nature there are, to which the teacher replies, "Sir *Isaac Newton* has laid down three."

Rather than make a further parade of all the authors who wrote of Newton's laws of motion as laws of nature, let me conclude by taking note that (as will be shown more fully in Chapter 5) John Adams's diary, which he kept while an undergraduate at Harvard College in 1754, recorded details concerning the lecture on "Experimental Phylosophy" which Adams attended on the 9th of April. The subject was "Sir Isaac Newtons three laws of nature proved and illustrated." In the lecture notes of Adams's teacher, Professor John Winthrop, we find the source of Adams's notes. The Newtonian three laws of motion were characterized as "the 3 Laws of Motion or Nature."

CONCLUSION   The evidence is thus overwhelming that, in the age of Jefferson, "laws of nature" meant what we would call "laws of science" and that Newton's three laws of motion commonly were called "laws of nature." This is the sense in which Alexander Pope referred to "Nature's laws" in his famous couplet about Newton:

> *Nature and Nature's laws lay hid in night.*
> *God said, Let Newton be, and all was light!*

Not only does this couplet refer to the laws of nature in the sense which we have just been exploring; these are obviously Newton's laws, since "all was light" only after God had let "Newton be."

In the face of such massive evidence, it is clear that whenever Jefferson considered a draft of the Declaration, the phrase "the laws of nature" would have had for him a specific Newtonian echo. A direct train of association would have led his conscious or unconscious line of thought to Newton's "axioms," the famous laws of motion found in Newton's *Principia,* one of the books that Jefferson so highly esteemed.

Lest any reader think it far-fetched to suppose that Newton's "axioms, or laws of motion" should be cited in a political context, let it be noted that a few years after the Declaration, John Adams would be citing Newton's third law of motion in the course of a political debate with Benjamin Franklin over the optimal form of the legislature (see Chapter 4). Furthermore, we have observed (in Chapter 1) that James Wilson introduced one of the definitions relating to Newton's first law in the context of a lecture on the principles of government. As we shall see, this Newtonian reading of the opening sentence of the Declaration reinforces the message which Jefferson obviously wished to convey.

The Declaration of Independence: "Self-Evident" Truths

For a full appreciation of the possible political significance of the Newtonian resonances of the first sentence of the Declaration, an examination of the second sentence is needed. Here, once again, the process of association suggests a Newtonian reading, one that is so closely related to the Newtonian sense of "laws of nature" that it serves as reinforcement for it. In the "original Rough draught"[125] Jefferson began this sentence with an assertion of belief. "We hold these truths," Jefferson wrote, "to be sacred and undeniable...."

This introduction will lead to a listing of the truths in question, including (in the final version) that "all men are created equal" and "that they are endowed by their Creator with certain unalienable Rights," among them the rights to "Life, Liberty and the pursuit of Happiness."[126]

Readers who remember the actual text of the Declaration will have observed that the sentence which I have cited originally contained some unfamiliar phrasing. In the final text, the one that is celebrated every year on the Fourth of July, Jefferson's truths are no longer "sacred and undeniable," as in his early draft, but have become "self-evident." The improvement is tremendous. Who can fail to be thrilled by the ringing cadence of the new phrase, "We hold these truths to be self-evident . . ."? Was this change merely an improvement in rhetoric? Carl Becker suggested that possibly it was not Jefferson but Benjamin Franklin who introduced this alteration.[127] Yet Julian Boyd, who probably knew Jefferson's handwriting better than any other scholar, concluded that there is no reason to suppose that it was not Jefferson himself who made this significant change.[128] For our purposes this problem need not detain us. Jefferson's final version was an expression of his belief, an improvement that he incorporated into his document and that remained through the final alterations and emendations. Our main concern is to try to discover what this phrase would have meant to Jefferson as he rewrote or reread the document and what it could have meant to the other signers and to readers of those days.

The canonical interpretation of the Declaration sees the words "self-evident" in relation to the doctrine of natural law in association with the Lockean theory of government or with ideas similar to Locke's but circulated by other writers, such as Burlamaqui. Thomas Jefferson and the signers of the Declaration were educated men; many had been trained in the law. They would have been acquainted with the principles of natural law, but they would also have studied Euclid. To anyone who has studied Euclid, the phrase "self-evident" immediately suggests "axiom." Over many years, I have tried a word-association game with students and colleagues, asking them to tell me what single word is suggested by association with "self-evident." In almost every instance, the reply has been immediate, without a moment's hesitation: "axiom."

John Harris's *Lexicon Technicum*, of which (as we have noted) there was a copy in Jefferson's library, makes a direct link between "self-evident" and "axiom," defining axiom as "such a common, plain, self-evident and received Notion, that it cannot be made more plain and evident by Demonstration." An example is "That a Thing cannot be and not be at the same time." Others are "That the Whole is greater than a Part" and "That where there is no Law there is no Transgression." Harris introduced this concept of axiom in his definition of "science." A "Science," he wrote, "is Knowledge founded upon, or acquired by clear, certain, and self-evident principles." It is obvious that in the age of Jefferson the words "self-evident" suggested axiom and made an association with science.

Axiom = Science

In the eighteenth century, Euclid was still a required subject of study. A commonly used textbook was the one prepared by Isaac Newton's teacher, Isaac Barrow, updated and reprinted again and again. Although Euclid is based on definitions, "Common Notions," and axioms, these had become known in the seventeenth and eighteenth centuries as "Definitions," "Postulates," and "Axioms." In geometry and in logic, self-evident statements or axioms are the foundations of knowledge. We know how highly Jefferson esteemed mathematics, a subject where there is universal agreement, where there are no grounds for individual dissenting opinions as to whether some statement is true or false. We have seen that in a letter to John Adams in 1812, after his retirement from political office, he declared that he was looking forward to reading his four favorite authors, one of whom was Euclid. As a man who loved the geometry of Euclid, Jefferson would have certainly appreciated the sentiment expressed in Edna St. Vincent Millay's line that "Euclid alone has looked on beauty bare." Part of the beauty of Euclidean geometry is the way in which the demonstrations lead from complex propositions back to the starkness of the primary axioms. This is apparently the sense in which John Adams later recalled that one of the benefits which he had gained from studying mathematics and science as a college student was patience, the need for perseverance in following through the stages of a difficult proof or a knotty problem.

The second of Jefferson's favorite scientific authors was Isaac Newton. Moreover, both the *Principia* and the *Opticks* were in Jefferson's library. Anyone who has actually looked at the two scientific treatises written by Newton is aware that, like Euclid's geometry, both Newton's *Opticks* and his *Principia* were founded on a set of "axioms." In Euclid and in Newton, two of his favorite authors and his two most cherished scientific authors, Jefferson would have found a system

based on axioms. The axioms in the books of Newton are different
in kind from the traditional axioms of geometry or of logic. This
nontraditional kind of axiom came into general usage during the Scientific Revolution. An early appearance of axiom in this new sense
is found in the *Commentariolus,* or "Little Commentary," written by
Nicolaus Copernicus in the sixteenth century. Copernicus is generally considered the author whose works mark the first stages of the
Scientific Revolution, and the *Commentariolus* is his first essay on the
heliocentric system that bears his name. Copernicus begins by
explaining that he will undertake to show why the old Earth-centered
models of the universe are wrong and why the center should be
located in an immovable Sun "if some postulates, which are called
axioms, are granted to me" ("si nobis aliquae petitiones, quas
axiomata vocant, concedantur").[129] Not one of these "axiomata" can
be considered universally self-evident in any ordinary sense. In fact,
each of them states a position counter to the tenets of the two systems
of the world generally accepted in those days, the Ptolemaic and the
Aristotelian-Eudoxian Earth-centered systems, and hence would
have seemed absurd to almost every reader of Copernicus's time.[130]

In 1611 Johannes Kepler used "axiom" in a very similar sense in
his *Dioptrice,* a short tract on the refraction of light, a work which
was studied by Newton when he first entered Trinity College as an
undergraduate. Here Kepler uses the word "axioma" in two ways. He
introduces "axioma opticum," or an "optical axiom," and also
"axioma physicum," or a "physical axiom," both of which assume
knowledge of science and obviously do not refer to propositions that
are universally self-evident. But there are also examples of an
"axioma" without any modifier; these too are not in any possible
sense self-evident to everyone. For example, axiom 6 reads: "The
refractions of crystal and glass are very nearly the same" ("Crystalli
& vitri refractiones sunt proximè eaedem"). Axiom 9 declares: "The
maximum refraction of crystal is about 48 degrees" ("Refractio Crystalli maxima est circiter 48. gradus"). Since these "axioms" depend
on the results of specific experiments, and could never be known
otherwise, they are like the axioms of Copernicus; they form the basis
for constructing a physical system and are not universally self-evident. Unlike the axioms of Copernicus, however, the axioms of
Kepler can be tested by direct experiment.

The eight "Axioms" on which Newton's *Opticks* is based are very
much like Kepler's axioms in that they are statements of physics,

known not by intuition or logic but by experiment. For example, axiom V states that the "Sine of Incidence is either accurately or very nearly in a given Ratio to the Sine of Refraction." This law was so non-evident that it defied the search for it until well into the seventeenth century. It was, for example, unknown to Kepler, who sought in vain for just such a law of refraction. Other axioms in Newton's *Opticks* are similarly known only as the result of experiment.

At the time when Jefferson was writing the Declaration, the dual usage of "axiom" had become so prominent that it was a topic of discussion in textbooks of natural philosophy. For example, the difference between the axioms of geometry and the axioms of physical science was made clear in John Keill's *Introduction to Natural Philosophy*. Keill could be assumed to speak with an authentic Newtonian voice since he was a close associate of Newton's and had been one of Newton's principal spokesmen in the dispute with Leibniz over priority in the invention of the calculus. Jefferson owned a copy of the 1730 Leyden edition of Keill's *Introductio ad Veram Physicam et Veram Astronomiam* and the English translation of the second part, *An Introduction to the True Astronomy* (6th ed.: London, 1769). Furthermore, Keill's textbooks were almost certainly studied by Jefferson when he was learning the principles of Newtonian science as an undergraduate at the College of William and Mary.

According to Keill's *Introduction to Natural Philosophy*, science or natural philosophy is based on "Philosophical Axioms." In Lecture VIII, Keill explains how natural philosophy differs from geometry. The "Objects of Natural Philosophy," he writes, "are Bodies and their Actions on one another" and these "are not so easily and distinctly conceived" as "those simple Species of Magnitudes, which are the Subject of Geometry." Hence, "in physical Matters" Keill would "not insist so much on a rigid method of Demonstrations" as to expect that "the Principles of Demonstrations, that is, the Axioms, [be] so clear and evident in themselves, as those that are delivered in the Elements of Geometry." The reason, he insists, is that "the Nature of the thing will not admit of such."

A similar position is taken by Henry Pemberton in his *View of Sir Isaac Newton's Philosophy* (London, 1728). Pemberton was also an authoritative interpreter of Newton, since he had edited the final version of the *Principia* under Newton's direction. In the opening chapter of his *View*, Pemberton contrasted the axiomatic principles of geometry and natural philosophy. Newton's laws, he wrote, "are

some universal affections and properties of matter drawn from experience." They "are made use of as axioms and evident principles in
all our arguings upon the motion of bodies." These Newtonian
axioms, he explained, are different from those used in geometry,
which are assumed "without exhibiting the proof of them." In natural
philosophy, "all our reasoning must be built upon some properties
of matter, first admitted as principles whereon to argue." In geometry, such principles or axioms are assumed "on account of their being
so evident as to make any proof in form needless." But in natural
philosophy, "no properties of bodies" can be considered "self-evident" in "this manner" of geometry. The reason why the properties
of bodies may not be regarded as self-evident in the manner of geometry, he explains, is that our knowledge of them does not depend on
"the nature and essence" of matter but rather on "experience." It is
to be observed that Pemberton, like other Newtonian commentators,
leaps from axioms and self-evident truths of geometry to their equivalents in natural philosophy, stressing the difference between the
two.

We may see how the two meanings of "axiom" were closely associated by examining Ephraim Chambers's *Cyclopaedia, or an Universal
Dictionary of Arts and Sciences* (London, 1752), one of the most useful
encyclopedic dictionaries of the eighteenth century. We can judge
the importance of this work from the simple fact that the celebrated
Encyclopédie of Diderot and d'Alembert began as a project to translate
Chambers's *Cyclopaedia* into French. Chambers devotes almost a
whole folio page of two columns to the discussion of "AXIOM." This
presentation has two separate parts. The first is concerned with an
axiom as "a self-evident truth," as "a proposition whose truth every
person perceives at first sight." The second part presents the usage
in science, namely, the way in which an axiom "is also an established
principle in some art or science." This is the sense, he explains, of an
"*Axiom* in physics," for example, "that nature discovers herself most
in the smallest subjects; also that nature does nothing in vain." Referring directly to Newton's *Opticks*, Chambers says that it "is an *Axiom*
in optics, that the angle of incidence is equal to the angle of reflection." It is similarly "an *Axiom* in medicine, &c., that there is no sin-
cere acid in the human body." He then explains that in "this sense the
general laws of motion are called *Axioms*," and he gives as examples
Newton's first and third laws. Any careful reader of this entry in

Chambers's *Cyclopaedia* would thus come away with a close mental association of self-evident truths and the axioms of physics.

The sequence of association that I have been suggesting—from "laws of nature" to axiom to Newton's laws of motion—is set forth specifically in the *Cyclopaedia*. Chambers states unambiguously that "*Laws of Nature* are axioms." Then he says that they are "general rules of motion and of rest, observed in natural bodies," to which he adds that "The *laws of nature*, and *of motion*, are in effect the same." In elaboration, Chambers explains that some "authors" use "laws of motion" for "particular cases of motion," but it is only the "more general, or catholic ones" which the authors "call *laws of nature*." From these fundamental laws, "as from axioms, the others are deduced." Chambers than proceeds to exhibit and discuss what he calls the three "*laws of nature*" which "Sir Isaac Newton has established," namely, the three laws of motion found in the *Principia*.[131]

When Jefferson was writing the Declaration, the most celebrated physical "axioms" were the laws of motion presented at the beginning of the *Principia*, which we have seen were boldly called "Axiomata, sive Leges Motus." It was obvious to Newton and to his readers that these "axioms, or laws of motion," could not in any real sense be considered universally self-evident. Consider the first axiom, which states what occurs in the case of a body on which there are no external unbalanced forces acting. Under these circumstances, according to Newton's first axiom or law, a body at rest will remain at rest, but a body in motion will persevere in moving straight forward at uniform speed. Newton was fully aware that throughout most of history, scientists and philosophers had believed in a quite different law, namely, that under these circumstances a body could never persevere in its motion but would come to rest. The second law was equally far from universally self-evident since it depends on the new concept of mass which Newton had just invented. The third law was also not universally self-evident, as may be seen in the simple fact that most people have never correctly understood it, wrongly believing that the third law is related to a condition of balanced forces and equilibrium. We shall see an example of this common error in John Adams's invocation of the third law in rebuttal of Franklin's concept of a unicameral legislature.

Even though Newton's "axioms" are not universally self-evident, they nevertheless could partake of the same degree of validity as the self-evident axioms of geometry, but only for those people who had learned to think about nature in the new manner. Every Newtonian was fully convinced that Newton's three laws of motion were the incontestable foundation of any true understanding of nature, just as the Euclidean axioms are the incontestable foundation of geometry. All believers in the "new philosophy," in the new science of the *Principia*, would assign the same level of truth value to Newton's axioms as to those of logic or geometry or any other certain knowledge. Should there be any lingering doubters, there was convincing evidence of the universal validity of these axioms in the text of the *Principia*, where the propositions that were derived from the "axioms, or laws of motion," served to explain the principal phenomena of the heavens and of our Earth.

This sense of "axiom," as a basic truth accepted by those who had been converted to a new and correct way of thinking, has an obvious transfer of meaning from the physical sciences to the realm of politics. Philosophers from Aquinas to Locke and Burlamaqui had always taken note that, especially in the context of natural law, the concept of "self-evident" does not mean evident to everyone. Morton White, in his study of the Declaration, discusses "an old philosophical tradition according to which a truth may be self-evident to some people but not to others, a tradition which goes at least as far back as Aquinas, who emphasized that a learned man can see as self-evident a truth which an ignorant and rude man cannot see as self-evident."[132] Jefferson was fully aware that the beliefs he had listed as self-evident truths were also not evident to everybody; they were not held by all people and were hardly accepted universally. Indeed, Jefferson does not write, "These truths are self-evident," but rather, "We hold these truths to be self-evident," thus suggesting that they are so within the universe of discourse represented by the Declaration of Independence, somewhat as Newton's axioms are self-evident in the context of the *Principia* and almost as though Jefferson is saying, "These are the axioms of this treatise." Remember that the "truths" in question were that all men are created equal and that they are endowed by their creator with the rights to life, liberty, and the pursuit of happiness. These truths were on a similar plane of belief to that of the laws of motion (that is, Newton's laws of nature, the axioms of Newton's *Principia*) in the sense that they were valid only for those who had adopted a new viewpoint. This kind of validity was the same with respect to physical science and to human rights and potentialities.

The train of association of ideas that I have suggested—from "self-evident" to "axiom" and from "axiom" in general to "axiom" in its new sense as it came to be used in physical science—may at first encounter seem forced. Such a sequence of ideas would be too extreme for easy credence if it depended only on a single expression, "self-evident." But, in the context of the Declaration, these words do not occur in isolation. Rather, they appear in the sentence immediately following Jefferson's reference to "the Laws of Nature." If a reference to laws of nature brings to mind an association with the axioms of Newton's *Principia,* it does not take a great leap of imagination to recognize that "self-evident" in the very next sentence has some reference to this same set of axioms. And this is all the more the case when we take account of the fact that this particular set of axioms was the foundation of what Jefferson and most of his contemporaries esteemed as the greatest creation of human reason, Newton's *Principia,* a book that Jefferson knew well.

Finally, lest any reader suppose it far-fetched to think that Jefferson used a logical or mathematical or scientific concept such as "axiom" in a political document, two examples will show that such usage was not uncommon. One of David Hume's essays bears the title, "That Politics May be Reduced to a Science."[133] Hume, like others of that day, meant by politics what we would call political science. In this essay, despite its title, Hume does not attempt to explore general principles of "politics" that prove it to be like physics or biology. Rather, he discusses some general and particular aspects of the British political system, together with the usual sprinkling of references to Greeks and Romans and an occasional nod to Machiavelli. The reduction "to a science" seems to consist in nothing more than the organized treatment of the problem. Assuming from the outset that politics *is* a science, Hume asserts that "politics admit[s] of general truths." One of the latter is "pronounced" by Hume to be "an universal axiom in politics."[134] It is "That an hereditary prince, a nobility without vassals, and a people voting by their representatives, form the best MONARCHY, ARISTOCRACY, and DEMOCRACY." Hume asserts that there are some "other principles of this science" which "seem to deserve" the "character" of "general truths" of this kind. Whatever we may think of Hume's axiom, the fact is that he considered it to be an axiom and believed that axioms of a sort have a place in political discourse.

A second example of the use of "axiom" in a political context occurs in *Federalist* No. 31, written by Alexander Hamilton. This essay, dated 1 January 1788, is of particular interest because it not only compares the "axioms" or "maxims" of geometry and of "politics," by which Hamilton, too, means political science, but also discusses the character of truths which are self-evident. Such truths are almost precisely designated by Hamilton as self-evident and are considered to be of two kinds. First, "there are certain primary truths, or first principles" that "contain an internal evidence which, antecedent to all reflection or combination, commands the assent of the mind."[135] These truths occur as "maxims" not only in geometry but also in ethics and politics. Second, in the "sciences" of ethics and politics, Hamilton finds "other truths" which, even "if they cannot pretend to rank in the class of axioms," are "direct inferences" from axioms and are "obvious in themselves." Hence "they challenge the assent of a sound and unbiased mind, with a degree of force and conviction almost equally irresistable" to that of proper axioms.

The "maxims in geometry" are presented through the customary examples: that "The whole is greater than its parts"; that things equal to the same are equal to one another; and that all right angles are equal to each other. The "maxims in ethics and politics" which are "of the same nature" include: "that there cannot be an effect without a cause; that the means ought to be proportioned to the end; that every power ought to be commensurate with its object; that there ought to be no limitation of a power destined to effect a purpose, which is itself incapable of limitation."[136] Truths of this kind must command the "assent" of anyone who is not prevented either by "some defect or disorder in the organs of perception" or by "the influence of some strong interest, or passion, or prejudice," while the truths which are almost "axioms" require an "assent" almost equal to that demanded by axioms. This leads Hamilton to a discussion of the reasons why there is a difference between the acceptance of axioms in geometry and the acceptance of truths, whether axioms or near-axioms, in ethics and politics.

The "objects of geometrical enquiry," according to Hamilton, are "intirely abstracted from those pursuits which stir up and put in motion the unruly passions of the human heart." This is the reason why we experience no difficulty in adopting "not only the more simple theorems of the science but even those abstruse paradoxes" which "are at variance with the natural conceptions" of the mind. In "the sciences of morals and politics," however, passion and prejudice may interfere with the "fair play" of people's "own understandings." Hamilton admits that "the principles of moral and political knowledge" do not "have in general the same degree of certainty with those of the mathematics." Even so, he argues, "they have much better claims in this respect" than we might suppose by judging "from the conduct of men in particular situations." In such cases, according to Hamilton, "the obscurity is much oftener in the passions and prejudices of the reasoner than in the subject." Of course, people are found to be "less tractable" in regard to "the sciences of morals and politics" than in regard to the science of geometry. This is a useful feature, however, because "caution and investigation" provide a "necessary armour" against "error and imposition." Yet Hamilton sounds a warning against carrying "this intractableness" too far, lest it "degenerate into obstinacy, perverseness or disingenuity."[137]

In considering Hamilton's lesson in logic, the point which may interest us the most is his discussion of the near-parallelism between the self-evident axioms characteristic of mathematics and their counterparts in political science. In particular, we may take note that Hamilton believed that there are axioms in the realm of political discourse just as there are axioms in geometry and other sciences. In fact, in *Federalist* No. 80, dated 28 May 1788, Hamilton introduces the concept of axiom with regard to "the propriety of the judicial power of a government being co-extensive with its legislature." He boldly asserts that "it is impossible by any argument or comment" to make this principle "clearer than it is in itself." Indeed, he argues, "if there are such things as political axioms, the propriety of the judicial power of a government being co-extensive with its legislative, may be ranked among the number." Even earlier, in *Federalist* No. 23, dated 18 December 1787, Hamilton declares with reference to another political point: "This is one of those truths, which to a correct and unprejudiced mind, carries its own evidence along with it; and may be obscured, but cannot be made plainer by argument or reasoning. It rests upon axioms as simple as they are universal."

Clearly Hamilton on more than one occasion introduced the concept of axioms, or truths closely related to axioms, which are self-evident if no mental or moral defect interferes with their recognition. He wrote of "axioms" and the concept of "self-evidence" in a political argument intended to win support for the new Constitution. There should, therefore, be little doubt that it would not have been unusual for Jefferson to have similarly been thinking of axioms in relation to self-evidence in the Declaration. As a matter of fact, Jefferson himself on one occasion introduced the concept of axioms in a somewhat related context, economics rather than politics. In a discussion of the "complicated" character of the "science" of political economy, Jefferson wrote of how "no one axiom can be laid down as wise and expedient for all times and circumstances."[138]

Conclusion

Jefferson's primary purpose in writing the Declaration was to produce a statement of principles and a list of grievances to justify a position for separation from Britain, for national independence. The philosophical introduction, however, goes beyond specific charges against the king and is an expression of the conscious and unconscious background of Jefferson's thought, drawing on his study of philosophy and political theory, his reading about natural law and natural right, his training and experience in civil law, and his knowledge of science. It is because Jefferson availed himself of such a rich and varied intellectual font that scholars have expended so much energy in interpreting the Declaration.

Whatever was in Jefferson's conscious mind when he wrote about "the laws of nature," these words had necessarily the sense of laws or principles of science. This was the meaning assigned by Jefferson himself when he used these very words in his *Notes on the State of Virginia*. It is also, for example, the meaning that appears in Thomas Pownall's book on the administration of the colonies, published in London in 1774, just two years before the Declaration. Pownall argued quite simply and straightforwardly that "the same laws of nature" apply in both the scientific and the political realms. In books that Jefferson had studied, "laws of nature" meant the axioms of Newton's *Principia*, the celebrated laws of motion. Whatever these words may have conveyed in the sense of natural law and implied natural rights, to Jefferson and to many readers of the Declaration the words "laws of nature" would have immediately started a train of association leading to Newton's laws of motion.

Whereas the train of thought from Jefferson's "laws of nature" is immediate and sure, the implications of "self-evident" are not so direct. Indeed, if these words were not being read in the context of "laws of nature," one might conclude only that they imply the notion of axiom. Moreover, if the sentence about "self-evident" had occurred without the preceding one about laws of nature and without the specification of truths which could not be self-evident in exactly the same way as geometric axioms, the train of association would have led to the most famous set of axioms of history, those found in Euclid's geometry. But "self-evident" does appear in the context of "laws of nature." Furthermore, there is a specification of three particular truths which plainly are not "self-evident" in exactly the same way as geometric axioms. Therefore, the self-evident "truths" are in context like the Newtonian "axioms, or laws of motion" and unlike the axioms of Euclid; they are plainly self-evident only in a particular way.

James Wilson, James Madison, John Adams, and other readers of the Declaration either had studied Newton's *Principia* or had learned their physics from one of the Newtonian textbooks of that day. For them, the concept of axiom, in the context of "laws of nature," would have initiated an immediate train of association from axioms in general to the particular set of axioms, Newton's laws of motion, which they knew as the "laws of nature." In taking note that these key phrases at the beginning of the Declaration would have had a scientific connotation for Jefferson and his contemporaries, we should keep in mind that it was Jefferson's ability as a scientist, his actual command of science, that was one of the three qualities that impressed his fellow delegates at the Congress. Certainly, they would not have been surprised to find some strong overtones of scientific thought in a document composed by someone whose intellectual life was permeated by his science.

I have mentioned a parallelism between the truth level or degree of acceptance of Newton's axioms and the self-evident truths of the Declaration. There is a further parallelism that would have been obvious to Jefferson and to anyone who had studied the Newtonian natural philosophy of the *Principia*. The truth of Newton's axioms was guaranteed by results, by their use in actual applications to Kepler's laws, planetary perturbations, the motions of the tides, the shape of the Earth and variation of weight with latitude, and the periodic return of comets. In 1776, when Jefferson composed the Declaration, the validity of the beliefs expressed in this document was still unproved, was still a matter to be decided by the future course of history. The Newtonian axioms had been validated by the correct retrodiction of past and present events and notably by the verification of the Newtonian prediction of the return of Halley's comet in 1758. Jefferson and his colleagues were convinced that the beliefs set forth in the Declaration would be similarly validated by events, by the future course of history. In this sense the establishment of the new nation was in fact a grand experiment, just as much a test in the realm of politics as the prediction of the return of a comet had been a test in the realm of nature.[139]

3

Benjamin Franklin: A Scientist in the World of Public Affairs

. . . . .

Some Aspects of Franklin's Political Thought[1]

Benjamin Franklin was unlike the other Founding Fathers in a number of major respects. First of all, he was the only one in 1775 who had won international fame for intellectual achievement. He was considerably older than the others, belonging to an entirely different generation. Born in 1706, he was twenty-six years older than George Washington, twenty-nine years older than John Adams, thirty-seven years older than Thomas Jefferson, and forty-five years older than James Madison. While in London as a young man, Franklin could have met Isaac Newton and almost did receive this honor, chronologically impossible for his younger colleagues. He was the oldest signer of the Declaration of Independence and the oldest delegate to the Constitutional Convention.

Franklin differed from Adams, Jefferson, and Madison in that he did not study at college, as they did. In fact, he finished his formal education at the age of ten and was apprenticed to his brother James, a Boston printer, when he was twelve. By any standard, however, Franklin was a well-educated man. In the fields of science, literature, philosophy, economics, political theory, and history, he was equal and often superior to colleagues and friends in America, in Britain, and in France. Franklin affords the proof that being highly cultured and even learned does not necessarily imply formal education—all the more so in the eighteenth century, when gentlemen attending college often learned very little.

Franklin also differed from Jefferson, Adams, and Madison in that he actually was a scientist, indeed, a very distinguished scientist, recognized and honored the world over for his contributions to our understanding of nature and its laws. At the time of the Revolution, Franklin was, in fact, one of the best known living scientists, not merely because of his sensational experiments on the discharge of lightning but for his discovery of many important new phenomena and his creation of the first satisfactory theory of electrical action.[2] Franklin's contemporaries were aware that the science of electricity had come into being in Franklin's time, and most of them would have agreed that Franklin was the primary creator of this new science. Not only had he produced a theory which enabled scientists successfully to predict the outcome of laboratory operations; he had also standardized the language of electricity by introducing in an electrical sense many basic terms that we still use: the word "battery" and the designation of charges as plus or positive and minus or negative. Critical historians of the twentieth century, writing about the development of electricity, share the high esteem in which Franklin was held by his contemporaries and still refer to the mid-eighteenth century as the Age of Franklin.[3]

An examination of science in relation to political thought among the Founding Fathers shows that the case of Franklin is different from the others in an even more fundamental respect, one which has been noted by almost every student of Franklin's political ideas. This feature has been succinctly set forth by Clinton Rossiter, one of the most astute scholars concerned with American political thought. Rossiter noted that the "pattern of Franklin's political theory is as perplexing as it is intriguing, as allusive as it is important."[4] In evidence Rossiter cited a fact immediately apparent to anyone who studies this topic: "The sum total of his strictly philosophical musings about government and politics would fill, quite literally, about two printed pages."

Franklin's political writings were primarily directed to issues, events, questions of the day, and problems of specific policy. He did not philosophize about the "nature and purpose of government." Rather, his arguments were set forth in terms of statistics, descriptions of conditions, considerations of outcome of policies, and direct propaganda, without "any appeal to fundamentals." Gerald Stourzh, who has made a thorough study of Franklin and policy, found that Franklin—unlike such other Founding Fathers as Jefferson, Madison, and Adams—did not write about separation of powers, nor did he use the authority of Montesquieu.[5] Nor was he under the "spell" of Harrington, as Adams was.[6] Rossiter found Franklin to have been "the one American patriot to write influentially about the events of 1763–1776 without calling upon natural law, the rights of man, and the social contract." This does not mean that Franklin had no guiding principles; we shall see, for example, that his views on population and the future of the British empire influenced much of his thought about the relations between America and the mother country. But, in general, such principles have to be inferred from his papers and are not the subject of specific essays on political philosophy.

Franklin has often been described as a political pragmatist, a man who was concerned with accomplishment rather than being the author of a political testament. Thus, unlike Adams, who wrote extensively (prolixly, in fact) about theoretical issues of government, or Jefferson, who developed theoretical ideas in the manner of a scholarly philosopher, Franklin never wrote an extensive theoretical political statement. Nor did Franklin produce a theoretical analysis of the structure and form of government of the kind produced by Madison and his collaborators in the *Federalist*.

Many of Franklin's political documents were letters to the press or tracts for the times, arguing the political advantage or disadvantage of a certain position, for example, the relative advantages to Britain and to colonial America of annexing Canada rather than Guadeloupe.[7] Even his essays on politics or on policy or on political economy were most often stimulated by his reading or were responses to practical issues such as restrictions imposed on American manufactures, the need for a colonial union, the British acquisition of territories in North America, the establishment of new colonies west of the Appalachians, international trade agreements, a unicameral versus a bicameral legislature, and representation in the Congress. The fact that most of his writings were directed to particular issues is shown in their titles: *A Modest Enquiry into the Nature and Necessity of a Paper Currency* (1729), *Plan for Settling Two Western Colonies in North America* (1754), *The Interest of Great Britain Considered with Regard to her Colonies and the Acquisitions of Canada and Guadaloupe* (1760), *Rules by Which a Great Empire May be Reduced to a Small One* (1773), *Queries and Remarks*

Respecting Alterations in the Constitution of Pennsylvania (1789).

Franklin did on occasion write essays on moral subjects and on human nature, and he did draw up rules of conduct, both for himself and for others. In fact, he became known as a "moral philosopher" long before he gained fame as a natural philosopher. In the category of moral philosophy are the maxims which appeared annually in his almanac, *Poor Richard*, begun in 1732, when he was twenty-six years of age.[8] His first published "independent" work, written at the age of nineteen, while he was working as a printer in London, was *A Dissertation on Liberty and Necessity, Pleasure and Pain* (1725), a philosophical argument that there could not be any free will or natural virtues and vices, since a good and omnipotent God permitted whatever was.[9]

Franklin's career embraced local issues of political action in Philadelphia and Pennsylvania, then decades of presenting American issues in London, followed by the glorious years as chief American representative in Paris during the years of the American Revolution. Although Franklin was author of some important political documents, such as the Albany Plan of Union, he was never particularly noted as the architect of political systems. His name, however, was associated with the principle of the unicameral legislature, especially as exemplified in the constitution of Pennsylvania, and he was always a warm defender of this idea.[10]

Franklin's Scientific Credentials

During the last half-century, historians have become aware that Franklin the scientist was more than a tinkerer or gadgeteer, more than an ingenious inventor. They have learned that he made many more scientific experiments than the one in which he flew a kite during a storm of thunder and lightning and that his science was not confined to the invention of the lightning rod.[11] Some historians have even come to appreciate the difference between Franklin's scientific discoveries about the lightning discharge and his invention of the lightning rod based on those discoveries. But most historians as well as the public at large would be hard put to define clearly and precisely the sense in which Franklin's contemporary and fellow scientist Joseph Priestley could liken Franklin's scientific achievements to those of Isaac Newton.

Even without a knowledge of science and its history, however, it is not difficult to discover Franklin's high place among the immortals of science, among those who have contributed in an important way to our fundamental knowledge of nature and its operations or who have enlarged the scope of science. By many different objective standards, Franklin was one of the foremost scientists of his age. His book *Experiments and Observations on Electricity,* in which he described his experiments and set forth his new theory of electrical action, ranks among the most notable books on science of that age and of any other age, as can easily be shown by a simple reference to its printing history. First published in London in three separate parts (1751–1753–1754), this book had two further editions (1754, 1760–1762–1765), after which Franklin revised the text and added a supplement dealing with many other topics of science.[12] This new version was published in 1769 and then revised and republished in 1774. That a book on an active branch of science should continue to interest the scientific public after two decades is an extraordinary tribute to both the significance of the contents and the style in which the science of Benjamin Franklin was presented. Long after Franklin's *Experiments and Observations* had ceased to be on the cutting edge of new science, it was recommended by the British chemist Humphry Davy, who told his students that in this book "science appears . . . in a dress wonderfully decorous, the best adapted to display her native loveliness." The "style and manner" of Franklin's "publication" on electricity, he wrote, "are almost as worthy as the doctrines it contains."[13] This compliment may be all the more esteemed when we take into account that Davy had made his reputation by the application of electricity to chemistry, culminating in the discovery of new chemical elements, and that Davy was himself a great stylist, celebrated for his poetry and even praised by Samuel Taylor Coleridge as one of the foremost literary lights of his age.

The fame of the book on electricity was not confined to readers familiar with the English language. A German translation appeared in 1758. The first French translation was published in 1752, preceded by a short history of electricity concluding with Franklin's achievements. A revised and enlarged edition of the French translation appeared a few years later, in 1756. In 1773, a few years before the American Revolution, a wholly new French translation appeared, now augmented by the supplementary material added by Franklin to the later English editions. A translation into Italian was published in 1774,[14] and a Latin version was finally undertaken, but never

[margin note: Book = worldwide fame]

completed, by Franklin's friend and scientific colleague Jan Ingen-housz.[15] It is hard to think of other scientific treatises of the eighteenth century, or of the nineteenth and twentieth centuries, that could rival this printing history of ten editions in four different languages.[16] To complete the publishing history, it may be noted that an American edition appeared in the mid-nineteenth century as volume six of Jared Sparks's edition of Franklin's works, that a scholarly critical edition was published in America in 1941, and that a Russian translation appeared in 1956, followed by a Spanish translation in 1992.

There are other gauges of the importance of Franklin's contributions to science. His achievement was recognized in April 1756 by election as Fellow of the Royal Society, the oldest scientific academy in the world and one of the most prestigious. In electing Franklin a Fellow, the Royal Society recognized the existence of electricity as a genuine branch of science. A few years earlier, in 1753, the Society had conferred on Franklin its highest award for scientific research, the Sir Godfrey Copley Gold Medal, and in that same year both Harvard and Yale had awarded him honorary degrees. Among other international scientific honors, only one other need be mentioned, perhaps the greatest scientific honor of them all, Franklin's election in 1772 as *associé étranger*, or foreign member, of the Académie Royale des Sciences of Paris. In order to appreciate the distinction which this election conferred, it must be noted that according to the rules of the Academy there could be only eight such "associés étrangers" at any one time.[17]

There are many other signs of Franklin's scientific distinction. When Denis Diderot was drawing up a curriculum for the university that Empress Catherine the Great of Russia was planning to establish, he recommended that Franklin's book on electricity (recently published in a French version) be required reading for all students as the highest expression of the experimenter's art.[18] In an oft-reprinted history of electricity, Joseph Priestley described how Franklin had become the most celebrated "electrician" (as electrical scientists were then called) of his age. He was widely known and respected throughout Europe, Priestley wrote, and his book "bid fair to be handed down to posterity as expressive of the true principles of electricity; just as the Newtonian philosophy is of the true system of nature in general."[19] Priestley also appreciated Franklin's style. "It is not easy to say," he observed, "whether we are most pleased with the simplicity and perspicuity" of Franklin's presentation and the "modesty with which the author proposes every hypothesis of his own" or "the noble frankness with which he relates his mistakes, when they were corrected by subsequent experiments."[20]

What were Franklin's chief contributions to science? On what scientific advances did his extraordinary fame rest? In the first place, a variety of different experiments led Franklin to invent the fundamental theory of electricity that was the hallmark of his scientific fame.[21] In this theory, all electrical effects arise from the action of an electrical "fluid." This was a fluid in the sense set forth in the supplemental part (or "Queries") of Newton's *Opticks,* a work which differed—as we have observed in Chapter 1—from Newton's more famous *Principia* in that it was written in English prose without a mathematical apparatus, that is, without developing the propositions by means of ratios and proportions, trigonometry, algebra, the calculus, or infinite series. Rather, the statement of each proposition was followed by a "proof by experiments."[22] The *Opticks* was thus a book that Franklin was able to read, whereas the *Principia,* with its chevaux-de-frise of difficult mathematics, forever remained closed to him. In the *Opticks,* Franklin and other scientists of his generation found a veritable handbook of the experimental art.

The electrical fluid postulated by Franklin was, in Newton's term, "elastic," that is, it would expand, spreading out on any conducting body. Franklin supposed that the potentiality of such expansion arose from the property that this fluid was made up of particles which repel one another, here using another suggestion made by Newton in the *Opticks.*[23] Introducing yet another Newtonian concept, that matter is particulate, Franklin supposed that all bodies have two compositional elements. One he called the particles of "ordinary matter," the other the particles of electrical fluid. Just as the particles of the electrical fluid repel one another, so are they attracted by the particles of ordinary matter.

When a body contains its "normal" quantity of electrical matter, Franklin supposed, it is electrically neutral; but when it loses or gains some electrical fluid, it becomes charged. A "plus" or "positive" charge results from a body's having gained some excess electrical fluid, more than its normal amount; similarly, a body acquires a "negative" or "minus" charge whenever it loses some of its normal amount of electrical fluid. In the language of Franklin's day, this fluid not only was invisible, without any odor, taste, or smell, but was also "subtle," in that it could penetrate and become lodged in the interstices between the particles of matter.

One extremely important consequence of Franklin's theory is that electrical charges arise from a redistribution of electrical fluid. Charges are not the result of something created by operations in the laboratory or in nature, as, for instance, the rubbing of a piece of amber with fur. Charges are not, in other words, creations of some new matter by the intervention of the experimenter's physical manipulation. It was a direct consequence of the theory, as Franklin proved by a number of different types of experiments, that charges must always appear in equal amounts of negative and positive electricity. Similarly, when charges are neutralized, they disappear in equal amounts of opposite sign. These two properties are generalized today under the name of law or principle of conservation of charge. The formulation of this principle and its proof and application constitute Franklin's primary claim to fame.

Since there is an attraction between ordinary matter and the electrical fluid, Franklin's theory easily explains why a negatively charged body (which has lost some of its "normal" supply of electrical fluid) will attract and be attracted by a positively charged body (with excess fluid). The theory also accounts for a large variety of phenomena which we classify today under the name of electrostatic induction.[24] Because Franklin's theory accurately predicts the outcome of operations in many electrical systems or laboratory situations, it won instant and wide acclaim. Although the theory eventually needed so many ad hoc assumptions that it was replaced, the new theories contained many elements of Franklin's original, while the many electrical phenomena he explored remain cornerstones of electrostatics.

With some modifications, we still use Franklin's kind of explanation in dealing with many problems of electrostatics. This aspect of Franklin's theory was clearly stated by J. J. Thomson, who discovered the electron, the fundamental particle in electric currents and one of the first of nature's fundamental particles to be identified and to have its physical properties established. Thomson said:

> The service which the one-fluid theory has rendered to the science of electricity, by suggesting and coordinating researches, can hardly be overestimated. It is still used by many of us when working in the laboratory. If we move a piece of brass and want to know whether that will increase or decrease the effect we are observing,

we do not fly to the higher mathematics, but use the simple conception of the electric fluid which would tell us as much as we wanted
to know in a few seconds.[25]

Robert A. Millikan, who measured the charge on the electron and
who provided direct evidence that there are particles of electricity,
went so far as to attribute to Franklin the credit for the electron,
on the grounds that Franklin had set forth the particulate nature of
the electrical fluid two centuries before either Thomson or himself.[26]

I have mentioned the fact that Franklin introduced basic terms
used in electricity, including positive and negative (or plus and
minus) charges. Other important discoveries and innovations relate
to the distinction between insulators and conductors, the effects of
grounding in electrical experiments, the difference between the
action of those conducting bodies that have blunt ends and those
that end in points, the mode of operation of the Leyden jar (or first
capacitor), and much else. Many of his contemporaries were particularly impressed by his analysis of the action of the Leyden jar, which,
until Franklin began his experiments, was one of the mysteries of the
mid-eighteenth century. The question was how a bottle filled with
water and held in an experimenter's hand could apparently "store"
so large a quantity of electricity; when discharged, there was a display
of seemingly unbelievable force or power. The chief inventor of the
Leyden jar, Pieter van Musschenbroek, was so impressed by Franklin's discoveries about the action of this jar that he dashed off a letter
to Franklin, stating that no one else had discovered so "many recon-
dite mysteries" of electricity.

Scientific investigation of the phenomena of lightning originates
with Benjamin Franklin. He accumulated analogical evidence
favoring the supposition that lightning must be an electrical discharge on a large scale. To test his hypothesis, he devised the sentry-
box experiment, which was first carried out successfully in France
and then repeated in other countries (see Figs. 22 and 23). He later
thought of an alternative experiment, the experiment of the kite,
which he seems to have designed and carried out before he had heard
the news of the success of the French experimenters in performing
the sentry-box experiment (see Fig. 24).

The importance of Franklin's experiments with the lightning discharge, the two just mentioned and many other later ones, can hardly

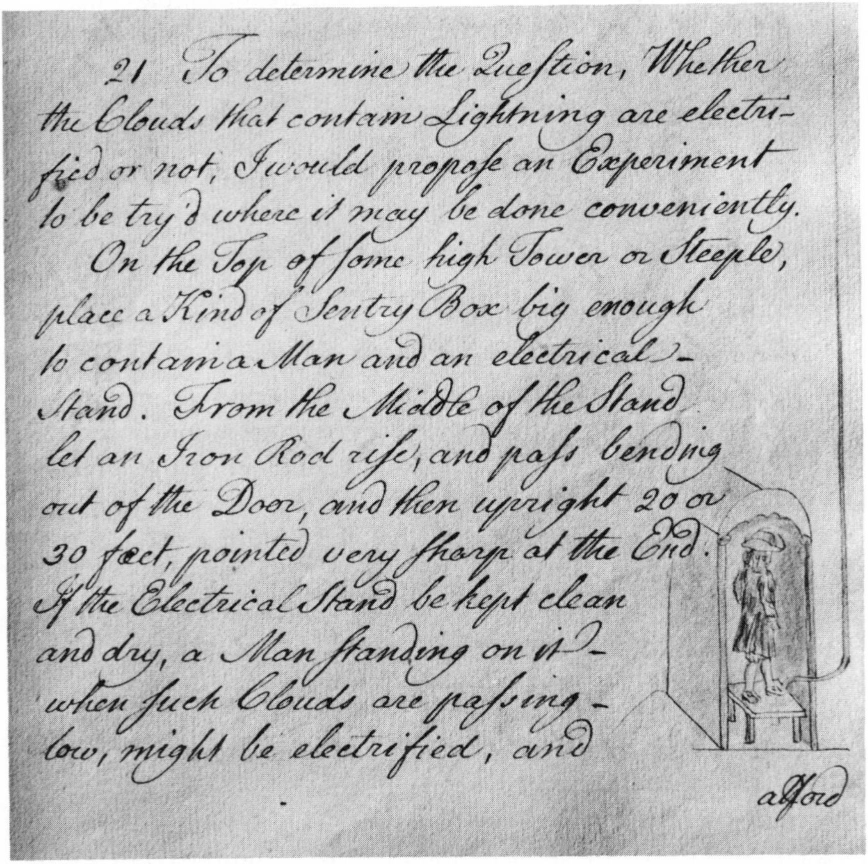
21 To determine the Question, Whether the Clouds that contain Lightning are electrified or not, I would propose an Experiment to be try'd where it may be done conveniently.
On the Top of some high Tower or Steeple, place a Kind of Sentry Box big enough to contain a Man and an electrical Stand. From the Middle of the Stand let an Iron Rod rise, and pass bending out of the Door, and then upright 20 or 30 feet, pointed very sharp at the End. If the Electrical Stand be kept clean and dry, a Man standing on it when such Clouds are passing low, might be electrified, and

afford

Fig. 22. Franklin's Sentry-Box Experiment. A wash drawing made under Franklin's direction was included in a pre-publication manuscript copy of his writings on electricity. Note that the experimenter stands on an insulating stool within the box, so as to be sheltered from the rain. A long pointed metal rod extends upward, intended to "draw" a charge from passing clouds and so prove that clouds are electrified—a consequence of which must be that the lightning discharge is an electrical phenomenon.

be overestimated. They not only showed that the new scientific subject of electricity was of importance, revealing properties of one of nature's most powerful forces; they also were a convincing demonstration that electrical phenomena—even large-scale electrical phenomena—occur spontaneously in nature. As Franklin's contemporary William Watson, the British physicist, put it, "The discoveries made in the summer of the year 1752 will make it memorable in the history of electricity." The lightning experiments, he explained,

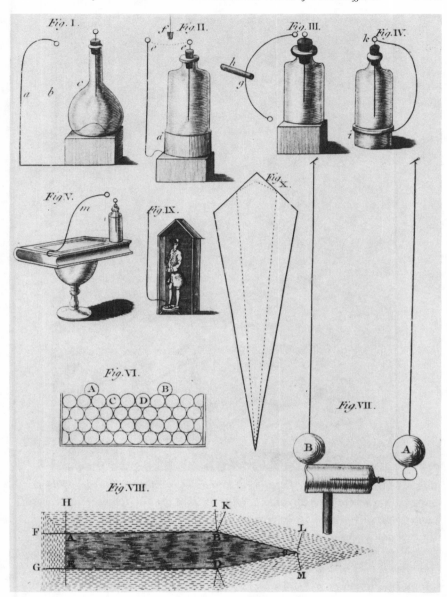

Fig. 23. Franklin's Fundamental Experiments on Electricity. The frontispiece to Franklin's Experiments and Observations on Electricity *(1751) shows, in the uppermost row, some of the experiments with the Leyden jar. The experiment at the extreme right demonstrates the principle of conservation of charge for the jar. The sentry-box experiment appears at the center.*

Fig. 24. An Electrical Kite Experiment Performed in Europe. The idea of the kite experiment came to Franklin after he had published the description of the proposed sentry-box experiment. This experiment was performed successfully many times in Europe and in America. The theoretical principles were identical to those of the sentry-box experiment. The experimenter is Jacques de Romas, who claimed he had invented the experiment after learning of the successful outcome of Franklin's sentry-box experiment but before hearing about Franklin's kite experiment.

"have opened a new field to philosophers, and have given them room to hope, that what they have learned before in their museums, they may apply, with more propriety than they hitherto could have done, in illustrating the nature and effects of thunder, a phaenomenon hitherto almost inaccessible to their inquiries."[27] Electricity, as the citation of the Copley Medal reads, now "appears to have a most surprizing share of power in Nature."[28] The consequence was that no physics of nature could thenceforth be considered complete if it did not encompass the science of electricity along with optics, mechanics, heat, and magnetism. Franklin's success in experimental physics also constituted a convincing proof that the approach to nature set forth in the "Queries" of Newton's *Opticks* was a fruitful guide for the creation of new experimental science.

As with all creative scientists, Franklin's curiosity and interests were not wholly limited to a single science. He made important discoveries and observations in areas other than electricity. For example, he found that northeast storms (that is, storms of wind blowing from the northeast) move toward the northeast, in the opposite direction of the flow of wind. He devised important experiments on heat conduction, including one still often performed in physics classes, in which a series of wax rings are placed at regular distances along rods of different metals. When the ends are heated, the rate of heat conduction is indicated by the speed with which the wax rings melt and drop off. Franklin measured the temperature in the ocean and was able to chart the path of the Gulf Stream. He was also interested in bioluminescence and studied the luminescence of seawater. He put forth arguments in favor of the wave theory of light more than a half-century before physicists found confirming evidence. He pioneered the study of thin films through his experiments on the effect of a thin surface layer of oil in stilling the waves of water.

In conclusion, it may be said that Franklin's legacy to scientific knowledge has three major components. His experiments led to the discovery of many new phenomena; the theory he invented brought order to a chaos of individual and apparently unrelated facts of observation and experiment; he proved that no system of natural philosophy could be considered complete if it did not encompass effects of electricity and magnetism. In addition, Franklin showed that experiments made in the laboratory, on the level of "toy physics," differed only in scale from the effects produced spontaneously by some of the most powerful forces in nature. His invention of the lightning rod was of special importance because it was a dramatic demonstration of the truth of the doctrine set forth by Francis Bacon and by René Descartes, that pure or disinterested scientific research of a fundamental kind would ultimately lead to inventions of use to humankind. Franklin also had the great distinction of having enlarged the Newtonian natural philosophy.

Science and Franklin's Political Career

One of the reasons why Franklin is often not considered to be a "proper" scientist, and so is relegated to the class of gadgeteers and inventors, is that he was not a university man. A careful examination of Franklin's preparation for scientific research, however, shows that he had studied science deeply in the textbooks of his age, available to him in his capacity as bookseller or on the shelves of the Library Company, of which he was the principal founder. Furthermore, the Junto, a mutual-aid association of young tradesmen and artisans, which Franklin founded, included the performance of scientific experiments among its activities. When the time came, Franklin was fully ready to embark on a career of scientific research because he had carefully studied the literature of experimental science and knew the work of such masters of experiment as J. T. Desaguliers and the methods of experiment set forth by Newton in his *Opticks*.

Of course, another reason why historians tend to stress Franklin's gadgets and inventions at the expense of his contributions to basic science is that they simply do not know enough physics to understand and to evaluate Franklin's contributions to fundamental theory. If any readers have doubts about the primary importance of Franklin's theoretical contributions to electrical science or about the importance of his laboratory explorations, they may turn for evidence to the technical histories of electrical science.[29]

A third reason for the neglect of Franklin's serious scientific career may be that its period of greatest intensity was relatively short. Franklin first encountered the subject of electricity when he was a successful businessman about forty years of age. Ready to retire, he would soon leave his printing establishment to be run by others. He quickly found that the study of electricity was congenial, and he devoted himself to it with force and alacrity. Learning that he was in advance of his European contemporaries provided additional stimulus. It is something of a puzzle, however, that after a few years of intensive research he entered more and more deeply into public life and never again devoted himself so wholeheartedly to the pursuit of science, although he continued to make experiments and observations and tried to emend his theory of electricity as new phenomena were uncovered. In any case, there is no question that his passion for science remained strong throughout the rest of his life, even though it was never again to become a primary pursuit. We shall see that at the very end of his career, on the occasion of a visit by Manasseh Cutler, he still showed his interest in scientific matters and his deep love of science.

Franklin himself suggested in his autobiography that he entered public life because of the consuming pressure of external affairs. No sooner was he fully launched on his career as experimenter than the threats of war appeared on the outskirts of Philadelphia, and Franklin, aware that there were no defenses, became active in the preservation of his city against possible enemy attack. In 1748, when the crisis had passed, he wrote to his friend and fellow scientist Cadwallader Colden, lieutenant governor of New York, that he had retired from active business affairs and looked forward to spending the next years "enjoying ... a great Happiness," that is, "Leisure to read, study, make Experiments, and ... produce something for the common Benefit of Mankind."[30] But in fact one thing led to another and he ended up as London agent for Pennsylvania in the struggle against the proprietors, later becoming agent also for Georgia, New Jersey, and Massachusetts. In 1750, at the moment of shifting into public life, Franklin sent an apologia to Colden. "Had Newton been Pilot but of a single common Ship," Franklin wrote, "the finest of his Discoveries would scarce have excused or attoned for his abandoning the Helm one Hour in Time of Danger." How "much less," he added, "if she had carried the Fate of the Commonwealth."[31]

Franklin never gave up being a scientist; what he did abandon was the career of being primarily a scientist, of being a full-time scientist. During the years in London and later in France, Franklin was in the midst of scientific activities of all sorts. He continued to make contributions to knowledge in a number of fields, including electricity, but extending to such areas as meteorology, oceanography, heat conduction, and even the life sciences. In London and in Paris he was invited to serve on the councils of science and to take part in dealing with important public issues that had scientific components, such as the investigation in 1784 of the claims of Franz Anton Mesmer.[32] In this sense, Franklin never abandoned the pursuit of science.

Of course, there is another reason why Franklin may have given up an intense full-time career as scientist. Perhaps he recognized that he had carried his contributions to the new science of electricity as far as his talents were able. After all, he had provided what was becoming the received theory; he had established a language of electrical discourse; he had discovered a large variety of new phenomena;

he had devised one of the most spectacular experiments ever to be performed. What higher achievement could possibly lie ahead?

Still it must be emphasized that for the rest of his life Franklin remained a passionate devotee of science. Science was, in the eloquent words of Carl Becker, "one activity which Franklin pursued without outward prompting, from some compelling inner impulse, one activity from which he never wished to retire, to which he would willingly have devoted his life."[33] Becker reminds us that science was the subject "to which he always gladly turned in every odd day or hour of leisure, even in the midst of the exacting duties and heavy responsibilities of his public career." Becker concludes that science "was after all the one mistress" to whom Franklin "gave himself without reserve and served neither from a sense of duty nor for any practical purpose."

Franklin's Use of Scientific Analogues and Metaphors in a Political Context

Franklin's political writings do not generally draw upon the concepts or principles of physics, not even the ideas of electricity, the science of which he was a master. Nor did he make much use of the principles of the biomedical sciences to buttress his political writings. His introduction of scientific ideas and theories in a political context is somewhat different from that of Jefferson, Adams, Madison, Hamilton, or any of the other Founding Fathers. One reason for this difference is that Franklin did not theorize about politics in the philosophical manner of others in America and Europe in the eighteenth century; accordingly his use of the images and concepts of science in a political context was not at all like that of his American contemporaries. But there were other reasons as well. By drawing on the phenomena or concepts or principles of science, Jefferson, Adams, and Madison could give their political ideas a special legitimacy by linking them to the domain of human thought most highly esteemed in the Age of Reason. An appeal to physics, astronomy, chemistry, or the biomedical sciences served not only to validate their ideas but to establish their credentials as citizens of the Enlightenment who understood scientific principles. We may readily understand why Jefferson, Adams, Madison, and their European contemporaries found the physical and biomedical sciences to be a rich source of analogues and metaphors, adding an exotic flavor of theoretical importance to their political prose.

But for Franklin, the case was quite different. His scientific credentials were well established by his discoveries and his recognition by peers. He had no need to display scientific learning out of its primary context. In addition, as he tells us in his *Autobiography*, he formed his prose style by copying and imitating great masters, primarily essayists who reached their peak before or during Franklin's youth, that is, mostly before the death of Isaac Newton. Franklin's style, accordingly, required no admixture of exotic elements or of references to the new science. Furthermore, since Franklin was writing for mass consumption, usually to gain a political point or to influence public opinion, his arguments would have been weakened by the introduction of esoteric references. Accordingly, whereas the physical and biomedical sciences provided a useful stock of metaphors to enhance the rhetoric of Jefferson, Adams, and Madison, there was no need for them in Franklin's writings. Of course, Franklin, like others writing in the 1750s and 1760s, could not help but introduce images from the latest news of science, especially the kind of science reported in almanacs, magazines, and newspapers and thus available to the general reader. This category would include reference to natural disasters, to the appearance of comets or meteors, and to new and exciting discoveries such as the peculiar properties of the polyp. Franklin also made use of fables about animals, constructing a kind of political bestiary in the great tradition of animal stories associated with Aesop. But his practice excluded such esoteric metaphors as the Newtonian laws of the universe, details of astronomical or physical laws and phenomena, new aspects of animal or plant physiology, and even the details of electrical experiments and theories.

Of course, from time to time, Franklin did introduce examples of scientific knowledge, just as any other learned person of that day might do. An example of Franklin's use of a biomedical analogue may be found in his *Cool Thoughts on the Present Situation* (1764), an attack on the proprietary system. The "miserable" situation in Pennsylvania, he wrote, does not have its cause "in the Depravity and Selfishness of human Minds." The "Cause is radical, interwoven in the Constitution, and so becomes of the very Nature, of Proprietary Governments."[34] This conclusion led him to a biological analogue. Some "physicians," he wrote, say that "every Animal Body brings into the World among its original Stamina, the Seeds of that Disease that shall finally produce its Dissolution." Similarly, he argued, the "Political Body of a Proprietary Government" contains "those convulsive Principles that will at length destroy it."

I have mentioned in Chapter 1 the vogue of interest in the polyp in mid-century and the fact that the properties of this curious entity would have been brought to the attention of Americans by Peter Collinson and also by an article appearing in the *Gentleman's Magazine*. Franklin referred to the properties of the polyp in a political context in one of his most important tracts, written in 1751, his *Observations Concerning the Increase of Mankind*, a major contribution to the nascent science of demography. In this politico-demographic study, Franklin aimed to demonstrate that there is "no Bound to the prolific Nature of Plants or Animals, but what is made by their crowding and interfering with each others Means of Subsistence."[35] He used this finding in part to argue that Britain should acquire sufficient territory for the needs of "the Increase of her People." Then he introduced the vivid simile. "A Nation well regulated," he wrote, "is like a Polypus." That is, if you "take away a Limb, its Place is soon supply'd; cut it in two, and each deficient Part shall speedily grow out of the Part remaining." This phenomenon of regeneration was so much in Franklin's mind that he referred to it in detail in *Poor Richard's Almanack* for that same year.

The discussion of the polyp in *Poor Richard* was part of an essay on the wonders of the microscope. Here the polyp was called the "most unaccountable of all Creatures."[36] I have already mentioned (in Chapter 1) some of the properties that made the polyp seem extraordinary. Here we may take note only of Franklin's description of what is so "wonderful, and almost beyond Belief," that "it will live and feed after it is turned inside out, and even when cut into a great many Pieces, each several Piece becomes a compleat Polype."

Franklin's analogy of the polyp with its power of regeneration led him to the following conclusions: "Thus if you have Room and Subsistence enough," you "may, by dividing, make ten Polypes out of one." So, he wrote, "you may of one make ten Nations, equally populous and powerful; or rather, increase a Nation ten fold in Numbers and Strength." But for such expansion à la polyp you must have adequate "Room and Subsistence."

A wholly different kind of biomedical metaphor was used by Franklin in his so-called Canada pamphlet of 1760,[37] which included as a kind of appendix virtually all of the essay of 1751 with its simile of the polyp. The second metaphor was based on an analogy between the state and the human body considered as a living organism, an analogy which had a long history and is known under the name of the body politic.[38] Franklin, as might be expected, gave this old analogy a new turn. "The human body and the political," he wrote, "differ in this, that the first is limited by nature to a certain stature, which, when attained, it cannot, ordinarily, exceed; the other by better government and more prudent policy, as well as by change of manners and other circumstances, often takes fresh starts of growth, after being long at a stand; and may add ten fold to the dimensions it had for ages been confined to." Nature determines the size of the human body, but there is no similar limitation to the size of a political body, to a state or nation. A "mother," of "full stature," Franklin wrote, "is in a few years equalled by a growing daughter." But, in "the case of a mother country and her colonies, it is quite different." In this case, according to Franklin, the "growth of the children tends to encrease the growth of the mother, and so the difference and superiority is longer preserved." The analogy suggests that mother England need not fear the expansion and growth of her daughter colonies in America, since the effect will be to "encrease" rather than decrease the relative growth of the mother. Franklin argues, in particular, that "this island" (England) by increasing its manufactures could "increase and multiply in proportion as the means and facility of gaining a livelihood increase," so that it is "capable of supporting ten times its present number of people" if "they could be employed."[39]

A celebrated example of Franklin's use of a biomedical analogy to make a political point occurs in a woodcut published by Franklin in the *Pennsylvania Gazette* on 9 May 1754. This was "probably drawn" by Franklin and appears to have been the "first American cartoon."[40] It is a crude picture of a curled snake, divided into eight pieces, marked with abbreviations of the names of seven states (New York, New Jersey, Pennsylvania, Maryland, Virginia, North Carolina, South Carolina) plus New England (see Fig. 25). It bears the caption: "Join, or Die." The significance of the cartoon was a supposed property of the fabled joint (or "joynt") snake. If broken apart, the snake would die; but if the broken piece were joined, it would continue to live. The political message was obvious.

Another snake that Franklin introduced in a political context was the snake with two heads. He was glad that the constitution adopted by Pennsylvania in 1776 called for a legislature composed of a single

Fig. 25. Franklin's "Joynt Snake."

chamber, a house of representatives directly elected by the people. A bicameral legislature, he remarked (and later wrote), reminded him of "the famous political Fable of the Snake, with two Heads and one Body." One day this snake "was going to a Brook to drink." Along the way she had "to pass through a Hedge, a Twig of which opposed her direct Course." Franklin wrote that "one Head chose to go on the right side of the Twig, the other on the left." The result was "that time was spent in the Contest, and, before the Decision was completed, the poor Snake died with thirst."[41]

In 1776 the two-headed snake was only part of a fable and not an item of Franklin's personal encounter with natural history. But a decade later, Franklin had just such a two-headed snake, preserved in a vial of spirit. He showed it to a Massachusetts visitor, Manasseh Cutler, who described a visit to Franklin in his home on Market Street, Philadelphia, on Friday, 13 July 1787. Cutler reported that he found Franklin "a short, fat, trunched old man, in a plain Quaker dress, bald pate, and short white locks, sitting without his hat under the tree." His "voice was low, but his countenance open, frank, and pleasing." Franklin introduced him "to the other gentlemen of the company, who were most of them members of the Convention."

Toward the end of the visit, Franklin showed Cutler "a curiosity he had just received, and with which he was much pleased. It was a snake with two heads, preserved in a large vial." Franklin informed him that this snake "was taken near the confluence of the Schuylkill with the Delaware, about four miles from this city. It was about ten inches long, well proportioned, the heads perfect, and united to the body about one-fourth of an inch below the extremities of the jaws. The snake was of a dark brown, approaching to black, and the back beautifully speckled (if beauty can be applied to a snake) with white; the belly was rather checkered with a reddish color and white." Franklin "supposed it to be full grown, which I think appears proba-ble, and thinks it must be a *sui generis* of that class of animals." Frank-lin assumed that this snake was not "an extraordinary production" of nature, "but a distinct genus," an opinion based on "the perfect form of the snake, the probability of its being of some age, and there hav-ing been found a snake entirely similar (of which the Doctor has a drawing, which he showed us) near Lake Champlain, in the time of the late war." According to Cutler, "the Doctor mentioned the situa-tion of this snake, if it was traveling among bushes, and one head should choose to go on one side of the stem of a bush and the other head should prefer the other side, and that neither of the heads would consent to come back or give way to the other." Franklin was on the point of mentioning "a humorous matter that had that day taken place in Convention, in consequence of his comparing the snake to America." Franklin "seemed to forget that every thing in Convention was to be kept a profound secret." But some member of the company present reminded him of "the secrecy of Convention matters," Cutler concluded, "and deprived me of the story he was going to tell."[42]

A final example is a fanciful tale about a whale, this time the bibli-cal whale rather than a whale of natural history. Soon after the sign-ing of the Declaration of Independence in July, Franklin was active in promoting a form of government for the new independent union of the colonies. He had long advocated a strong association of the colonies and had been the chief architect of the Albany Plan of Union in 1754. In 1775 he drew up a proposed "Articles of Confederation and Perpetual Union" for what he designated as "The United Colo-nies of North America." While some approved of his plan, others, according to Jefferson, "were revolted at it." Many features were incorporated into a new plan, adopted in 1776, and were later included in the first federal constitution.[43]

On 1 August 1776, about a month after the signing of the Declaration of Independence, during discussions of the form of organization to be adopted, Franklin proposed that each state be represented in the new Congress by a number of delegates and votes proportional to the relative size of its population. John Adams reported that Franklin, in support of his motion, tried to assuage the fear that "the great Colonies will swallow up the less." Franklin dismissed any such argument, according to Adams's report, by remarking that "Scotland said the same thing at the union."[44] Jefferson's two accounts[45] are somewhat more complete than Adams's summary and give us a sample of the characteristic wit with which Franklin argued his cause. "At the time of the union" of England and Scotland, Franklin said, as reported by Jefferson, "the Duke of Argyle was most violently opposed to that measure, and among other things predicted that, as the whale had swallowed Jonah, so Scotland would be swallowed by England." And yet, Franklin is reported to have continued, "when Lord Bute came into the government, he soon brought into its administration so many of his countrymen, that it was found in event that Jonah swallowed the whale." Jefferson concluded the anecdote by commenting: "This little story produced a *general* laugh, and restored good humor, and the article of difficulty was passed."[46]

Demography: A Science of Use for Policy

Although Franklin's writing on science tended to be separate from his writing on politics and although his political writing tended to be practical and directed to immediate issues rather than theoretical and devoted to philosophical argument, in at least two publications he combined the development of important scientific ideas with the development of his political thought about the future of America in relation to the British empire. In two pamphlets treating this question of policy, he was writing in response to specific political issues, but these pamphlets differed from his other political tracts in that they were based explicitly on theoretical scientific notions. They may even be regarded as essays on population, the first presenting Franklin's ideas and the second giving them further development and application.

The earlier pamphlet was *Observations Concerning the Increase of Mankind,* written in 1751 and circulated in manuscript until 1755, when it was published anonymously, first in Boston and then in London, as an appendix to William Clarke's *Observations on the Late and Present Conduct of the French.* Franklin's essay, with various excisions, appeared subsequently in a number of publications. It was attributed to Franklin by name when it was reprinted by the *Gentleman's Magazine* in 1755 and by the *Scots Magazine* in 1756. It then appeared in 1760 as a supplement to Franklin's *Interest of Great Britain Considered* and in the *London Chronicle* and in Burke's *Annual Register.* It was further published in 1769 in a supplement to the fourth English edition of Franklin's *Experiments and Observations on Electricity* and again in 1779 in Franklin's *Political, Miscellaneous, and Philosophical Pieces,* which was read by Thomas Malthus. Among Franklin's younger contemporaries whose thought was influential, Adam Smith is known to have had two copies in his library, and Turgot read the essay in the French translation published in 1769.

Students of Franklin have pointed out that the *Observations Concerning the Increase of Mankind* was written, as other of Franklin's works tended to be, in response to a particular political event, the British Iron Act of 1750, which restricted the manufacture of iron in the American colonies.[47] The essay was, however, primarily an important theoretical contribution to the new science of demography.[48]

Franklin's demographic presentation had the special virtue of being based squarely on numerical data. This was an age when the new statistical sciences were coming into existence and when it began to be recognized that questions of polity could be decided on numerical or quantitative grounds. One of Franklin's intellectual mentors, William Petty, from whom Franklin learned the labor theory of value and the associated principle of the division of labor, had been so fully convinced of the importance of numerical or statistical considerations for government that he had called one of his major publications *Political Arithmetick.* Newton's contemporary Leibniz, mathematician and philosopher, had even assumed that a time would come when "there will be no more need for disputation between two philosophers than between two accountants." It "will be enough for them to take their pens in their hands and sit down to their sums and say to each other (calling in a friend if they wish): 'Let us calculate.' "[49]

The call for numerical analysis was set forth most plainly in Franklin's day by Stephen Hales, the founder of modern plant physiology, who reduced many aspects of plant growth and physiology to numbers. It was Hales who, as secretary of the Royal Society, notified Franklin that the lightning experiments had been carried through to a successful conclusion in France. In his *Vegetable Staticks*, published in 1727, Hales wrote that "the all-wise Creator has observed the most exact proportions *of number, weight and measure, in the make of all things.*" Therefore, he concluded, "the most likely way" to "get any insight into the nature of those parts of the creation, which come within our observation, must in all reason be to number, weigh and measure." Like Jefferson, Madison, and others of that time, Franklin was fascinated by numbers, although he did not suffer from Jefferson's mania for numerology. He took great joy in creating magic squares and even invented a kind of magic circle of numbers which was reproduced in the later editions of his book on electricity. Convinced of the power of numerical persuasion, he also used numbers in his later campaign to encourage inoculation as a preventive measure against smallpox.

The very first paragraph of *Observations* clearly and even abruptly gave the signal that this work was based primarily on an analysis of comparative statistical data. "Tables of the Proportion of Marriages to Births, of Deaths to Births, of Marriages to the Numbers of Inhabitants, &c. formed on Observations made upon the Bills of Mortality, Christnings, &c. of populous Cities, will not suit Countries," Franklin wrote, "nor will Tables formed on Observations made on full settled old Countries, as Europe, suit new Countries, as America."[50] This opening statement suggests that Franklin was well acquainted with the statistics of various European urban centers. He had read William Petty's *Treatise of Taxes and Contributions* before 1728[51] and possibly knew other works by Petty. He seems to have known Edmond Halley's classic work on *The Degrees of the Mortality of Mankind*, based on "Tables of the Births and Funerals of the City of Breslaw."[52] Another likely source of information was Thomas Short's *New Observations, Natural, Moral, Civil, Political, and Medical on City, Town and Country Bills of Mortality*, which was published in London in 1750.

One of Franklin's important contributions to demography was his estimate that, under the American conditions which provided unchecked growth, the population would double every twenty or twenty-five years.[53] Since the population of one million colonists would double in about twenty-five years, Franklin predicted, the next century would see more Englishmen "on this Side the Water" than in England.[54] Let Franklin speak for himself about the consequences:

> What an Accession of Power to the British Empire by Sea as well as Land! What Increase of Trade and Navigation! What Numbers of Ships and Seamen! We have been here but little more than 100 Years, and yet the Force of our Privateers in the late War, united, was greater, both in Men and Guns, than that of the whole British Navy in Queen Elizabeth's Time. How important an Affair then to Britain, is the present Treaty for settling the Bounds between her Colonies and the French, and how careful should she be to secure Room enough, since on the Room depends so much the Increase of her People.[55]

We can see, therefore, why Franklin concluded that an expansionist policy in North America was a necessity and that British America was destined to become the most populous and the most important part of the British system. Gerald Stourzh has shown that after 1751 the "increase of mankind" became for Franklin "almost an obsession." Stourzh found that this aspect of Franklin's study of population "is the central point of his expansionism and the very core of his faith in the inescapable growth of American power, either within the framework of the British Empire or without and even against it."[56] It also demonstrates more generally the way in which practical political considerations are implicated in Franklin's theoretical work on demography.

Franklin's contributions to demography have further earned him the title of being one of the principal founders of the modern scientific study of population. Long before Malthus, he wrote of the tendency of any species to increase in numbers, thus doubling itself periodically, or increasing in what Malthus called a geometric proportion and what we call an exponential increase. Like Malthus, Franklin was aware that any such expansion of a population is limited by subsistence. Even though populations tend in theory to increase indefinitely at an exponential rate, Franklin recognized, they can never in the real world exceed the limit set by the available food supply. Franklin's principle was a general one, not limited to human populations, as may be seen in the following dramatic presentation of his ideas in the *Observations*:

> There is . . . no Bound to the prolific Nature of Plants or Animals, but what is made by their crowding and interfering with each others Means of Subsistence. Was the Face of the Earth vacant of other Plants, it might be gradually sowed and overspread with one Kind only; as, for Instance, with Fennel; and were it empty of other Inhabitants, it might in a few Ages be replenished from one Nation only; as, for Instance, with Englishmen.[57]

Moreover, like Malthus a half-century later, Franklin was aware that there are a number of factors that tend to check the unlimited growth of populations. These include not only the diminution of the available sustenance or food supply but also the various results of wars and disease, the loss of trade, bad government, owning slaves, and especially factors causing postponement of marriage, such as love of "Foreign Luxuries and needless Manufactures" or want of "industrious Education."[58]

The principle that populations increase exponentially and so tend to outrun their available food supply was stated by Malthus in his *Essay on the Principle of Population,* published in 1798. Malthus added the idea that any increase in the food supply can only be arithmetical. It is well known that Darwin made use of Malthus's two laws to argue that there must be a competition for survival among individuals in any species, that there must be a process of natural selection. In thus setting forth his theory of evolution, Darwin claimed that he was making a gigantic leap from Malthus's laws for human populations to similar laws for all kinds of populations, flora as well as fauna. In the *Origin of Species,* published in 1859, he twice stated explicitly that the "Struggle for Existence" was "the doctrine of Malthus applied . . . to the whole animal and vegetable kingdoms."[59] Actually, however, both Malthus and, before him, Franklin had anticipated Darwin's generalization; the differences, which are, of course, notable, include particularly the emphases and the directions taken by the conclusions.

Malthus had not read Franklin's *Observations* while writing the first *Essay on the Principle of Population,* but in the preface to the second edition, published in 1803, he refers to "Dr. Franklin" as one of those who have treated the subject of "increase of population."[60] Near the opening of the first chapter of this edition Malthus supports his thesis by noting:

It is observed by Dr. Franklin, that there is no bound to the prolific nature of plants or animals, but what is made by their crowding and interfering with each others means of subsistence. Were the face of the earth, he says, vacant of other plants, it might be gradually sowed and overspread with one kind only; as, for instance, with fennel: and were it empty of other inhabitants, it might in a few ages be replenished from one nation only; as, for instance, with Englishmen.[61]

We have seen that these remarks occur in Franklin's *Observations Concerning the Increase of Mankind,* which was read by Malthus in Franklin's *Political, Miscellaneous, and Philosophical Pieces.*[62] It was a tribute both to Franklin's eminence and to his power of expression that he was included by Malthus "among our own writers"[63] and that his phrasing was used to begin and bolster an argument which Malthus had already made quite clearly in 1798, before reading Franklin's essay.

Although Franklin and Malthus differ in the direction of the ideas which they share on the increase of population, since Franklin presents the growth of population chiefly as a source of good and Malthus as a source of evil, Franklin's thinking anticipates and perhaps even influences that of Malthus in a number of specific respects beyond the basic notion of a prolific increase that will be limited by the means of subsistence. For example, in the first edition of his *Essay on the Principle of Population,* Malthus set forth two causes, misery and vice, which explain why populations do not increase as they would if there were no checks. In the second edition he added a third such cause, "moral restraint," which involves in particular a virtuous postponement of marriage,[64] and in later editions he clarified this idea through a pithy definition: "By moral restraint I would be understood to mean a restraint from marriage, from prudential motives, with a conduct strictly moral during the period of this restraint."[65] We do not know for certain whether in this matter Malthus was influenced by Franklin, but Franklin did suggest a similar notion, in relation to being "cautious . . . of marriage."[66] In the second edition of the *Essay* Malthus referred explicitly to Franklin in relation to the "great extent of territory required for the support of the hunter" and also in relation to "the drains of [Africa's] population" effected by the spread of slavery.[67] Franklin certainly anticipated Malthus's state-

ment "that population, when unchecked, goes on doubling itself
every twenty five years,"[68] and, although Malthus used this estimate
in the first edition of the *Essay,* before he had read Franklin's discussion, his source is likely to have been one that depended upon
Franklin.[69]

The second important publication in which Franklin expounded
his ideas on population is the so-called Canada pamphlet, actually
entitled *The Interest of Great Britain Considered, with regard to Her Colonies, and the Acquisitions of Canada and Guadaloupe.* This pamphlet was
published in 1760, while Franklin was serving as a colonial agent in
London. In 1759, when Franklin began to compose this pamphlet,
the British victories over the French in both the North American continent and the West Indies had convinced many observers in Britain
that there would soon be an end to what Americans know as the
French and Indian War (1754–1763) but what is known in Europe as
the Seven Years' War because of a new war that started there in 1756,
merged with the colonial war, and ended with it in 1763. It was
apparent, therefore, that the British spoils of victory would very likely
include a choice between annexing Canada or Guadeloupe. Besides
including the *Observations* as a supplement to the new publication,
Franklin argued for the acquisition of Canada on much the same
grounds set forth in the essay written almost ten years earlier. The
result, he proposed, would be to provide new regions into which a
naturally expanding population would spread, thus peopling with
British colonials the lands south of the Great Lakes and east of the
Mississippi as well as Canada proper. Protection from French foes
and their Indian allies and the availability of cheap land would guarantee a natural increase of population. The consequence would be
an ever-expanding market for British manufactured goods. That is,
since "people increase and multiply in proportion as the means and
facility of gaining a livelihood increase," and since Britain "is capable
of supporting ten times its present number of people" (if only "they
could be employed"), it would follow that "in proportion . . . as the
demand increases for the manufactures of Britain, by the increase of
people in her colonies, the numbers of her people at home will
increase, and with them the strength as well as the wealth of the
nation."[70]

Suppose now, Franklin argued, that Britain chose Guadeloupe
and that there was no room for natural expansion in British North
America. In this case, the colonists would be "confined within the

mountains"[71] and the natural increase of population would cause the population density to increase until it became as great as that of Britain. The cost of land would rise and wages would fall. The extractive industries—agriculture and hunting—would, under these circumstances, no longer be as profitable as before and the colonists would be forced to turn to manufacturing. As a result, Americans would be rivals of Britain rather than consumers. Hence Americans would become increasingly free from any dependence on the mother country. Franklin's eloquent argument speaks for itself:

> A people spread thro' the whole tract of country on this side the Mississippi, and secured by Canada in our hands, would probably for some centuries find employment in agriculture, and thereby free us at home effectually from our fears of American manufactures. Unprejudiced men well know that all the penal and prohibitory laws that ever were thought on, will not be sufficient to prevent manufactures in a country whose inhabitants surpass the number that can subsist by the husbandry of it. That this will be the case in America soon, if our people remain confined within the mountains, and almost as soon should it be unsafe for them to live beyond, tho' the country be ceded to us, no man acquainted with political and commercial history can doubt. Manufactures are founded in poverty. It is the multitude of poor without land in a country, and who must work for others at low wages or starve, that enables undertakers to carry on a manufacture, and afford it cheap enough to prevent the importation of the same kind from abroad, and to bear the expence of its own exportation. But no man who can have a piece of land of his own, sufficient by his labour to subsist his family in plenty, is poor enough to be a manufacturer and work for a master. Hence while there is land enough in America for our people, there can never be manufactures to any amount or value.[72]

Franklin concluded that the "extended population" that would prove to be "most advantageous to Great Britain" would "be best effected, because only effectually secured by our possession of Canada."[73]

Thus Franklin's Canada pamphlet was double-barreled. From one point of view it was a political argument for acquiring Canada and the lands stretching out westward to the Pacific Ocean; from another point of view it was an essay on population providing the theoretical or scientific grounds for choosing Canada rather than Guadeloupe. In accomplishing this twofold program, Franklin was invoking the authority of numerical science in support of a political decision. Furthermore, he again stressed one of the important demographic principles that had been announced in the *Observations*. This was that the density of a population determines the major occupation of inhabitants.[74] When he repeated this idea in the Canada pamphlet, Franklin may even have been alluding humorously to his own earlier essay by referring to "a striking observation of a very *able pen*," which he now paraphrased as having presented the idea

> that the natural livelihood of the thin inhabitants of a forest country, is hunting; that of a greater number, pasturage; that of a middling population, agriculture; and that of the greatest, manufactures; which last must subsist the bulk of the people in a full country, or they must be subsisted by charity, or perish.[75]

In any case, Franklin not only reprinted the *Observations* as a supplement to the Canada pamphlet but also included in the later text itself various applications and restatements and expansions of arguments which make him one of the important early contributors to the science of demography.[76]

Political Implications of Franklin's Lightning Rod

Franklin's invention of the lightning rod was a significant event in the history of technology, not least because it showed how pure science or research aimed primarily at elucidating nature's secrets could yield applications useful to human life. We shall note below that Franklin's studies of the lightning discharge and his invention of the lightning rod were hailed as a victory in the warfare of reason against superstition. But there were a number of cases in which superstitious fear opposed the introduction of "Franklin rods," as they were sometimes called. This opposition took many forms. Some people were concerned that dire effects would follow from attempts to interfere with the forces of nature. Others believed that lightning storms were signs of a divine wrath, from which humankind could not escape by clever technology.

There are many examples of the popular superstitious fear of using lightning rods. When a Czech Catholic priest, Father Procopius

Diviš, erected a lightning rod of his own invention in 1754, a superstition arose that the device had caused a drought and in 1760 it was torn down by the people of the village in which it had been installed.[77] Some decades later, when the Comte de St. Omer erected a Franklin rod, there was the same kind of popular pressure to have it removed. This time, since the count refused, the case was taken to court. Expert witnesses from the Royal Academy of Sciences were brought from Paris to testify in favor of the lightning rods. After hearing testimony from both sides, the magistrate ruled in favor of the rods. The Comte de St. Omer was so well pleased by the outcome of the case that he had the lawyer's brief printed and widely distributed. The lawyer was young Robespierre, who was launched on his public career by this success.[78]

All of the controversies concerning lightning rods did not center on prejudice against their use. In England in the 1770s Franklin's invention became the subject of a violent political dispute, centering on Franklin's own scientific discoveries. At issue was the question of the mode of action of lightning rods and the consequent determination of their optimum shape. Franklin had argued, from analogy with laboratory experiments, that the best shape was one with pointed ends. The reason was twofold. First of all, Franklin had found in the laboratory that if a pointed grounded conductor, such as a needle held in the experimenter's hand, was brought anywhere near an insulated charged body, such as a metal globe mounted on a glass stand, then—even though there was no contact between the end of the needle and the charged object—the needle would "silently" draw off the charge of that neighboring body.[79] This phenomenon would occur, Franklin discovered, only if the grounded metal object that was brought near the charged conductor was pointed, not if the end was blunt. Accordingly, when he first thought of the lightning rod, he conceived that its principal function would be to act like the grounded needle in his laboratory experiments and thus "silently" draw off the charge of passing electrified clouds and in this way prevent a discharge. For this purpose, analogy suggested that the grounded rod should end in a sharp point. So sure was Franklin of the analogy between the lightning discharge and the electrical discharges produced in his laboratory that he published a description of the lightning rod in *Poor Richard* for 1753, even before he had made an experimental test to prove that the lightning discharge is an electrical phenomenon.[80] In a letter published in England before the first lightning experiments were performed, Franklin explained how he thought a lightning rod would work. Here is his own presentation:

There is something however in the experiments of points, sending off, or drawing on, the electrical fire, which has not been fully explained, and which I intend to supply in my next. For the doctrine of points is very curious, and the effects of them truly wonderfull; and, from what I have observed on experiments, I am of opinion, that houses, ships, and even towns and churches may be effectually secured from the stroke of lightening by their means; for if, instead of the round balls of wood or metal, which are commonly placed on the tops of the weather-cocks, vanes or spindles of churches, spires, or masts, there should be put a rod of iron 8 or 10 feet in length, sharpened gradually to a point like a needle, and gilt to prevent rusting, or divided into a number of points, which would be better—the electrical fire would, I think, be drawn out of a cloud silently, before it could come near enough to strike; only a light would be seen at the point, like the sailors corpusants.[81]

When Franklin's sentry-box experiment was performed in France (Fig. 26; also see Fig. 27), the result did more than merely confirm his hypothesis that the lightning discharge is an electrical phenomenon. The French experimenters found additionally that a grounded pointed rod would attract an actual stroke of lightning and conduct it safely into the ground. Franklin himself soon confirmed this finding, and he came to appreciate that a primary function of the rods would be their ability to conduct a stroke of lightning safely into the ground. A certain amount of confusion then arose over whether a Franklin rod might increase danger precisely because it attracted the lightning; there was always a possibility that all of the discharge would not be safely conducted into the ground. Perhaps, therefore, some scientists argued, the better course would be to make the rods terminate in knobs, or have blunt ends, so as not to attract a stroke of lightning unnecessarily and dangerously.

In 1772, while Franklin was in England, the Royal Society appointed a committee to report on the best shape of lightning rods to be used to protect the powder magazine at Purfleet.[82] Franklin was a member of the committee, as was a rival "electrician" named Benjamin Wilson, who was also a distinguished portrait painter and

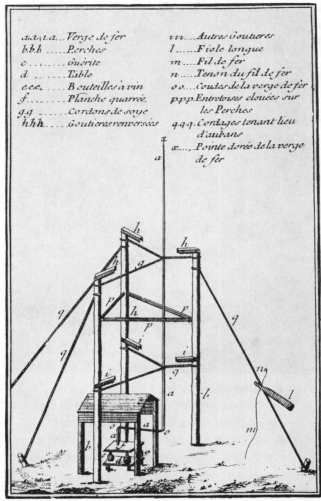

Fig. 26. The French Version of the Sentry-Box Experiment. The second edition of the French translation of Franklin's Experiments and Observations on Electricity *(1756) contains this plate showing the apparatus for the successful test of Franklin's theory of the electrification of clouds and the conclusion that lightning is an electrical discharge, as well as a validation of the principles of the lightning rod. Instead of erecting the test rod on the top of a high building, the rod is placed on a structure in an open field.*

Fig. 27. A Version of Franklin's Sentry-Box Experiment Performed in Siberia in 1761. The engraving shows an insulated vertical rod erected in the middle of an open field. The charge is collected in a Leyden jar within the shed. The scientist within the shed is making observations and measuring the altitude of the discharge, surrounded by a group of apparently overawed and frightened soldiers and peasants.

whose portrayal of Franklin is believed by many Franklinists to be the best representation of him in the middle years of his life. Wilson dissented from the report, publishing his disagreement with that part only which recommended that each conductor should "terminate in a *point*."[83]

Wilson's argument is as follows: "Dr. Franklin, in his conjectures, that lightning and electricity were one and the same fluid, considered how he should *invite*, or *bring down and collect the lightning*, so as to make experiments upon it." For this end, a pointed rod served well. But, Wilson continued, "when *curiosity*, which I apprehend was one of the first motives for introducing points to invite the lightning, was *satisfied;* and *experience* had taught us, that we had it in our *power* to *collect* that fluid which occasions it; and when the *principle* of its action was from experiments thus investigated and *ascertained,* this manner of *invitation,* viz. by using points, ought, in my opinion to have *ceased;* because a greater quantity of lightning, than we have yet experienced, may chance to attack us." Pointed ends are unsafe, Wilson concluded, because of "their great readiness to *collect the lightning in too powerful a manner,*" and he accordingly recommended that the conductors have knobbed ends. The rest of the committee, however, having read Wilson's dissenting paper, responded, "We find no reason to change our opinion."

Franklin believed that Wilson did not properly understand the action of lightning rods and wrote that "some electricians . . . recommend knobs on the upper ends of the rods, from a supposition that the points invite the stroke."[84] It is true, he added, "that points draw electricity at greater distances in the gradual, silent way; but knobs will draw at greater distances a stroke." The Society as a whole agreed with Franklin's conclusions and recommended the use of pointed conductors. Wilson continued the attack, however, publishing articles and pamphlets. In 1812 the historian of the Royal Society sadly recorded that Wilson, "by his obstinacy and improper conduct," introduced "those unhappy divisions which had so unfortunate an effect upon the Royal Society, and were so disgraceful to the cause of science and philosophy."[85] Indeed, we know from Franklin's own writings how bitter the controversy became. Franklin was pressed to "make some Answer," but in the end he did not do so because he had "an extreme Aversion to Public Altercation on Philosophic Points."

With the Declaration of Independence and the onset of the Revolution, the controversy shifted from a purely scientific disagreement and acquired political connotations. An opinion as to the "blunts" and the "points" could even be taken as an index of a person's views with respect to the case of the Americans against the crown. George III then entered the lists and, from a purely political rather than a scientific point of view, ordered blunt conductors set upon the royal palace. The king is said to have gone even further than this, however, entreating the president of the Royal Society, Sir John Pringle, to make use of his influence in order to have the Royal Society reverse its decision concerning the most appropriate shape of lightning rods. Pringle's reply noted that His Majesty might change the laws of the

Fig. 28. Destruction of an Unprotected Church. This mid-eighteenth-century illustration proves how lightning rods can protect churches and other buildings from lightning. Without a pointed and grounded lightning rod, a church (on the left side) is destroyed by the fire resulting from a stroke of lightning. In order to demonstrate the efficacy of the lightning rod, a model of a church is made with collapsible roof and walls, with a small charge of gunpowder placed inside. When the experimenter (whose hand is shown at the bottom of the picture) discharges the Leyden jar, the charge is absorbed without damage if the model has a grounded and pointed rod. In the second stage of the experiment, the situation being portrayed, a tiny knob is placed over the point. Now the discharge produces a spark (simulating a stroke of lightning) which blows up the church.

land but could not reverse or alter the laws of nature (see Fig. 28). So bitter was the reaction to Pringle's stand that he was forced from office and was replaced by Sir Joseph Banks.[86]

Franklin was fully aware of this politically oriented scientific controversy. He wrote to a friend in 1777 that Wilson "seems as much heated about this *one point,* as the Jansenists and Molenists were about the *five.*" He explained that he had "never entered into any controversy" in defense of his "philosophical opinions," which—he believed—ought to "take their chance in the world." If "they are *right,* truth and experience will support them; if *wrong* they ought to be refuted and rejected."[87] He did note, however, that he had "no private interest in the reception" of his inventions, "having never made, nor proposed to make, the least profit by any of them." He then shifted roles, writing as a politician rather than a scientist when he observed: "The King's changing his *pointed* conductors for *blunt* ones is . . . a matter of small importance to me." In fact, Franklin added, if he had a wish it would be that the king had rejected lightning rods "altogether as ineffectual." It was, Franklin held, only since George III "thought himself and family safe from the thunder of Heaven, that he dared to use his own thunder in destroying his innocent subjects."

Most observers of this political controversy, centering around Franklin's discoveries, would have agreed that the wits carried the day. A little verse, circulating in England at the time, is more eloquent than the anti-Franklin polemics.

> *While you, great George, for safety hunt,*
> *And sharp conductors change for blunt,*
> > *The nation's out of joint:*
> *Franklin a wiser course pursues,*
> *And all your thunder fearless views,*
> > *By keeping to the point.*

The Scientist as Diplomatist

Historians generally agree that Franklin's greatest service to his country was as commissioner and then as "minister," or ambassador, to France during the years of the Revolution.[88] Toward the end of 1776 Benjamin Franklin, Silas Deane, and Arthur Lee were chosen to be the three commissioners representing Congress in France. Then, toward the end of 1778, after Deane had been replaced by John Adams, Franklin was made sole minister to the Court of Versailles. He remained in this post until the middle of 1785. His achievements were monumental.

One of the outstanding needs of the American army was the help of skilled military engineers. As early as December 1775 the Continental Congress had resolved that

the Committee of Correspondence be directed to use their endeavours to find out and engage in the service of the united colonies skilful engineers not exceeding four, on the best terms they can, and that the said Committee be authorised to assure such able and skilful engineers as will engage in this service, that they shall receive such pay and appointments as shall be equal to what they have received in any former service.[89]

Franklin, a member of the Committee of Correspondence, accordingly wrote about a week later, on 9 December 1775, to his friend Charles-Guillaume-Frédéric Dumas, who on Franklin's recommendation was appointed secret agent for the American, cause in the Netherlands. After noting that the Americans had "hitherto applied to no foreign power," Franklin mentioned some pressing needs, including one that was especially relevant to the recent resolution:

> We are in great want of good engineers, and wish you could engage and send us two able ones, in time for the next campaign, one acquainted with field service, sieges, &c. and the other with fortifying of sea-ports. They will, if well recommended, be made very welcome, and have honourable appointments, besides the expenses of their voyage hither. . . .[90]

When Franklin himself went to France as commissioner in 1776, the instructions from the Continental Congress which he carried with him included the duty of engaging "a few good Engineers in the Service of the United states."[91]

Franklin was fortunate in being able to enlist the services of a French military engineer of extraordinary talents, a member of the celebrated "Corps de Génie," Louis Lebègue de Presle Duportail, an unsung hero of the Revolution. Not only did Duportail serve Washington as chief engineer during the war years, but once the war was over he drew up plans for a school to train future military engineers for the United States. His ideas contain the seeds of the later United States Military Academy at West Point. Duportail was also responsible for the creation of a national society intended to foster the creative use of science in both the military arts and peaceful pursuits such as agriculture, manufacturing, commerce, and communications.[92] The organization included both army personnel and civilians and bore the proud name of the United States Military Philosophical Society. As is shown in Figure 29, the engraved certificate of membership indicates the various pursuits of peace and war in which science and

Fig. 29. Thomas Jefferson's Certificate of Membership in the United States Military Philosophical Society. Conceived by the French military engineer Louis Duportail, this society was a national organization composed of the corps of army engineers, outstanding civilian scientists and inventors, and public officials. The motto "Scientia in Bello Pax" combines two ideas, "Science in War" and "Peace." Note the symbols of science in use for peaceful pursuits such as agriculture and fishing as well as for military purposes.

its applications might be useful. Although short-lived, this society was the first truly national scientific or engineering society to come into being in America.

In addition to engineers, the Revolutionary army needed ammunition, arms, uniforms, and other supplies,[93] together with the securing of loans to pay for them. It is difficult today to conceive the magnitude of the military needs of the new nation, cut off from all normal sources of supply in Britain. In 1776 Congress had specified these needs—to be supplied by France—as uniforms and arms for 25,000 men, 100 pieces of field artillery, and "goods to help win over the American Indians."[94] In early 1777, this demand was upped to 40,000 uniforms (and sufficient cloth to make 40,000 more), 80,000 blankets, 100,000 pairs of stockings, 1 million flints, and 200 tons of lead to make bullets and cannonballs.[95] The uniforms were manufactured and sent to America along with 20,000 muskets. Of some 3 million livres' worth of equipment, all but some 200,000 livres' worth arrived safely.[96] The costs for these items had to be met with loans, negotiated with the French government. A request was made for £2 million sterling (at 6 percent), the equivalent of some $10 million in old dollars or 47 million livres, supplementing a grant of 2 million livres from the French government plus 1 million livres as an advance from the French tax-collection agency, the Fermiers-Généraux. We may gain some idea of the magnitude of the sum by taking note that it represented about 10 percent of the annual budget of the French government.

Franklin and his fellow commissioners were also given the assignment of obtaining the services of military officers, of whom the most famous are Lafayette, Von Steuben, and Pulaski. In addition, they had the very important missions of gaining recognition for the United States as a sovereign nation and winning support in the form of military aid through the actual sending of soldiers to engage in combat against the British forces and through military action by the French navy. Another major assignment was to negotiate a true alliance with France and to gain treaties of trade and commerce.

Here we are not concerned to describe how Franklin, with the aid of his fellow commissioners, achieved these ends. That story has been presented with great insight and in fine detail by Jonathan Dull. Our study is limited to an analysis of the role which Franklin's science may have had in his extraordinary success as a diplomatist.[97] We shall concentrate on the question of whether or not his scientific skills and great scientific reputation were of significance in this achievement.

Franklin's preparation for his mission to France included two assignments in Britain during the two decades before the Revolution. The first began with his appointment by the Pennsylvania Assembly in 1757 to negotiate the differences that had arisen between the Assembly and the proprietors of the colony, the Penn family. The second began in 1764 and ended with the coming of the Revolution. This time he was colonial agent not only for Pennsylvania but also for Georgia (after 1768), New Jersey (after 1769), and Massachusetts (after 1770). His activities in behalf of the colonies involved lobbying for their causes and advocating legislation favorable to America. Although these earlier efforts ended in failure, they developed certain skills which enabled Franklin to serve his country in France with an ability and a success that surpassed the potentialities of any rival candidate.

Franklin's qualifications for the French post were neatly summed up by John Adams in a later letter to Mercy Warren, on 8 August 1807. "Who, in the name of astonishment, in all America, at that time," he asked, "had a knowledge of courts?" Franklin "alone," he wrote, "had resided in England as . . . agent at the Court of St. James's" though "despised and scorned." In "address and good breeding, he was excelled by very few Americans." In France and in Holland, Adams added, "I know that his manners, address, learning, knowledge, and good sense were acknowledged by all who conversed with him." Furthermore, Adams explained, "if by 'address' you mean graceful attitudes and elegant motions and gestures, he had received as genteel an education as any man in America; if you mean a civil and polite conversation, he was, at least, equal to any American then in Europe."[98] This statement is all the more significant in light of the many problems that arose between Adams and Franklin as fellow commissioners in France. Indeed, on an earlier occasion, on 20 September 1779, in a letter to Thomas McKean, Adams described Franklin as "a wit and a humorist," remarking, "he may be a philosopher for what I know," but "he is not a sufficient statesman for all the business he is in." According to Adams, Franklin "knows too little of American affairs, of the politics of Europe, and takes too little pains to inform himself of either to be sufficient for all these things—to be ambassador, secretary, admiral, consular agent, etc." On the other hand, Adams wrote, "such is his name on both sides of the water, that it is best, perhaps, that he should be left there; but a secretary or consuls should be appointed to do the business. . . ."[99] Adams's earlier statement was, of course, conditioned by the pressure of events, whereas the later judgment, made at some remove, is more in accord with the historical record.

Among the skills which were to serve Franklin as diplomatist in France were "a remarkable flexibility of mind, a negotiator's temperament, and an extraordinary breadth of vision."[100] Jonathan Dull has pithily summarized Paul Conner's suggestion that Franklin's "suppleness and power" of mind had a quality of "order underlying its freedom" and that this feature was connected with "the years of working with racks of printer's type" and with his "exposure to the structure of eighteenth-century science" as also with the "social experience of living and succeeding in the most orderly (yet socially and intellectually open) of American cities."[101] We have seen how, after the French and Indian War of 1754–1763, Franklin developed a vision of British America expanding to greatness and even drew on his theory of population to show that the future of the British empire would be centered in America.

Franklin's diplomatic style has been discussed in the many biographies and in various books on his activities in France. He developed a quiet and seemingly passive strategy which may be contrasted with the methods of John Adams, who seems to have been impatient for results and accordingly made peremptory demands. Adams's conduct in this regard was so offensive that the Comte de Vergennes, Louis XVI's foreign minister, refused at a certain point to treat with him at all.[102] We may well understand why, at the end of hostilities, Congress decided not to have Adams be the sole representative in peace negotiations and, instead, created a group of five commissioners which included both him and Franklin.[103]

In the early years of the Revolution, although Franklin and his fellow commissioners successfully negotiated the procurement of uniforms, arms and ammunition, and various supplies, and dealt with the problem of a safe haven for American privateers, they were not at first able to obtain a treaty of alliance, trade agreements, or a formal recognition of the new country. Yet Franklin and his fellow commissioners "enjoyed two successes." By refraining from "issuing any categorical demands," they avoided any "rupture with France." In particular, by "cultivating French public opinion," they "helped prepare for the later smooth functioning of the alliance."[104]

Franklin's efforts to win public opinion to his side differed from the campaign of pamphlets and letters to the press which had been part of the failed mission to England in the years before the Revolution. In France, Franklin seems to have chosen "the proper medium for his message," recognizing that "it was at the dinner table rather than the breakfast table that the influential in France formed their opinions."[105] Franklin was not only a guest at innumerable grand dinners but became a conspicuous figure in the salons where public opinion was being informed and molded. Thus while he was apparently having a good time in French society he was actually working in the best possible way to advance the American cause.

We must be careful lest we exaggerate the force of Franklin's personal influence. Yet Franklin's cultivation of the nobility and of the wealthy bourgeois was a recognition that these groups "were the only section of the French public to have influence at court." The members of these two groups "did not make government policy" but "could block it quickly,"[106] as Franklin's friend Turgot, the former finance minister, learned to his sorrow.

We may agree that "Franklin was the perfect revolutionary for the purpose of reassuring the French privileged class" that the Americans did not threaten the established order in Europe. He was "different enough to be interesting but familiar enough not to be frightening."[107] There was good reason for the French to be uneasy on this score, and Louis XVI would naturally be reluctant to enter a contest in which subjects were in revolt against their king. The problem of colonies versus crown must have seemed especially distressing since Englishmen had twice before shown a disrespect for the divine right and constituted authority of the crown: once in 1649, when Charles I was beheaded, and again in 1688, when James II was forced to abandon his crown and flee for safety to exile in France. Was the revolt of Englishmen in America against the authority of the crown a repetition of events, suggesting a cycle or revolution in affairs that might justify the dread of a similar revolution in France? Franklin himself, in his own person, could allay such fears. To the British and to many Americans he may have been a thorough revolutionary,[108] but, on account of his demeanor and manners and the policies that he openly espoused, Franklin and the cause he stood for did not appear to the French as threatening the established order but rather seemed to epitomize reform and reconstruction. Although a colonial by birth and education, Franklin had years of experience at the court of St. James and knew how to charm his French friends and wide circle of acquaintances. He achieved a popularity that has lasted in France until this day and that has been shared by no other American. His reputation helped turn French opinion toward war with England as an ally of America, fanning the flames of an existing Anglophobia. He himself, in his own person, aided this campaign by assuring the French that the American revolutionaries were not campaigning to undermine monarchical government in general and that they were no threat to the French monarchy in particular.

Franklin was fully aware that a central component of Vergennes's foreign policy was to lessen British influence, but he apparently did not understand the degree to which "the underlying reason was to permit France to cope with the powers of eastern Europe."[109] The French were still smarting from the wounds of their defeat in the war with England some years earlier, when they had lost Canada and had seen the eclipse of their influence in the New World. The wish to retaliate against England was a significant part of French policy. On the other hand, the expense of a war with England was likely to be a cause of bankruptcy, and there was no easy source of new revenue. There was, of course, the prospect of garnering the trade with the United States that had formerly been the exclusive commercial terrain of England. Hence Vergennes's policy at first was to encourage the Americans by providing them with means to continue the war, while not allowing France an active military role. But by 1778 more positive intervention could be offered. The chief reason, according to the traditional interpretation, is that the French did not know whether the American war was a lost cause until the defeat of Burgoyne at Saratoga so impressed them that they feared American negotiation of a separate accord with Britain. Jonathan Dull has shown, however, that there is no documentary foundation for this opinion and that the French decision to give open assistance and make treaties of commerce and alliance with the Americans coincided with the rebuilding of the French navy, which was essential if France was to engage in war with England.[110] In any case, France agreed to the alliance and became an active partner in the war, recognizing the new nation. After Vergennes had sent a minister plenipotentiary to America, Congress recognized that America must make an equal diplomatic response. It was at this time that Franklin was elevated from one of a group of commissioners to the rank of minister, the equivalent of a modern ambassador.

Franklin's posture in France has been called a "strategy of humility."[111] The diplomatist likened America's position to that of a courted virgin. "While we are asking Aids," he wrote, "it is necessary to gratify the desires, and in some Sort comply with the Humours, of those we apply to," since our "Business now is to carry our Point." But, he insisted, "I have never yet changed the opinion I gave in Congress, that a Virgin State should preserve the Virgin Character." Such a state should "not go about suitoring for Alliances, but wait with decent Dignity for the Applications of others." And so, with decent dignity, Franklin kept some constant pressure on Vergennes, obtaining various supplies for the armies and funds to pay for them, dealing with the questions of American privateers and other vital matters, until the right moment came and it was to Vergennes's advantage to conclude a military alliance and to develop treaties of trade and commerce.

Of course, in the final analysis, French intervention was predicated primarily on French interests and the national desire to humiliate Britain. In addition, there must have been a feeling that America was the proper stage on which the French should deliver a decisive blow to Britain, since it was in America that—only fifteen years earlier—France had been humiliated by being forced to cede to Britain her major territory in the New World. But there was no consensus on support for the American cause. Turgot, although a friend of Franklin's, was against active intervention, "predicting that its costs would be so overwhelming that they would postpone, perhaps forever, any attempt at necessary reform. He even went so far as to suggest that the fate of the monarchy might hinge on this fateful decision."[112] From an economic point of view, it was not even clear that the loss of the North American colonies would make a significant dent in the British economy. While it is true that the colonies were a market for British manufactured goods and supplied Britain with raw materials, notably sugar and tobacco, on balance the British were pouring more money into America than they were taking out.

Simon Schama, the astute historian of the French Revolution, has wisely observed in reference to the part played by Lafayette that it "would be naive to imagine that popularity alone could have pushed France down the road to a more aggressive intervention in the American war, had not Vergennes and Maurepas, the King's ministers, decided upon that course for reasons wholly unconnected with 'Liberty' or any other fancy modern notions." Yet, even admitting that the foreign policy of France was primarily determined by its own national political and economic interests, we may also recognize that the eventual foreign policy was made popular in France notably because of the extraordinary personality and reputation of Benjamin Franklin. Schama, in studying the background of the French Revolution, found Franklin's popularity to have been "so widespread that it does not seem exaggerated to call it a mania." Franklin was "mobbed wherever he went, and especially whenever he set foot outside his house in Passy." Indeed, he "was probably better known by sight than the King, and his likeness could be found on engraved glass, painted porcelain, printed cottons, snuff boxes and inkwells, as well as the more predictable productions of popular prints issuing from the rue Saint-Jacques in Paris."[113] Franklin was fully aware of this profusion of likenesses and wrote on 3 June 1779 to his daughter that "your father's face" is "as well known as that of the moon." And then, never able to resist a pun, he added that "it is said by learned etymologists that the name *Doll*, for the images children play with, is derived from the word IDOL; from the number of *dolls* now made of him, he may be truly said, *in that sense*, to be *i-doll-ized* in this country."[114]

In assessing the role of this popular cult of Franklin, we should particularly take note that—unlike John Adams or Silas Deane or any other member of the American delegation—Franklin was in fact an internationally famous scientist and sage, someone to be reckoned with. He came to France with an enormous scientific reputation. Only a couple of years earlier he had been elected one of the eight *associés étrangers* of the French Academy of Sciences and his book on electricity and natural philosophy had recently been issued in a sumptuous new French translation in two volumes.

In France, according to John Adams, Franklin's "reputation was more universal than that of Leibnitz or Newton, Frederick or Voltaire, and his character more beloved and esteemed than any or all of them."[115] In a similar vein, William Pitt the Elder had earlier referred to Franklin as a man "whom all Europe held in high Estimation for his Knowledge and Wisdom, and ranked with our Boyles and Newtons; who was an Honour not to the English Nation only but to Human Nature."[116] We should take note that in both of these encomiums Franklin is being associated with Isaac Newton, the supreme icon of the Age of Reason.

Not only was Franklin world famous for his scientific achievement; he was considered one of the sages of the New World. In February of 1778 there occurred a touching and widely reported scene in which, when Franklin was paying a visit to Voltaire, the French philosopher at Franklin's request gave the blessing of liberty to Franklin's grandson. In April of the same year, according to eyewitness accounts, Voltaire and Franklin met at a public session of the French Academy of Sciences and there, to "noisy acclamation," embraced. To the French public Voltaire was known as a great poet and sage, and Franklin as a great scientist and sage. Franklin's science became "a vital feature of his appeal because it seemed to be as much the work of the heart as the head: it was wisdom moralized."[117] It was this image of the scientific sage that was presented in the French version of the sayings of Poor Richard, given to the French as *La Science du Bonhomme Richard,* which became a best-seller on its publication in 1773 and which was reissued in a number of editions. This was an age of great general interest in science, a subject reported regularly in the newspapers and brought to the attention of the curious through popular lectures and demonstrations. The French public was infatuated by the image of Franklin as a scientific wizard who preached the virtues of dignity and liberty.

Benjamin Franklin personified the primary virtues and talents that were celebrated in the Age of Reason. Born and brought up in America, he seemed to symbolize the "new man" whose mind, character, and habits were formed in the rural conditions of the New World, where nature could influence intellectual and moral development untrammeled by the artificial environment of urban civilization as found in the Old World. Franklin, of course, was in no sense a "child of nature"; he was not reared in a rural environment far from the artificial influence of urban civilization. Quite the contrary! He was born and spent his early years in Boston, then a thriving metropolis, and he grew to maturity and spent the next decades living in Philadelphia, which in the 1760s was possibly "larger than any city in England except London itself."[118]

In actual fact, only on rare occasions did Franklin spend any real time in the country. Yet in France, Franklin's plain dress and fur cap symbolized the frontier. He wore no wig, but proudly exhibited his bald pate and gray curls. He was admired for his prose style, for his self-education, for his achievement in rising from poverty to a condition of success in business, for his "homely" wisdom, for his adroitness in political realms, and for his sense of humor. He was also a skilled printer who set up a private press in his Paris residence at Passy. And he had all these qualities in addition to his renown for scientific achievement.

Franklin's popularity was obviously taken into account by the French government when making certain decisions relevant to his activity. During the first part of the mission to France, when the French policy involved a combination of secret aid to America and official neutrality toward Britain, Franklin "had been officially instructed not to show himself at public assemblies." The "celebrity-seekers who had attended the April [1777] public meeting of the Academy of Sciences had been disappointed by his absence." It is significant that "royal permission and approbation of the dedication to Franklin of a scholarly book had been revoked."[119] But during 1778, when France's support became overt, Franklin was allowed and even perhaps encouraged to make public appearances, as at the Academy of Sciences or at the opera, and to receive the acclamation which was enthusiastically manifested on such occasions. He was also able to utilize this liberty and this popularity to speak openly about the American cause.[120] In the light of these conditions the special character of his renown becomes even more worthy of analysis.

As a scientist, Franklin's fame was widespread to a degree that by far exceeded that of his scientific contemporaries. The reason is that one part of his accomplishment could be easily understood. Most people do not have sufficient scientific training to understand and appreciate the niceties of important discoveries in theoretical physics. The average man and woman in Franklin's day did not comprehend the complexities of electrostatic experiments or the theories of electrical action which were the basis of Franklin's fame among scientists and the reason for his election to the Academy of Sciences. But everyone knew the terrible force of lightning and the frightening effects of its destructive power. Even if ordinary men and women could not understand Franklin's analysis of the discharge, they could easily appreciate that Franklin had shown that the immense force of lightning is merely a manifestation on a large scale of the same processes of nature that produce sparks during laboratory experiments and that lightning is the result of the same natural causes that produce sparks and crackling when we walk on deep rugs or comb our hair on a dry winter's day.

The lightning experiments enjoyed a particular popular appeal for yet other reasons. Not only had Franklin proved by experiment that the lightning discharge is an ordinary electrical phenomenon; he had gone a step further and used his own discoveries to invent the lightning rod, a device that effectively tamed this terrifying force of nature (see Fig. 30). Dr. Burney, father of the novelist Fanny, reflected popular attitudes toward Franklin's contribution when he reported a night of fear, a time of "real danger" during a major lightning storm in Bavaria. "I lay on the mattress, as far as I could from my sword, pistols, watch-chain, and everything that might serve as a conductor." He "never was much frightened by lightning before,"

Fig. 30. Models to Demonstrate the Efficacy of Lightning Rods. These three models were used at eighteenth-century Harvard to demonstrate the use of lightning rods. They also served the same purpose in popular lectures on electricity. The two steeples are made of blocks that come apart as the result of a spark discharge. The little collapsible church is just like the one shown in Fig. 28.

he concluded, but that night he longed for one of Doctor Franklin's insulated beds, "suspended by silk cords in the middle of a large room." Nothing, he wrote, could persuade the Bavarians "to put up [lightning] conductors to their public buildings," even though the lightning in Bavaria was "so mischievous, that last year, no less than thirteen churches were destroyed by it."[121]

Franklin's study of the lightning discharge had a special component that gave his work a natural appeal in the Age of Reason. It was an axiom of that time that the exercise of reason—expressed in its highest form in the sciences—should conquer superstition. Before Franklin, the lightning discharge was a source of terror. Many people believed it to be a thunderbolt hurled from heaven as the action of an angry God against sinners down on Earth. Others believed that lightning must be a manifestation of the forces of darkness. There was a custom in those days that, during times of storm, church bells would be rung in order to dissipate the harmful effects. To this end, the bells themselves often bore a Latin inscription:

Vivos voco, mortuos plango,
Deum laudo, fulgura frango.

That is, "I summon the living, I mourn the dead, I praise God, I shatter the lightning." Franklin took special note of how "lightning seems to strike steeples of choice, and that at the very time the bells are ringing."[122] Dr. Burney's experience of a night spent in fear of the lightning provoked him to observe "that the people of Bavaria were, at least, 300 years behind the rest of Europe in philosophy, and useful knowledge." Thus, he reported, "nothing can cure them of the folly of ringing the bells whenever it thunders." During the storm, Burney reported, "the bells in the town of Freising were jingling the whole night," but this noise "had not the effect of an opiate upon me," rather reminding him of the fears the storm had aroused and "the real danger" he was in.[123]

Franklin's dual role as tamer of the lightning and fighter against tyranny was the subject of a popular epigram composed by Turgot, friend of Franklin and one-time finance minister of France. "Eripuit coelo fulmen," he wrote, "sceptrumque tyrannis." Franklin "snatched the lightning from the sky and the scepter from the tyrants." The sentiment here expressed became well known when it was used as the caption for various portrait representations of Franklin.[124] One of the most famous of these is the allegorical engraving by Marguerite Gérard after a design by Jean-Honoré Fragonard, showing the aged Franklin seated on a throne, guarded by Minerva holding a shield against a bolt of lightning, with a symbolic Britain lying stricken at his feet (see Fig. 31). The association of Franklin with lightning and politics took many forms. John Adams recorded that it "is universally believed in France, England and all Europe, that his electric wand has accomplished all this revolution."[125] There seems no evidence, however, that Franklin was ever called "the electric ambassador," as some historians have maintained.

Various engravings and medallions of Franklin with Turgot's motto, along with porcelain figures and printed cloth, flooded the Paris market. For many there was the obvious implication of a link between nature and liberty, Franklin overcoming simultaneously the forces of nature and of monarchical tyranny. We may well understand that Louis XVI, who was nobody's fool, recognized the implications and was annoyed to the extreme by seeing this motto everywhere, notably as a caption to the likenesses of Franklin dis-

Fig. 31. Fragonard's Celebration of Franklin. Seated on a heavenly throne, Franklin uses his left hand to help support the shield of Minerva to ward off the lightning he has learned to conquer, while with his other hand he directs Mars to conquer avarice and tyranny. America, holding a bunch of "fasces" or sticks, symbols of a republic, sits at his left. Britain and all enemies are shown in disarray. At the bottom of the etching is the title given by Fragonard, "To the Genius of Franklin," together with the celebrated motto coined by Turgot, "Eripuit coelo fulmen sceptrumque tyrannis," that is, "He snatched the lightning from the sky and the scepter from the tyrants." The print was etched by Mlle. Gérard, the sister-in-law of the artist, Jean-Honoré Fragonard.

played for sale at Versailles. Madame Campan, first lady-in-waiting to Marie Antoinette, wrote in her memoirs that "even in the palace of Versailles, at the exhibition of Sèvres porcelains, they were selling, under the King's eyes, medallions of Franklin bearing the legend: *Eripuit coelo fulmen, sceptrumque tyrannis.*" Madame Campan has further recorded for us an index of "Louis XVI's secret feelings" in response to this situation. The king ordered from the porcelain factory at Sèvres a chamber pot in which the medallion was set at the bottom of the bowl and which he sent as a New Year's present to his mistress, Countess Diane of Poitiers, who shared the general infatuation with Franklin.[126]

There is additional evidence, should any more be needed, that Franklin's scientific achievement and reputation were very important ingredients of his public image in France. Madame Campan recorded that Franklin came with "the renown of one of the ablest scientists," to which were added "the patriotic virtues which had made him embrace the noble role of apostle of liberty."[127] This primacy of Franklin the scientist, the natural philosopher, appears also in Thomas Jefferson's observation that when "Dr. Franklin went to France, on his revolutionary mission, his eminence as a philosopher, his venerable appearance, and the cause on which he was sent, rendered him extremely popular."[128] Note that the first of Franklin's qualities listed by Jefferson is "eminence as a [natural] philosopher" or as a scientist. It is significant in this regard that in France Franklin was almost always referred to as "Dr. Franklin" and not "given the title *Monsieur*."[129]

Franklin's combination of roles—public figure, scientist, sage, and politician—was officially recognized when it was decided to halt the Mesmer craze which had gained a strong foothold in Paris and was potentially a force subversive of orthodoxy or authority on many levels: scientific, medical, and political. A committee of the most distinguished scientists in France was set up by the Academy of Sciences to investigate the cures produced by Mesmerism and the effects of the so-called Mesmeric fluid. The members of this committee included, among others, the historian-astronomer, and later mayor of Paris, Jean-Sylvain Bailly; Antoine-Laurent Lavoisier, the world's most distinguished chemist, then in the throes of reforming the whole subject of chemistry; Doctor Joseph Ignace Guillotin, a leading figure in the world of medicine, author of a "humane" invention for beheading criminals which still bears his name; and Benjamin Franklin. When the report was issued and Mesmerism discredited, popular prints celebrated the occasion. In one of them (see Fig. 32), Franklin is shown reading the commission's negative report, which he holds in his hands, while Mesmer (with the head of an ass) and his associates fly off discredited. Clearly, the report was given authority because the spokesman for the commission which produced it was the famous

Fig. 32. Franklin Confounds the Mesmerists. Benjamin Franklin is shown at the lower left, holding the report of the French commission of scientists and medical doctors to investigate the claims of cures produced by Franz Anton Mesmer and his disciples. The report concluded that there is no such thing as a "Mesmeric fluid" to cure patients' ills and that Mesmer was a charlatan. Faced with the report, the Mesmerists are shown flying off from a scene of disarray, Mesmer himself being portrayed with the head of an ass.

American. Franklin was the senior member of the commission, and his name appeared in the final *Rapport* before all the others.

It is certainly true that foreign policy in the age of Franklin or at any other time has not been made by scientists. The Fellows of the Royal Society of London and members of the Royal Academy of Sciences were not consulted in their capacity as scientific experts in determining the actions of either Britain or France with respect to the American colonies in their war for independence. It is also true that Franklin was known in France not only as a scientist but as a wit who could make his mark in Paris, as a man of popular wisdom who was author of widely read and admired maxims, as a noble sage who could assure the French that the support of America was a respectable undertaking, as a representative of rationalism who could symbolize to the French that the American cause had a great and serious meaning. No one could match his popularity, a factor which assured general public support for the cause he represented and thus made it easier for the government to support active intervention. But the chief foundation of this popularity was his enormous scientific fame. Furthermore, the reason for this scientific fame was not only that he had made electricity into a science—in fact, a new science, one not even known to Isaac Newton—and that he had discovered a scientific explanation of the lightning discharge but that he was the liberator who had freed men and women from the chains of superstitious beliefs about lightning, who had shown how to protect buildings, animals, and human beings from lightning's dreadful effects. There is hardly a contemporaneous account of Franklin's renown in France that does not stress his having been a scientific celebrity. It cannot be doubted, therefore, that his years of scientific research proved to be of primary importance in his greatest political service to his country.

Constitution and Emancipation

By the time of the Constitutional Convention, Franklin was an "elder statesman," respected for his long services to his city, state, and nation. At the age of eighty-one, he was considerably older than his fellow delegates. He also belonged to an older school of liberal thought and held notions about the form of government which were not acceptable to the framers of the Constitution. For example, he believed in a unicameral legislature, a form of government adopted only in Pennsylvania, where it did not last. He advocated rotation of public office and even supported the choice of a plural executive rather than a single president. He was convinced that executives should serve without pay. Such ideas were not favorably received by the delegates to the Constitutional Convention.

Franklin was not certain that his failing health would permit him to serve as a delegate, but in fact he attended the sessions regularly for four months. He had never been good at oratory, his forte having been the pen. He was more effective in committees than in public assemblies. At the Convention he was too frail to make any but very brief speeches; his longer speeches were read for him. Yet, in character sketches of delegates to the Convention, William Pierce reported of Franklin not only that "all the operations of nature he seems to understand," not only that he "tells a story in a style more engaging than anything I ever heard," but that he "is 82 years old, and possesses an activity of mind equal to a youth of 25 years of age."[130] Above all, he was an important force in effecting compromise on certain crucial issues. When there was a need to overcome dissension, it was Franklin who suggested that the sessions begin with prayers. His chief service was proposing the "Great Compromise," the plan in which the smaller states would have equal representation with the larger ones in the Senate, while all states would have representation according to their population in the House. This was a reversal of his earlier position, held when he was involved with the Albany Plan of Union and with the Articles of Confederation, of which he was the primary author. He had then believed—as he still did—in a direct democracy based on proportional representation in the legislative body. Many years earlier, at Albany, he had argued that the smaller states need have no fear that the larger states would "swallow" them up. At that time, he had tried to assuage the fears of the smaller states by introducing an anecdote from British history centering around a variation of the story of Jonah and the whale. At the Convention, however, he recognized the reality of the problem of the smaller states and became the spokesman for compromise, the guarantee that there would be equal representation for all states in the Senate.

On the last day of the Convention, 17 September 1787, Franklin gave an address which became known as "Franklin's Final Speech."[131] It is a masterpiece of political wisdom. Admitting that there were aspects of the new Constitution of which he did not approve, he countered that he was "not sure" that he "would never approve" them. He continued by indicating succinctly what old age could offer to a person who had lived in accord with reason and an experimental temper: "I have experienced many instances of being obliged, by better information or fuller consideration, to change my opinions even on important subjects." Furthermore, he thought that "there is no *form* of government but what may be a blessing to the people, if well administered." On the basis of these and other reflections, chief of which was the importance of adopting "a general Government" at that time, he declared his consent "to this Constitution, because I expect no better, and because I am not sure that it is not the best."

Although there is no direct evidence, I believe that for Franklin one of the important features of the Constitution was the provision for amendments.[132] Good empiricist that he was, and a successful scientific experimenter, he was all too aware that the "best laid plans" may prove on trial to be faulty, just as the nicest theories may prove to be wanting. In 1747, when he was in the midst of creating his theory of electricity, he expressed this point of view feelingly. After "further Experiments," he wrote on 14 August, "I have observed a Phenomenon or two that I cannot at present account for on the Principles laid down" in earlier writings. "In going on with these Experiments," he continued, "how many pretty Systems do we build, which we soon find ourselves obliged to destroy!" Moreover, this experience of learning by actually making trials produced yet another effect, a moral one: "If there is no other Use discovered of Electricity, this, however, is something considerable, that it may *help to make a vain Man humble.*"[133] It was both as scientist and as man of political experience that he wrote to Jonathan Shipley on 24 February 1786, fifteen months before the Convention: "You seem desirous of knowing what Progress we make here in improving our Governments. We are, I think, in the right Road of Improvement, for we are making Experiments"[134]

It was characteristic of Franklin that his last public act was in the service of his fellow human beings and that the last letter he wrote was in the service of his country. The former was connected with his continuing efforts to promote both the abolition of slavery in America and the provision of education for African-Americans.[135] During his early years, in the 1730s, Franklin, like others of that day, had accepted the institution of slavery without considering the moral implications. In his shop he was what Carl Van Doren called "a kind of general trader," dealing not only in the hiring of indentured servants, male and female, but also in the buying and selling of male and female slaves.[136] In his own household, although his servants were generally white, he also kept black slaves.[137]

In his essay on population, which was written in 1751 although first published only in 1755, Franklin argued that slaves were not an economical investment. In 1751 he was employing this insight as part of his demonstration that America would never be able to compete with Britain in manufacturing through the use of slave labor and that Britain therefore did not need to fear or prevent the growth of American manufacturing.[138] Franklin pointed out that the economic disadvantages of slavery were many, not the least being that having slaves as workers resulted in a "Neglect of Business." He stressed the cost of "Interest of the first Purchase of a Slave, the Insurance or Risque on his Life, his Cloathing and Diet, Expences in his Sickness and Loss of Time," plus the "Expence of a Driver to keep him at Work." In the version published in 1755 he expressed the belief that slaves will tend to engage in "Pilfering" because "almost every Slave" is *"by Nature* a Thief."[139] When he published a revised text in 1769, he clarified his attitude by altering *"by Nature"* to read "from the nature of slavery."[140] There are many indications that from 1756 to 1762, Franklin and his wife, Deborah, and his son William owned slaves.[141]

Yet, during the years of his stay in England between 1757 and 1762, Franklin gave much encouragement and counsel to Dr. Bray's Associates for Founding Clerical Libraries and Supporting Negro Schools in America.[142] In part because of this collaboration, a school for black children was opened in Philadelphia on 20 November 1758, and on 9 August 1759 Deborah Franklin wrote from Philadelphia that she had decided to send her slave Othello to the school.[143] In 1760 Franklin was elected member and then chairman of the Associates, and in these capacities he was able to give detailed advice and other assistance which was very helpful in the establishment of several schools for black children in American cities.[144] Before or at the beginning of the Revolution these schools all closed, but afterward Franklin's cooperation was a factor in the reestablishment of the school in Philadelphia.[145]

As the years passed, Franklin's commitment to the provision of education for African-Americans endured, and he gradually adopted a commitment to the abolition of slavery. In the spring of 1761 he was elected for a second year to the chairmanship of the Associates of Dr. Bray,[146] and after his return to America he gave attention to the affairs of various schools that had been or could be established.[147] But the most impressive account of his attitude and of his ability to change his attitude is found in a letter of 17 December 1763, written from Philadelphia to John Waring, secretary of the Associates of Dr. Bray. Here Franklin frankly admits the error of his former beliefs and argues persuasively about the intellectual ability of black people. He had just paid a visit to the Negro school and reported that he was "on the whole much pleased." On the basis of what he "saw," he had "conceived a higher Opinion of the natural Capacities of the black

Race" than he had "ever before entertained." He found their "Apprehension" to be "as quick, their Memory as strong, and their Docility in every Respect equal to that of white Children." His correspondent might "wonder perhaps that I should ever doubt it." On this score, Franklin said, he would "not undertake to justify all my Prejudices, nor to account for them." Franklin's conclusion is remarkable, as the editors of his *Papers* observe, for being "one of the earliest, if not the very first, of all statements by distinguished Americans of a belief, based on personal observation, that Negro children's intellectual capacity fully equals that of white children."[148]

In a "conversation" on slavery, published in 1770, Franklin rejoiced that "many Thousands" of Americans "abhor the Slave Trade." They not only "conscientiously avoid being concerned with it," he continued; they "do every Thing in their Power to abolish it." After remarking that British traders "bring the Slaves to us, and tempt us to purchase them," he confessed, "I do not justify our falling into the Temptation."[149] Yet this publication reads enough like an apologia for the American situation that it has been labeled "a defense of American slaveholders."[150] And later in the same year, as agent for the Georgia Commons House of Assembly, Franklin worked in Britain to secure assent to the slave code of Georgia.[151]

Two years later, in 1772, partly under the influence of correspondence with the anti-slavery Quaker Anthony Benezet[152] and in response to the freeing in England of the runaway slave James Sommersett, Franklin published in the *London Chronicle* "his first attack on the institution [of slavery] itself."[153] He began by proposing initial steps toward liberation:

> It is said that some generous humane persons subscribed to the expence of obtaining liberty by law for Somerset the Negro. It is to be wished that the same humanity may extend itself among numbers; if not to the procuring liberty for those that remain in our Colonies, at least to obtain a law for abolishing the African commerce in Slaves, and declaring the children of present Slaves free after they become of age.

To this calmly presented and moderate proposal, he added a description of the horrors suffered by slaves and then posed the outraged rhetorical questions:

Can sweetening our tea, &c. with sugar, be a circumstance of such absolute necessity? Can the petty pleasure thence arising to the taste, compensate for so much misery produced among our fellow creatures, and such a constant butchery of the human species by this pestilential detestable traffic in the bodies and souls of men?[154]

Almost two years later, on 20 March 1774, Franklin noted briefly his opinion of the capabilities of African-Americans living in freedom. The occasion was a letter to the Marquis de Condorcet answering a number of questions which Condorcet had asked, in a scientific context, about conditions in America. Observing that the free Negroes in the colonies were "generally improvident and poor," Franklin explained that they were not "deficient in natural Understanding." The basic reason for their apparent failings and failures was rather that "they have not the Advantage of Education."[155]

During the years of the American Revolution Franklin was engaged with promoting the American cause as a diplomat in France, but after his return to America he was able to turn his attention once again to the serious problem of slavery existing in the newly free nation. In 1787 he became president of the Pennsylvania Society for Promoting the Abolition of Slavery, and the Relief of Free Negroes Unlawfully Held in Bondage. This society, the first such organization to be established in the New World, had been founded by Quakers in 1775 but was inactive during the Revolution.[156] Under Franklin's presidency it would resume and extend its activity. Moreover, during 1788, in his last will and testament, Franklin left a special bequest to his son-in-law Richard Bache, forgiving him a debt of over two thousand pounds, with the request that in return "he would immediately after my decease manumit and set free his negro man Bob."[157] And during the next two years, which were the last two years of his life, he was intimately connected with four documents devoted to the prohibition or at least the discouragement of slavery and to the provision of education and employment for African-Americans who had already received emancipation.

The first of these documents was a memorial signed by Franklin as president of the Society for Promoting the Abolition of Slavery and presented to the House of Representatives on 12 February 1789. The signing of this memorial, which urged the new government to do everything within its power to discourage the traffic in human beings, turned out to be Franklin's "last public act."[158] But two related acts were to follow.

On 9 November 1789, Franklin signed, again as president of the Abolition Society, "An Address to the Public" which he may also have composed.[159] This appeal for support is connected with an "annexed plan," obviously identical or similar to the "Plan for Improving the Condition of the Free Blacks" which may also have been drawn up by Franklin sometime between 1787 and 1789.[160] The program included the establishment of a subcommittee for education to "superintend the school instruction of the children and youth of the free blacks" in two possible ways: "They may either influence them to attend regularly the schools already established in this city, or form others with this view."[161] Another subcommittee was charged not only with providing job training but also with procuring "constant employment for those free negroes who are able to work."[162]

The "Plan" is written for the most part in a style that is businesslike and dry, as befits its purpose of describing the functions of the committee and subcommittees formed to carry it out. The "Address to the Public," on the other hand, appeals to the ethics and the emotions of potential contributors to the support of the project, which extends the activity of the society beyond the goal of abolition to the further step of promoting the education and employment of those who have gained freedom. "Slavery is . . . an atrocious debasement of human nature," Franklin declared. It is so base an institution, he reminded his readers, "that its very extirpation, if not performed with solicitous care, may sometimes open a source of serious evils."[163]

Franklin could not help but be aware that, in the world he knew, black men and women appeared to be "inferior" to whites. As a scientist, however, Franklin understood that experience must always be interpreted by reason. Many years before, he had been led to realize that blacks are not by nature "deficient in natural Understanding."[164] Accordingly, there had to be a reason why they seemed inferior to whites. Their condition was not caused by some racial quality but was rather a result of the wretched conditions under which they had lived. "The unhappy man, who has long been treated as a brute animal," he declared, "too frequently sinks beneath the common standard of the human species." More precisely, the "galling chains, that bind his body, do also fetter his intellectual faculties, and impair the social affections of his heart."[165] Franklin had also come to believe that free Negroes suffer in white society because they have lacked "the Advantage of Education."[166] As always, Franklin held, "Let the experiment be made!" And so he advocated a program to help black people in order that they might not only be "restored to freedom" but prepared "for the exercise and enjoyment of civil liberty." Such a program would not only "furnish them with employments suited to their age, sex, talents, and other circumstances" but also "procure their children an education calculated for their future situation in life." It was the belief of the Abolition Society that this plan would "essentially promote the public good, and the happiness of these our hitherto too much neglected fellow-creatures."[167]

Three weeks before his death, Franklin wrote his last attack on slavery. It was also his last hoax, presented in the form of a defense by one Sidi Mehemet Ibrahim of the right and duty of Algerians to own and to sell Christian slaves.[168] Parodying the arguments made in his own day in favor of slavery, Franklin applied them in spirited rhetorical fashion to the pretended justification for enslaving Europeans a century earlier. The satire reached its audience through publication in the *Federal Gazette* in March 1790.

Nine days before his death, on 8 April 1790, Franklin wrote his last letter—addressed to Thomas Jefferson. Jefferson had asked him whether he could remember, or had any documents to help him to recall, whether the treaty of peace set the boundary of the United States between Maine and Nova Scotia at the western or the eastern "river of the Bay of Passamaquoddy." Franklin found the original map, with "that part of the boundary traced." He was ill, he wrote, and had not been able to answer Jefferson's letter at once. He now sent the needed information, adding these magnificent words, "I am perfectly clear in the remembrance."[169]

Toward the end of his life Franklin expressed his hope that France, undergoing revolution, might "by the Struggle" obtain and secure "Liberty, and a good Constitution" as the "Blessings" that would follow and "amply repair" the difficulties of the present. "God grant," Franklin wrote to David Hartley on 4 December 1789, "that not only the Love of Liberty, but a thorough Knowledge of the Rights of Man, may pervade all the Nations of the Earth, so that a Philosopher may set his Foot anywhere on its Surface, and say, 'This is my Country.' "[170]

Brilliant

4

Science and Politics: Some Aspects
of the Thought and Career
of John Adams

·····

Science in the Career of John Adams

John Adams, vice-president of the United States under George
Washington, and second president, was not a man of science.
Unlike Jefferson, for whom science was a ruling passion, Adams was
interested in science only as an educated citizen of the Enlightenment
who saw in the physical and biological sciences the highest form of
human knowledge based on reason and experience.[1] Adams did,
however, advance the sciences to the degree that he was a primary
founder of the American Academy of Arts and Sciences in Boston.
In this endeavor, he was motivated to establish a Boston-based "philosophical society" on the model of Franklin's American Philosophical
Society, with which he had contact when he was in Philadelphia in
1776 as a delegate to the Continental Congress.[2] While in Europe in
1778–1779 "in the commission to the King of France, with Dr. Franklin and Mr. Arthur Lee," Adams later wrote, he had noted the praise
bestowed on the Philadelphia organization and its publications and
he vowed to establish a similar society in Boston, "where I knew there
was as much love of science, and as many gentlemen capable of pursuing it," as in Philadelphia or "any other city of its size."[3]

On a number of important occasions, Adams's political thought
invoked scientific concepts and principles. For example, we shall see

that in a public debate with Franklin on whether a legislature should be bicameral or unicameral, Adams tried to overwhelm his political rival by citing the authority of Isaac Newton and the third of the Newtonian axioms or laws of motion. He even tried to undermine Franklin's political position by an example drawn from Franklin's own discoveries in electricity, which he had studied while an undergraduate at Harvard College. The mere fact that science appears in Adams's thought, even though it was not a principal focus of his intellectual concerns, makes him all the more an effective witness for the importance of science in the world outlook of the Founding Fathers.

John Adams received as good an education in science as was possible in America at that time. His physics teacher, Harvard's Professor John Winthrop, Fellow of the Royal Society, was an astronomer, as fully scientifically qualified as any college or university teacher of his day in Europe or in America. A later pupil of Winthrop's, Count Rumford, is reputed to have called Winthrop a "happy teacher." When Adams entered Harvard College at the age of fifteen, as a member of the class of 1755, he was fortunate in having a science professor of Winthrop's stature and ability. Winthrop's teaching went beyond mere book learning, since he was able to use Harvard's first-rate collection of scientific instruments for making demonstration experiments.[4] These were the gift of Thomas Hollis, who also endowed Winthrop's professorship, the oldest endowed chair in the sciences in the Americas, the Hollis Professorship of Mathematics and Natural Philosophy. This professorship, established in 1727, was actually the second one that Hollis endowed at Harvard, the first having been the Hollis Professorship of Divinity (1721).

In 1812, more than half a century after being graduated from Harvard College, Adams lamented that his knowledge of science was extremely limited, that he had devoted so much of his time and energy to the study of philosophy (and political philosophy) rather than to the more useful area of science. The occasion was a letter from Thomas Jefferson, informing Adams that he had abandoned politics and had ceased to read the newspapers, preferring to dedicate his intellectual energies to Tacitus and Thucydides and to Newton and Euclid.[5] Adams's reply expressed his envy of Jefferson, a man who could make an exchange of newspapers "for Newton!" He told Jefferson how sorry he was that he could not, as Jefferson had done, rise "from the lower deep of the lowest deep of Dulness and Bathos to the Contemplation of the Heavens and the heavens of

Heavens." He regretted that he himself had not "devoted to Newton and his Fellows" the "time which I fear has been wasted on Plato and Aristotle, Bacon, (Nat) Acherly, Bolin[g]broke, De Lolme, Harrington, Sidney, Hobbes, Plato Redivivus, Marchmont, Nedham," and "twenty others," on "Subjects which Mankind is determined never to Understand." He could not help but add that "those who do Understand them are resolved never to practice, or countenance [them]."[6]

Adams's Education in Science

When Adams entered Harvard, the college was—by American standards—quite venerable, having been founded more than a century earlier. Almost from the start, the Harvard curriculum had included the study of science. In 1672, President Leonard Hoar wrote a letter to the Honorable Robert Boyle, one of the leading scientific figures of the day, known worldwide for the law of gases named after him, about Harvard's scientific needs. Believing that "reading or notions only are but husky provender," Hoar announced his plan to construct an "ergasterium for mechanick fancies" and a "laboratory chemical for those philosophers, that by their senses would culture their understandings."[7] When the Hollis Professorship was established, the college was provided with funds to purchase apparatus so that the teaching of science would include experimental as well as theoretical natural philosophy—the basic principles of physics and astronomy, with some chemistry.

The science curriculum, as specified by Hollis, comprehended "Pneumaticks, Hydrostaticks, Mechanicks, Staticks, Opticks," and other parts of physical science plus algebra, geometry, and "plain & Spherical Trigonometry." In addition, students were taught "the general Principles of Mensuration, Plains and Solids," and the "Principles of Astronomy & Geography." There was instruction in "the Motions of the Heavenly Bodies according to the different Hypotheses of Ptolemy, Tycho Brahe & Copernicus."[8] By today's standards, when the science requirement in college education has been "watered down" to a bare minimum of a single course of the "general education" or the "core curriculum" variety, this was quite a dose for the ordinary students, more nearly equivalent in its intellectual content to what we would teach today in high-level introductory courses for science majors. In addition, of course, these students were required to learn mathematics on a level beyond that attained by many college freshmen of our day.

John Adams entered Harvard College in August 1750. He studied the customary college subjects: Latin and Greek, logic and metaphysics, ethics, natural philosophy (or physics), geography, astronomy, and mathematics. Much of the instruction was provided by "tutors," and seems to have consisted of memorizing and reciting.[9] Just a few years before Adams entered Harvard, in 1744, a student published a poetic "Lament" about the subjects he had to learn. It reads in part:

> *Now algebra, geometry,*
> *Arithmetick, astronomy,*
> *Opticks, chronology, and staticks,*
> *All tiresome parts of mathematicks,*
> *With twenty harder names than these*
> *Disturb my brains, and break my peace.*

This student recalled that

> *We're told how planets roll on high,*
> *How large their orbits, and how nigh.*

In a somewhat more facetious mood, the student declared,

> *If I should confidently write,*
> *This ink is black, this paper white,*
> *They'd contradict it, and perplex one*
> *With motion, light, and its reflection,*
> *And solve th'apparent falsehood by*
> *The curious structure of the eye.*[10]

Although the tutors were not scholars of real distinction, the Hollis Professor was a first-rate scientist. Appointed to his post in January 1739, seven years after receiving his A.B. from Harvard, Winthrop gained scientific fame as an astronomer, but his scientific endeavors embraced meteorology, mathematics, and geology. His observations of the transit of Mercury and of a lunar eclipse in 1740 produced his first communication to the Royal Society, of which he was later to become a Fellow (1766). A member of the American Philosophical Society (1768), he was awarded an honorary LL.D. by the University of Edinburgh (1771) and another by his own college (1773). In 1761 he led a college expedition to Newfoundland to observe the transit of Venus. Winthrop was thus a practicing scientist, recognized for his researches, and not a mere tutor or lecturer.[11] He was an ideal teacher for Adams, whose greatest intellectual interests as an entering student were science and mathematics.

Lecture Notes

There are several sources that enable us to reconstruct Adams's learning experience as a student of natural philosophy or science. First, we know the contents of the lectures and scientific demonstrations given by Winthrop, which Adams attended, because Winthrop's lecture notes have been preserved.[12] We also have a detailed list of the scientific instruments in the Harvard collection which Winthrop used to illustrate and explain the abstract principles he was teaching,[13] some of which would have been used by Adams and other qualified students. An additional source, which was discovered too late to be used by his biographers, is a diary which Adams kept from June 1753 to April 1754, covering Adams's junior year as an undergraduate student[14] and including notes about Winthrop's lectures. Thus we know not only the topics which Winthrop presented but the detailed contents of the lectures and the nature of the scientific demonstrations which Winthrop performed to illustrate the main points. These notes also enable us to ascertain the degree of Adams's comprehension of the principles of physics which Winthrop had been presenting.[15] The students in Winthrop's course would also have studied parts of the standard textbooks of natural philosophy of those days, for example, the two-volume work by W. J. 'sGravesande.

Some of the topics of Winthrop's lectures as recorded in Winthrop's own notes appear in Adams's later political discourse. One was the "Mechanical Powers," that is, the simple machines such as the balance, the pulley, the lever, and so on. Another was "The Laws of Motion" which "Sir Isaac Newton has laid down," that is, a discussion of the "3 Laws of motion, by which every thing that belongs to Motion may be Explained." Each of the three laws was presented by Winthrop and then elucidated and illustrated, concluding with the observation that by means of "the 3 Laws of Motion or Nature," all of nature's "Phaenomena may be Solved." Winthrop told his students, "Indeed all Mechanicks are nothing but Different Applications of These Laws."

Another primary subject was "Gravity." Here Winthrop explained that gravity is defined as "that power by which all bodies tend toward one another." This chapter concludes: "What is here Said of the Earth may be Said of all the Planets. Hence, 1st it may be Inferred that Gravity affects every particle of Matter, & 2d that the Satellites of Saturn & Jupiter Gravitate to their Primarys; as the Moon (which is our Satellite) Doth towards the Earth; & as all the Planets & Comets Do toward the Sun. But Whether the Power of Gravity is Infinitly Extended or not is not yet known; though it is known to extend 5 times the Distance of Saturn's Orbit, by the Eccentricity of Some of the Comet's Orbit's. 3dly it follows that Gravity is the Same in all bodies whatever be the constitution & texture of their Parts, & 4thly whatever be the Cause of Gravity it Penetrates through all Sorts of Bodies, & that even to the very centre."

Winthrop's course was very up-to-date. That is, he included the latest subjects of scientific research, such as electricity, which was then being developed into a science by Benjamin Franklin. Winthrop later became a correspondent and friend of Franklin's. It was through Franklin's influence that Winthrop was awarded an honorary degree by the University of Edinburgh. In the course which Adams was attending, Winthrop gave only eight lectures because he was away from Cambridge on a trip to Philadelphia, where he met Franklin for the first time.[16]

"This Electricity," Winthrop told his students, "Since the Year 1743 has made a Considerable noise in the World." The year 1743 marked the discovery of the Leyden jar, the first condenser or capacitor, which greatly increased the scale of electrical experiments. It is supposed, Winthrop said, that several "of the (at present hidden) Phaenomina of Nature" depend on electricity. We shall see below that Adams became interested in this new scientific subject and made notes about the recently invented lightning rod. We know that Adams was familiar with some parts of Franklin's book, *Experiments and Observations on Electricity* (London: 1751–1753–1754 and later editions), which took the form of a collection of letters and articles, since later in life, while a delegate to the second Continental Congress in 1775, Adams recalled Franklin's theory of storms of thunder and lightning, referring to the book as "Franklin's Letters on Electricity." I have not been able to find out whether Adams became acquainted with Franklin's book while still an undergraduate or later.

The text of Winthrop's lecture notes shows that he carefully explained each of Newton's three laws of motion. He spent more time on the third law than on the first two, since he was aware that students might misinterpret it. The law, as stated by Newton in the *Principia,* reads: "To any action there is always an opposite and equal reaction; in other words, the actions of two bodies upon each other are always equal and always opposite in direction." In explanation, Newton wrote: "Whatever presses or draws something else is pressed or drawn just as much by it." He gave two examples, of which the first reads: "If anyone presses a stone with a finger, the finger is also pressed by the stone." According to the second example, "If a horse draws a stone tied to a rope, the horse will (so to speak) also be drawn back equally toward the stone, for the rope, stretched out at both ends, will urge the horse toward the stone and the stone toward the horse by one and the same endeavor to go slack and will impede the forward motion of the one as much as it promotes the forward motion of the other."

Winthrop went to some pains to explain to his students that the third law does not relate to equilibrium since the forces of "action" and "reaction" are not applied to the same body. The reader can easily understand the difference between conditions of equilibrium and of the third law by considering an example. Let there be a case of equilibrium in which two forces act on one and the same body; for equilibrium to result, the two forces must be equal in their magnitudes but exactly opposite in the directions in which they act. An example would be a ten-pound object suspended by a spring. Two opposite forces act on the ball. One is the force of the weight arising from Earth's gravity pulling the ball downward, the other the upward force exerted on the ball by the tension of the stretched spring. Both forces will be of the same magnitude, so as to "balance" each other. Both forces act on the same body, the ball.

Winthrop's notes for his lectures show how, in example after example, he illustrated the ways in which action and reaction forces are not exerted on a single body, but rather on separate bodies. One such example was a stretched cord, in which the cord "Bears as much against the body that Stretched it, as the body Bears against the Chord." He discussed how a man in a boat "pulls another to him," in which case he is "as much" pulled toward the other boat as the other boat is pulled toward him. Other examples included gravitational attractions and the action of a magnet on a piece of iron. Winthrop noted that, in the latter case, "it is proved by Experiment that the Iron Draws the Load-Stone as much as the Load-Stone Draws the Iron." That is, in accordance with Newton's third law, the iron must pull on the lodestone or magnet with exactly the same amount of force that is being exerted in the opposite direction by the magnet. In this case, however, the two forces are not acting on one and the same body, since the magnet acts on (or draws) the iron and the iron acts on the magnet. Because there are no equal and opposite forces acting on the same body, there is no equilibrium and the magnet will actually draw the iron to it. It is to be observed that in the example given by Newton of a horse pulling a stone, the horse moves forward dragging the stone along with it. That is, the third motion clearly does not imply a condition of being at "rest," with all the forces acting on a single body being in equilibrium.

The third law leads to important consequences for Newton's system of the world based on a force of universal gravity. As Newton explained in the *Principia,* the Sun keeps the Earth in an orbital path by a force that pulls the Earth inward toward the center, toward the Sun. But, according to Newton's third law, the Earth must also pull the Sun toward itself with a force of equal magnitude. In an extreme illustration, the third law requires that an apple must pull the Earth upward with exactly the same amount of force with which the Earth pulls the apple downward.

I have stressed the meaning of Newton's third law because this is the law later invoked by Adams in a political debate with Benjamin Franklin. As we shall see, Adams introduced Newton's law of "action equals reaction" in order to overwhelm Franklin's argument for a unicameral rather than a bicameral legislature. We shall thus have occasion to find out whether Adams any longer had in mind the clear explanation given him by Professor Winthrop when he was a Harvard undergraduate.

On 9 April 1754, Adams recorded in his diary that "Sir Isaac Newtons three laws of nature . . . together with the application of them to the planets," were "proved and illustrated" in Professor Winthrop's lecture. The teacher had shown that the planets "are kept in their orbits by two forces acting on them, viz that of gravity and that which is call'd their Centrifugal force whereby they strives to recede from the Center of their orbits and fly off in tangents."[17] In an earlier note, entered on 1 April 1754, Adams recorded Winthrop's general preliminary discussion of the principles of science and the laws of nature. He there referred to Winthrop's presentation of a "motion [which] is subject to Certain laws which he explained and I have forgot." But "thus much" he did remember: "that motion, produced by gravity, was universally in right lines, from the body acted on by gravity, to the Center of gravity, as the Center of the earth for instance, or the like." Adams's diary, unfortunately, does not contain any specific information concerning his comprehension of Newton's third law. Since he was a bright student, especially interested in physics and mathematics, however, we may—I believe—safely assume that under Winthrop's expert tutelage he learned the correct form of the law and its applications.

Equilibrium: The Study of Forces

Since the notion of balance was of real significance in Adams's political thought, we may take special note that this part of physics, the subject generally known as statics, was developed at length in Winthrop's lectures. In the world of nature and in common experience, there is rarely an equilibrium produced by only two forces. And so we may understand why Winthrop and other teachers or writers of textbooks in that age stressed the equilibrium produced by three or more forces. Indeed, in the cases where only two forces are acting, they usually are not exactly equal. When the forces do happen to be equal, they tend not to be exerted exactly along the same line of action and so there is no balance, no equilibrium.

Of course, in nature there is one condition in which two forces occur that are exactly equal and opposite: the case of forces of "action" and "reaction" in accord with Newton's third law. Thus the Sun's gravitational force on the Earth has an exactly equal and opposite counterpart in the gravitational force of the Earth on the Sun. But, as we have seen, in this case the forces act on different bodies (one on the Sun, the other on the Earth) and thus do not balance each other and cannot produce a condition of equilibrium.[18]

In nature and in common experience, and in examples studied in courses on natural philosophy or physics, students have traditionally encountered cases in which three or more forces are applied to one and the same object to produce equilibrium. A classic example is that of a sliding object (such as a sled) being held in place on a frictionless or slippery surface (such as an icy slope) by a cord which is parallel to the slope. This is the famous problem of the inclined plane, a subject studied in elementary natural philosophy or physics ever since the Scientific Revolution. A classic analysis of the forces on an inclined

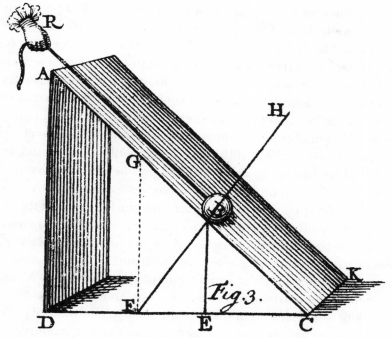

Fig. 33. Analysis of Forces on an Inclined Plane.

plane is found in John Keill's textbook on natural philosophy, which was reprinted again and again (both in the Latin original and in English translation) during the first half of the eighteenth century. (A diagram of this analysis of forces, taken from Keill's book, is reproduced in Fig. 33.) Keill's book was based on lectures to students at Oxford, where he was a professor, and it took on a quality of high authenticity from the fact that Keill had been a close associate of Isaac Newton.

Keill analyzed the forces acting on a small cart or ball or other object (marked in the diagram by the letter B), which is held in place on an inclined plane by a cord BR (parallel to the inclined plane AC), held in the experimenter's hand (R). There are, Keill explained, three forces in equilibrium. The "first is the Force of Gravity acting according to the direction BE"; this is the weight-force which always acts straight downward or vertically. The "second is the Power R drawing the Body according to the Direction BR parallel to AC," that is, the force exerted by the experimenter's hand pulling on the string. The "place of the third Power is supplied by the Resistance of the

Plane [or the reaction] acting according to the Line BH perpendicular to it." Keill showed the students how to use a "force diagram" in order to find both the force exerted by the string and the "reaction" of the surface for a ball or cart of known mass on a plane of any given inclination. If the weight of the cart is known, then the geometry of the situation enables the student to compute both the force of reaction of the surface and the force exerted through the string.

In a variation of this experiment, the string would pass over a pulley at the top of the plane; different weights could then be hung from the end of the string. In this way an experimenter could verify the correctness of the computed value of the force exerted by the string. Sometimes the apparatus would be set up in such a way that the angle of inclination of the plane could be varied. Such demonstrations offered the students convincing proofs of the correctness of the principles of equilibrium of three non-collinear forces.[19]

Another example of the equilibrium of three forces made use of an ingenious experimental device. Known as a "force table," this instrument was discussed in many of the textbooks of natural philosophy in the middle of the eighteenth century. For example, it is described W. J. 'sGravesande's popular and often reprinted work, *Elements of Natural Philosophy, confirm'd by Experiments,* which bore the subtitle, *An Introduction to Sir Isaac Newton's Philosophy.* This device consisted of "a round Board of about 8 Inches Diameter," in "a horizontal position," mounted on a pedestal (see Fig. 34). "Round the edge of it, within the Thickness of the Wood, is a Groove whereby pulleys are applied at Pleasure to any Part of the Circumference." Each pulley has "a brass Plate perpendicular to it, which fits into the Groove." One of the experiments performed with this machine involves three pulleys, set at any given angle to one another. Three threads are knotted together and then each of the free ends is passed over one of the pulleys. Weights are now attached to the ends of two of the threads and the experimenter then tries different weights at the end of the third thread until equilibrium is obtained. A variant of the experiment would begin with three weights and then have the pulleys moved around with respect to one another until there is a condition of equilibrium (see Figs. 35 and 36). Equilibrium occurs when the knot is free at the center and not pulled toward one or the other pulley. In interpreting the results, a force diagram is drawn, just as in the case of the inclined plane. Lines are drawn having the same angles between them as the angles between the cords (from

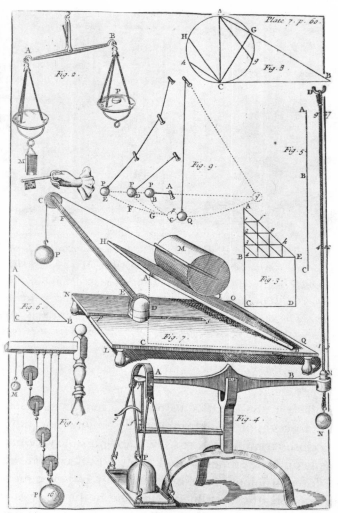

Fig. 34. Demonstrating the Equilibrium of Forces on an Inclined Plane. A plate from W. J. 'sGravesande's popular Mathematical Elements of Natural Philosophy *shows a variety of problems of forces. At the center is an apparatus consisting of a smooth plane (AHOQ) which can be adjusted at any angle. A roller (M) is held in place (or can be pulled upward along the plane) by a string which passes over a pulley C and has a known weight (P) attached to it.*

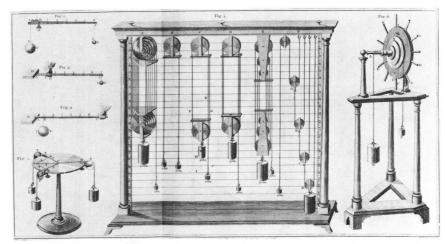

*Fig. 35. Equilibrium of Forces in the Mid-1700s. Another illustration from 'sGraves-
ande shows various combinations of forces in equilibrium. At the bottom, lower left, a
circular force table consists of three pulleys, which can be set at any angle to each
other along the rim. Three strings are knotted together at the center; each string then
passes over one of the pulleys and has a weight attached at the end. In science courses
at the time when John Adams was a college student, this device was used in an experimental test of the mathematical analysis of the equilibrium of three forces.*

knot to pulley) in the force table, and of a length representing the
weights or forces. For equilibrium, these lengths and angles must
produce a closed triangle. There were many possible variants of these
experiments and the student could easily advance from the case of
the three-force problem to one involving four and even more forces.

In the example of the inclined plane and in the study of the force
table, the subject is statics or the equilibrium of balanced forces acting
on a body at rest. The Newtonian physics of the *Principia* was concerned with a quite different subject, namely, dynamics, the physics
of unbalanced forces acting to produce accelerations or changes in
the state of motion or of rest of a body. The goal of the *Principia,*
unfolded progressively in three "books," was the study of forces and
motions, the different kinds of paths or trajectories that bodies would
follow under the action of unbalanced forces. Newton did treat of
statics—the resolution and composition of static forces and the action
of simple machines—but only briefly, in a single scholium following
the laws of motion. This subject is not further explored in the main

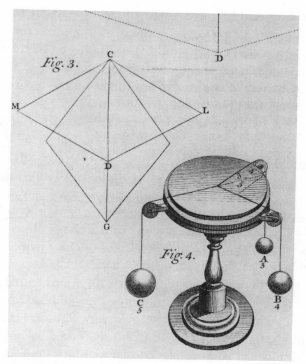

Fig. 36. Demonstration of the Equilibrium of Three Forces. An eighteenth-century textbook shows the force table in perspective. Students in Adams's day would try various combinations of weights and angles in order to find the conditions of equilibrium empirically as well as to check the results of theoretical computation. The use of a force table would produce a vivid impression of the way in which three (and even four) forces produce a condition of equilibrium.

body of the text and is not a principal part of the physics of the *Principia*.

We should take note that historians have tended to stress that part of physics which Newton most advanced and to which the *Principia* is almost exclusively devoted, that is, the Newtonian laws of motion and their applications in studying the accelerations produced by various kinds or conditions of forces, plus the elaboration of the Newtonian system of the world according to the law of universal gravity. But in Adams's day, the subject of statics was an important part of natural philosophy, providing an example of the application of rational principles to the phenomena of nature and the use of mathematics in physical situations. The naked logic of the analysis of forces on the inclined plane was an impressive exercise of the mind for students in the eighteenth century as it would continue to be to their counterparts in the nineteenth and early twentieth centuries. I can vividly remember the thrill of the analysis of forces on the inclined plane as presented in my own freshman physics course.

The analysis of problems involving three forces would certainly have made a deep impression on young John Adams, all the more so in that it would have had a strong appeal to his appreciation of logical argument. Later in life, as I have mentioned, Adams developed his political theory around the concept of a balance of power, stressing that stable equilibrium conditions require more than two forces. He founded his "science of government" on mechanical principles, not those of Newton's *Principia*, not dynamics or the physics of celestial forces and motions, but statics—the science of balanced forces, of equilibrium. If there are two powers or forces, he argued, one must always be stronger and will prevail unless checked by a third force.[20] In physics, a struggle between two powers will always continue "till one is swallowed up or annihilated, and the other becomes absolute master."[21] Similarly, if there are two legislatures, "one or the other will be most powerful, and whichever it is, will continually scramble till it gets the whole."[22] A third power is thus always needed to "preserve or restore the equilibrium" by throwing "weights into the lightest scale."[23] In short, "there can be, in the nature of things, no balance without three powers."[24] Thus, by analogy with physics, a stable society requires three "orders," with "a third to aid the weakest," a system of "three different orders of men *in equilibrio*" or of "three orders forming a mutual balance," producing the "only natural" and "complete," that is, "triple" or "tripartite" balance.[25] Similarly, the government itself, the sovereign power, requires a division into "three equiponderant, independent branches," because "in this way, and in no other, can an equilibrium be formed."[26] In reading Adams's statements about forces and powers in his *Defence of the Constitutions of Government of the United States of America* and other writings, one is sometimes hard put to discern whether he is writing about politics and social issues or about the science of statics. There is no possible doubt that his study of the equilibrium of forces under John Winthrop had left an abiding impression on his thoughts and that he made good use of the examples from the mathematical physics of statics which he had studied as an undergraduate.

Adams and Franklin's Science

Adams's later hostility toward Benjamin Franklin became intense during their association in Paris as members of the American mission during the years of the Revolution. In part, Adams's feelings concerning Franklin must have had a component of jealousy. Adams's journals and letters express his frustration and anger on being constantly eclipsed by the general esteem Franklin attracted as a world-famous scientist and as a leading public figure. He was constantly made aware that Franklin's writings were being published in French translation, and he saw portrait prints of Franklin for sale in the shops. He finally believed that Franklin was paying too much attention to social affairs, while he himself was doing the hard work, and that Franklin was taking on to himself too much credit for the great events in the New World that had brought them both to Europe.

About a dozen years later, in 1790, Adams repeated this expression of paranoid jealousy of Franklin's universal fame, once again involving a political parody of Franklin's electrical science. In a letter to Benjamin Rush, the Philadelphia physician and pioneer in the treatment of mental disorders, Adams expressed his disappointment that his country had not given him recognition for his services. Adams's lament is all the more interesting to the degree that it suggests by implication that Franklin's importance during the years of the Revolution was a distortion of history caused by Franklin's enormous scientific reputation as founder of the science of electricity.

The occasion for the outburst was the furious public antagonism to Adams aroused by the printing of his *Discourses on Davila,* an attempt to prove to his countrymen that they were making a false and misleading analogy between the principles of their own revolution and those of the French Revolution. Adams was pilloried for his conservative views, especially those concerning natural hierarchies or orders in society. As his biographer has noted, "the furor that the *Discourses* aroused" was so dismaying to Adams that it "activated his paranoia once again." He felt that his services to his country had never been appreciated, and he came to conclude, as he wrote to Benjamin Rush, that the "history of our Revolution will be one continued lie from one end to the other." He continued: "The essence of the whole will be *that Dr. Franklin's electrical rod smote the earth and out sprung General Washington. That Franklin electrized him with his rod— and henceforward these two conducted all the policy, negotiation, legislation and war.*" These lines, he remarked, "contain the whole fable, plot, and catastrophe." Today's reader of these lines will be in the position that Adams then referred to, when he wrote about what a "reader will say" if "this letter should be preserved and read a hundred years hence." Such a reader, according to Adams, "will say, 'The envy of . . . JA could not bear to think of the truth,' " and will conclude that Adams scribbled to Rush the "blasphemy that he dared not speak when he lived."[27]

John Adams's first contact with Franklin must have been in 1753, while he was still an undergraduate. In that year Harvard College bestowed on Franklin an honorary A.M., the first true honorary degree to be awarded by an American college. Adams would have been introduced to Franklin's important researches in electricity as part of the lectures and demonstrations on that subject by Professor Winthrop, who was a scientific colleague of Franklin's.[28] Adams would have learned, in particular, about Franklin's spectacular experiments proving that clouds are electrified and that the lightning discharge is an electrical phenomenon. Some forty years later, as we shall see in a moment, Adams would make use of some remembered aspects of Franklin's discoveries in his efforts to confute Franklin's own plan for a legislature consisting of a house of representatives without a senate.

Not long after Adams's graduation, Franklin's invention of the lightning rod became a subject of controversy in the greater Boston area, generating a pamphlet literature of controversy which attracted Adams's attention. At issue was a theological controversy over the propriety of using lightning rods. Franklin's invention was rather generally esteemed because it enabled people to protect their structures—homes, barns, churches, and other public buildings—from destruction by lightning. But in Boston the invention became the subject of theological debate when a local preacher, the Reverend Thomas Prince, pastor of the Old South Church in Boston, began to attack its use in his sermons.[29] Prince was stimulated to oppose the use of lightning rods following an earthquake that occurred in Boston on 18 November 1755, just a few months after Adams was graduated from Harvard College. This event followed by just seventeen days the great earthquake which wreaked enormous destruction in Lisbon and which was the occasion for Voltaire's fierce satire *Candide*. The news of the Lisbon earthquake had not as yet reached Boston, but—as has been remarked—the "people of New England" were "frightened enough without it."

We can gain some measure of the severity of the earthquake from the description recorded by John Adams. He was jolted to such a degree that he recorded the event in a diary. "We had a very severe shock of an earthquake," he wrote, that "continued near four minutes." He was then staying at his father's house in Braintree and was awakened from his sleep by the force of the tremors. "The house seemed to rock and reel and crack," he recorded, "as if it would fall in ruins about us." The severity can be gauged by Adams's comment that chimneys "were shattered by it within one mile of my father's house."[30]

A week later, on 25 November, Prince published *An Improvement of the Doctrine of Earthquakes,* declaring that such phenomena are divine providences, messages sent as warnings to earthly sinners. On the next day, 26 November, Professor Winthrop read a lecture in the Harvard chapel on the physics of earthquakes, explaining the recent event as a natural phenomenon. In order to help quell any superstitious interpretations, Winthrop decided to give his interpretation as wide a circulation as possible. Accordingly, he sent his text to the printer so that its message could be disseminated beyond the confines of the Harvard College Yard. While Winthrop's lecture was still being readied for publication in the printer's shop, Prince brought back into print a sermon he had published in 1727, on the occasion of an earlier earthquake. The title suggests the nature of Prince's message, *Earthquakes the Works of God.* A postscript bearing the date 5 December 1755 introduced an electrical component into the theory of earthquakes. That is, he went beyond the mere declaration that the earthquake had been sent from heaven as a warning to the inhabitants of Earth of their sinful condition, as a sign for them to repent and reform. He now adopted the theory that God produces earthquakes by means of the electrical substance and he declared that the recent earthquakes could be directly attributed to the introduction of the lightning rods invented a few years earlier by "the sagacious Mr. *Franklin.*"[31]

Before condemning Prince for such a fanciful theory we should take note that at that time a number of the leading members of the scientific establishment, the accepted scientific community, believed that earthquakes can be caused by the electrical fluid. Franklin's electrical discoveries, including experiments in which there were violent discharges, lent credence to this theory. The notion that earthquakes are caused by the accumulation of electrical fluid in the earth was espoused by such scientific luminaries as Stephen Hales, the founder of plant physiology and foreign secretary of the Royal Society.[32]

Prince's argument was simple and straightforward. The more lightning rods are erected, he wrote, the more the earth must become "charged" with "the *Electrical Substance*." Any region of the earth that becomes "fuller of *this terrible Substance*" may become "more exposed to *more shocking Earthquakes*." In short, he argued, by conducting the lightning safely into the earth, Franklin's lightning rods acted so as to circumvent the action of God's wrath. But of course, Prince argued, "there is no getting out of the mighty Hand of God!" If we avoid God's wrath "in the *Air*," we cannot do so "in the *Earth*," where "it may grow more fatal."

When Winthrop read this postscript, he hastened to the printer to add a seven-page appendix to his *Lecture*. There is no record that Adams ever read Prince's sermon, but he did obtain a copy of the rebuttal by his old professor, John Winthrop, and entered some very perspicacious annotations in the margins. Winthrop countered Prince's earthquake theory with a non-electrical explanation. He was particularly concerned that Prince's postscript would "fill with unnecessary terror the minds of many people" and—even more important—"discourage the use" of lightning rods, an invention "which, by the blessing of God, might be a means of preventing many of those mischievous and sorrowful accidents, which we have so often seen to follow upon thunder storms."[33]

Adams's copy of the pamphlet by his former professor, with his annotations, may still be read in the copy preserved in Adams's library, now in the Boston Public Library. From these handwritten remarks we learn that many people agreed with Prince, considering "Thunder and Lightning as well as Earthquakes, only as Judgments, Warnings etc. and have no Conception of any Uses they can have in Nature." Adams had heard "some Persons of the Highest Rank" say that "they really thought, the Erection of Iron Points was an impious Attempt to robb the Almighty of his Thunder, to wrest the Bolt of Vengeance out of his Hand."[34]

On the final blank leaf of the pamphlet, Adams recorded some general thoughts about innovations. "The Invention of Iron Points to prevent the Danger of Thunder," he wrote, "has met with all that

Opposition from the Superstitions, Affectations of Piety, and Jealousy of New Inventions, that Inoculation to prevent the Danger of the Small Pox, and all other usefull Discoveries, have met with in all Ages of the World." In this early stage of his life Adams sided with Benjamin Franklin and did not show any sign of the antipathy that was to develop after their close personal contact in France.

Franklin's science came to Adams's mind in the summer of 1777, while he was in Philadelphia as a delegate from Massachusetts to the Congress. The heat in August was greater than anyone could remember for the last twenty years. Barely able to sleep, he was "enfeebled and irritated," as he wrote to his wife, Abigail, complaining that the "horrid hot weather melts the marrow within my bones and makes me faint away almost." With the temperature standing at over 100 degrees in the shade, Adams could draw small comfort from the thought of Admiral Howe and the British fleet sweltering on the ocean. When the heat spell finally broke, it did so in a thunderstorm of enormous proportions. He recorded the splendid spectacle of the "world . . . all of a blaze with lightning," with "grand rolls of thunder" shaking "the very chamber" in which he was staying. "The windows jar," he reported, while "the shutters clatter, and the floor trembles." Observing this grand spectacle of nature's great power, Adams recalled his reading about these phenomena and took note that "Dr. Franklin in his letters on electricity," that is, in his book entitled *Experiments and Observations on Electricity,* had "explained the philosophy [that is, the physics] of it." Perhaps because he remembered Franklin's scientific explanation, Adams appears not to have been frightened by the thunder and lightning but was simply overjoyed to behold the cooling rain fall upon the city "most delightfully."

Adams and the Concept of Political Balance

Like many other political thinkers of the founding age of the American republic, Adams seems to have been obsessed with the idea of a balance. Adams's discussion of the balance in relation to the American system of government invokes a feature of the Constitution generally known as the separation of powers and is directly related to the doctrine of "checks and balances." Every American schoolchild learns of the balance among the executive, the legislative, and the judicial branches of government. The early discussion of this notion, as in the *Federalist* papers, which are considered in the next chapter, centered not so much on this tripartite balance as on an opposition of the powers of the two branches of the Congress, the Senate and the House of Representatives. Often the concept of political balance is linked to the physical action of a machine or mechanical device and is even said to be based on an analogy from physics, specifically the Newtonian system of physics. Two of the major exponents of this latter view have been Woodrow Wilson, while still president of Princeton University, and the British science writer and historian J. G. Crowther, who has written of "the Newtonian notions of checks and balances, and mechanical equilibrium, . . . introduced into the Constitution."[35] These allegedly Newtonian interpretations are presented in the next chapter. There we shall see that the discussions and debates concerning the Constitution did indeed often invoke principles of science, but these were generally not Newtonian.

The balance of forces, equilibrium or equipoise, is a part of physics known as statics, the science of forces at rest. Newtonian physics, as has been pointed out, is concerned with a different subject, dynamics, the physics of forces and accelerations, rather than statics. The subjects studied in the *Principia* are examples of non-equilibrium, of unbalanced forces that produce accelerated motion. As I have noted in Chapter 1, equilibrium or balance of forces implies, according to Newton's first law of motion, that there must be a condition of rest or of uniform motion—that is, no acceleration, and hence no gravity, no Newtonian dynamics. (On the principles of Newtonian dynamics as expounded in the *Principia,* see Supplement 2.)

In any event, the concept of political "balance," and notably "balance of power," long antedates the scientific work of Isaac Newton. According to the *Oxford English Dictionary,* an appearance in English of this phrase occurred in 1579, about a hundred years before the publication of Newton's *Principia,* in a translation by Sir Geoffrey Fenton of Francesco Guicciardini's *The Historie of Guicciardini Containing the Warres of Italie,* written before 1540. The concept of a balance of power, in the sense in which it became current in the eighteenth century and in which we still use it today, was traced by John Adams to the writings of Niccolò Machiavelli in the sixteenth century. As Adams was aware, the concept of balance of power came into prominence in the middle of the seventeenth century. It did not arise in the direct context of physics, however, in relation to machines or to the balance of statical forces. Rather it came into circulation in a political context through the writings of John Harrington, primarily his *Oceana*. This work, displaying a kind of Utopian system of government, was published in 1656, more than thirty years before the appearance of Newton's *Principia*. Harrington's *Oceana* was not only chronologically pre-Newtonian; it was ideologically committed to the thesis that the physical sciences are of absolutely no use as a source of analogies for political discourse. Harrington pilloried Thomas Hobbes for having drawn on physics and astronomy in constructing his political system.

It has already been mentioned that Harrington drew heavily on the concepts and theories of William Harvey, that his political writings are replete with analogies drawn from the Harveyan physiology. So squarely did he base his political system on the principles of the new science of biology that those who discussed and applied his ideas during the days of the Constitutional Convention could not help but be aware that the notion of balance of power had been set in the physiology of Harvey rather than the physics of Newton.

Harrington was a remarkable thinker who has been esteemed as the founder of the economic interpretation of history.[36] The reader may be reminded that British and American historians have long been aware that Harrington exerted a very important influence on the framers of the American system of government. This subject was developed in a scholarly article by Theodore Dwight in 1887 and was taken even further by H. F. Russell Smith in 1914. More recently, the impress of Harrington's ideas on the American Constitution has been a feature of a monograph by Samuel Beer, published in 1993.[37]

Although Harrington's political ideas were set forth in a general framework of Harveyan biology, even being considered a form of "political anatomy," the particular concept of balance of power was not developed in *Oceana* in biological analogies. Rather, the principle of the balance was advanced by Harrington in economic terms, as part of his general belief that economic forces must affect politics, that political power cannot be considered separately from its economic base. This presentation not only was a prominent feature of the opening sections of Harrington's *Oceana*, but was stressed in John Toland's biographical introduction to Harrington's works in all the editions available to readers of the eighteenth century, published in London in 1700, 1737, 1747, and 1771. John Adams had a copy of both the 1747 and the 1771 editions in his personal library.[38]

Toland put Harrington's principle simply and straightforwardly;

it is that *"empire follows the balance of property,* whether lodg'd in one, in a few, or in many hands."[39] To use Harrington's own set of examples: If a king owns or controls three-quarters of the land in his realm, there is a balance between his monarchical power and his property. But if the king's property was only one-quarter, there would be no balance and any absolute monarchical system would be unstable. Similarly, if "the few or a nobility, or a nobility with the clergy," were the landlords, or should "overbalance the people unto the like proportion," the result would be a "Gothic balance," and "the empire" would be a "mixed monarchy." Finally, there is the case in which "the whole people be landlords, or hold the lands so divided among them, that no one man, or number of men, within the compass of the few or aristocracy, overbalance them." In this event, "the empire (without the interposition of force) is a commonwealth."[40]

During the years of the American Revolution and the Constitutional Conventions, many American statesmen were aware that the concept of "balance" in a sociopolitical context could be traced to James Harrington's *Oceana.* Thus, John Adams wrote in his *Defence of the Constitutions* that this political concept was Harrington's discovery and that Harrington was as much entitled to credit for it as Harvey was for the discovery of the circulation of the blood.[41] In this sentiment Adams was echoing the praise given by John Toland, in his edition of Harrington's works. Incidentally, we may take note that Adams here associated Harrington's name with Harvey and the greatest discovery in the life sciences made during the Scientific Revolution, the circulation of the blood. There can be no doubt that Adams was aware that the concept of balance had come into current political thought in a biological and not a physical context, well before the creation of Newtonian science.

Adams and the Physical Principles of the Balance

The notion of a balance is historically associated with a device for weighing, in which a horizontal beam, with a scale (or pan) hanging from each end, is poised so as to move about a central pivot. The oldest form of such a balance is the equal-arm balance, in which the horizontal beam is suspended or pivoted at its exact midpoint, so that there is the same distance from the pivot to the point at each of the two ends from which the scales are suspended. When equal weights are placed in both pans, the bar remains in its horizontal position, in a condition of equilibrium; but when one pan weighs more than the other, it "tips the balance" and the bar will incline in that direction. This kind of weighing device is used by chemists, goldsmiths, and others who need exact weights of small quantities. For accurate weight, there must be an exact counterpoise between the two sides of the balance, between the two halves of the beam and the two pans and their supports. In practice such a perfect counterpoise is not usually possible and so there is a tiny "rider" or sliding weight which is used for "fine tuning."

The unequal-arm balance is the one we encounter most often, as in a doctor's scale for determining our weight, a steelyard, or the scales commonly used in grocery stores (although many of the latter are controlled by springs). In all of these scales, a large weight close to the pivot is balanced by a smaller one farther away from the pivot, according to the rule that there is an equality between the product of each of the weights multiplied by its respective "turning arm." In such an unequal-arm balance—for example, a steelyard—a pound weight ten inches from the pivot will balance a twenty-pound weight suspended on the other side, one-half inch from the pivot.

Balances appear in ancient Egyptian pictures and also in one of the signs of the zodiac, Libra. The Latin word "libra" has a number of meanings, including the device for weighing, a weight, and the Latin pound. In late Latin, a two-scaled balance was described by the use of the compound adjective "bilanx, bilancis" (meaning "having two scales" or "two-scaled"—a combination of "bi" meaning "two" and "lanx" meaning a plate or a dish or a scale). It is from the second part of the Latin name "libra bilanx" (or two-scaled weighing device) that our word "balance" is derived. The associated word "equilibrium," signifying equal weight, comes from the combination of "aequus" (or "equal") and "libra." It is obvious that both "balance" and "equilibrium" can easily be used metaphorically as well as for the designation of a physical object.

John Adams encountered the physical principles of equilibrium and the balance in the physics course that he took from John Winthrop at Harvard. Winthrop's outline of his lectures on natural philosophy (which Adams attended) shows that he discussed the balance in some detail and even showed the students various types of balances from the Harvard collection of scientific apparatus. A large portion of Adams's notes on Winthrop's lecture of 3 April 1754 is devoted to the balance, which is included among discussions of center of gravity, the lever, the pulley, and what was known as "the axis in peritrocheo." The latter was a device essentially consisting of a wheel and axle. In common use in Adam's day, the axis would lie horizontally, supported at each end, with a wheel fastened to it. Usually, there were holes in the wheel, into which staves or rods could be inserted, as in a windlass or capstan, so that the axle could be turned easily. A rope wound round the axle would draw up a weight.[42] Thomas Jefferson has given us a record of the use of such a device, which he encountered in 1788 and described (in his "Notes of a Tour through Holland and the Rhine Valley") as a "machine for drawing light *empty* boats over a dam at Amsterdam." The device consists, Jefferson wrote, of "an Axis in peritrocheo fixed on the dam. From the dam each way is a sloping stage. The boat is presented to this, the rope of the axis made fast to it, and it is drawn up. The water [on one] side of the dam is about 4. f. higher than on the other"[43] (see Fig. 37).

In Winthrop's lecture, according to Adams, "the Ballance was principally insisted on." The theory of its action "was fully explained and the method of weighing, viz. the distances of the Bodys from the Center of motion must be precisely in a reciprocall proportion of their quantitys of matter or weights." Winthrop added that this relation was theoretical, that in practice one had to make allowance for such factors as "the weight of the Beam on which they [the weights] are suspended, as well as friction, and the falsity of the supposition that radii proceeding from the center of the earth are parrellel."[44] Two days later, according to Adams's record, Winthrop went more deeply into the "theory of the Ballance, scales, steel-yard etc. and the 3 species of lever's." Winthrop generalized the principle of the lever, to which "he referred allmost all the instruments in life, and universally."

Winthrop set forth a general rule for all of these devices, which Adams entered into his notebook. To "make a aequilibrium," he wrote, "the product of the quantity of matter in the weight multiplyed into its distance from the Center of motion, must be equal to the quantity of matter in the power, multiplyed into it's distance from said Center." This is, as we have seen, the exact form of the law of moments for the lever or any form of balance.

A draft of a letter written a few years later, in 1758, three years after graduation from Harvard, shows how obsessed the young

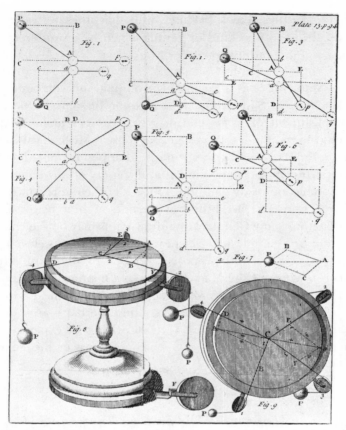

Fig. 37. Equilibrium in Three Simple Machines. An eighteenth-century plate demonstrates the balance of forces in various kinds of pulley systems, in three forms of levers, and in the "axis in peritrocheo" or wheel and axle (of the sort described by Thomas Jefferson). At the lower left is shown a force table for the study of the equilibrium of three forces.

Adams was with the interactions of machines. No doubt his interest in machines, notably their interlocking parts and operational principles, would have been greatly stimulated by the working models of machines in the Harvard collection of scientific apparatus with which Winthrop demonstrated the principles of physics to his students. In this letter, Adams asked himself the question, "What are the Proofs, the Characteristicks of Genius?" His answer was the "Invention of new Systems or Combinations of old ideas." The "first and greatest" such "Proof of Genius" was "Invention of new Wheels, Characters,

Experiments, Rules, Laws." Wheels and machines were first in his hierarchy, and the first kind of genius which he considered was the gifted mechanic. Let him speak for himself:

> The Man, who has a faculty of inventing and combining into one Machine, or System, for the Execution of some Purpose and Accomplishment of some End, a great Number and Variety of Wheels, Levers, Pullies, Ropes &c. has a great Mechanical Genius. And the Proofs of his Genius, (unless it happen by mere luck) will be proportionably to the Number, and Variety of Movements, the Nice Connection of the, and the Efficacy of the entire Machine to answer its End. The last, I think at present, ought to be considered in [estimating?] any Genius. For altho Genius may be shewn in the Invention of a complicated Machine, which may be useless, or too expensive, for the End proposed, yet one of the most difficult Points is to contrive the Machine in such a manner, as to shorten, facilitate, and cheapen, any Manufacture &c. For to this End a Man will be obliged to revolve in his Mind perhaps an hundred Machines, which are possible but too unwieldly or expensive, and to select from all of them, one, which will answer the Purposes mentioned.[45]

Scientific Metaphors Used by John Taylor in Rebuttal of Adams's Views

Later in life, Adams developed ideas concerning the system of checks and balances in the American system of government in which he made use of the concept of a physical balance and the notion of a giant machine. Historians have often cited his controversy with John Taylor, during which Adams discussed the concept of balance in relation to the Constitution. Taylor was a Virginia planter,[46] author of *An Inquiry into the Principles and Policy of the Government of the United States* (published in Richmond in 1814).[47] This book was, in large measure, an attack on Adams's political and social philosophy, expressly some ideas put forth in Adams's "Defence" of the American state constitutions.[48] Taylor was particularly disturbed by Adams's arguments in favor of a "natural aristocracy," Adams's opinion "that aristocracy is natural, and therefore unavoidable." Taylor concentrated on the word "natural" and posed the contrary of Adams's pronouncement by saying that aristocracy is "artificial or factitious, and therefore avoidable."[49] He cited Adams's position that "there are only three generical forms of government; monarchy, aristocracy and democracy, of which all other forms are mixtures; and that every society naturally produces an order of men, which it is impossible to confine to an equality of rights."[50] The resemblance of the American system of government to any of these forms, Taylor observed, "depends upon the resemblance of a president or a governor to a monarch; of an American senate, to an hereditary order; and of a house of representatives, to a legislating nation."[51] Taylor alleged that Adams had used "this threefold resemblance" in order to "bring the political system of America within the pale of the English system of checks and balances," that he had modified "our temporary, elective, responsible governors, into monarchs; our senates into aristocratical orders; and our representatives, into a nation personally exercising the functions of government."[52]

In a study of Adams's political ideas, we need not follow Taylor through the details of his counterarguments. But we may take note that he made use of a biological analogy in order to score a point against his political rival. He observed, first of all, that "the terms 'monarchy, aristocracy and democracy,' or the one, the few, and the many," may be no more than numerical designations of the locus of power. Or, he continued, introducing a bit of plant science, perhaps these three forms of government are "characteristic" of the organization of the state in the sense that "the calyx, petal and stamina" are characteristic "of plants." Here we see an example of the use of biological analogies or metaphors which was so common in the age of the Founding Fathers. But in this case Adams would respond sarcastically: "Linnaeus is upon my shelf, very near me, but I will not take him down to consult him about calyx, petal, and stamina, because we are not now upon gardening, agriculture, or natural history," but "politics."

Taylor was concerned with the concept of political balance. He noted, accordingly, that the above-mentioned three terms—monarchy, aristocracy, democracy—may possibly be "complicated with the idea of a balance." Even so, he concluded, these terms "have never yet singly or collectively been used to describe a government, deduced from good moral principles."

Before considering Adams's reply, we may note an additional reference to the sciences introduced by Taylor. For example, he compared the history of politics and the history of astronomy. Taylor argued that the American form of government displayed no features of resemblance to traditional definitions of monarchy, aristocracy, and democracy and that "our form of government" must "be rooted in some other political element." He concluded that Adams, and others of like opinion, had made an error by using "a numerical or exterior classification" rather than one "founded in moral principles." This error has put a restraint on "political volition," confining attention to the three forms of government without giving a place to "the moral efforts of mankind, towards political improvement." It was Taylor's thesis that "these three generical principles of government, or a mixture of them, have been universally allowed to comprise the whole extent of political volition." The result has been that "whilst the liberty enjoyed by the other sciences, has produced a series of wonderful discoveries," the science of "politics" has been "circumscribed by an universal opinion," just "as astronomy was for centuries," and thus politics "remained stationary from the earliest ages, to the American revolution."[53] A little later on, Taylor contemplated a possible conflict between "the democracy of England on one side, and the nobility on the other," with the "warfare" being "moral or physical." He likened this conflict to a "combat . . . between the universe and an atom."[54]

In his use of analogies, Taylor even mixed mythology and mathematics. He introduced the example of the mythical hydra with its many heads—not the polyp, used by Franklin in his essay on population—in order to illustrate how victory "over enemies is followed by defeat from friends." An "enemy destroyed abroad," he observed, "is only the head of an hydra," since it "produces two at home."[55] Then, shifting from mythology to mathematics, Taylor concluded that a belief that "aristocracy can only exist in the form of a hereditary order, or a hierarchy, is equivalent to the opinion, that the science of geometry can only be illustrated by a square or a triangle."[56]

In another passage, Taylor introduced a medical analogy, poking fun at the notion of a balance by referring to "doctor Balance." He argues that political doctors might each find the single limb or organ that he was examining to be in perfect health, so that their "aggregate" opinion is soundness of the organism, even though it may be obvious to all of them that the patient is dying. Taylor complained that "mankind are distracted by an host of political doctors." There is agreement, he noted, "that the British policy is afflicted with some inveterate distemper, but each doctor asserts his favorite limb to be sound." Thus, while "the aggregate by one opinion pronounces it to be in the agonies of death, the same aggregate by many opinions announces it to be in perfect health":

> Funding, banking, patronage, charter, mercenary armies and partial bounties, are each admired as a panacea by some one; even corruption is defended as a happy expedient for managing the house of commons; and doctor Balance, venerable with the rest of antiquity, excites universal astonishment by declaring with unaffected gravity, that a nobility endowed with enormous wealth, virtue and talents, is only wanting [in order] to renovate it throughout.[57]

Taylor was probably referring to Adams and his idea of a balance when he added: "Such doctors are labouring to patch up a policy for the United States, out of the self-same limbs, with an animal thus compounded, lying in [mortal] convulsions before their eyes."

Adams's Rebuttal of Taylor: The Separation of Powers and Adams's Ideas concerning Political Balance

Adams's rebuttal of Taylor took the form of a series of letters, written by Adams seriatim as he plowed his way through Taylor's book.[58] These letters are notable for Adams's disagreement with one of the central themes of Taylor's book, that the Constitution of the United States had been founded on (Taylor used the word "deduced" from) sound moral principles.[59] Adams also addressed one of the points introduced by Taylor, whether the Constitution of the United States was "complicated with the idea of a balance."[60]

Is there, he asked, "a constitution upon record more complicated with balances than ours?" His reply embodies eight different acts of balance. First of all, "eighteen states and some territories are balanced against the national government"; second, the House of Representatives "is balanced against" the Senate, while the Senate is balanced against the House. Third, "the executive authority is, in some degree, balanced against the legislative"; fourth, "the judiciary power is balanced against the house, the senate, the executive power, and the state governments." Fifth, "the senate is balanced against the president in all appointments to office, and in all treaties," which Adams

Balances [margin annotation]

believed to be "not merely a useless, but a very pernicious balance." Sixth, "the people hold in their own hands the balance against their own representatives, by biennial, which I wish had been annual elections"; seventh, the "legislatures of the several states are balanced against the senate by sextennial elections." Finally, eighth, "the electors are balanced against the people in the choice of the president," which Adams believed to be "a complication and refinement of balances, which, for any thing I recollect, is an invention of our own, and peculiar to us."[61]

And, yet, Adams went on, "all this complication of machinery, all these wheels within wheels, these *imperia* within *imperiis* have not been sufficient to satisfy the people." And so they "have invented a balance to all balances in their caucuses." The result is that there are now "congressional caucuses, state caucuses, county caucuses, city caucuses, district caucuses, town caucuses, parish caucuses, and Sunday caucuses at church doors." It is "in these aristocratical caucuses," he concluded, that *"elections are decided."*

Adams feared that "another balance to all these balances" would "soon be invented and introduced," an "over balance of all 'moral liberty' " and of "every moral principle and feeling." He had in mind "the balance of corruption." Adams saw this "corruption" active in the "spirit of party," in the "spirit of faction," and even in "the spirit of banking," which he regarded as prevailing in America "to a greater degree" than in any other country. And this led him to conclude, in some agreement with another of Taylor's theses,[62] that there is not "any aristocratical institution by which the property of the many is more manifestly sacrificed to the profit of the few" than by the "spirit of banking." These are harsh words from a political conservative like Adams.

In these extracts we may see how Adams has used a concept of balance that is not directly connected with the economic origins of balance of power which he found in Harrington's *Oceana*. He employs the notion of balanced machinery, of wheels within wheels. In introducing this set of physical analogies, however, Adams was not, as is often supposed, introducing a Newtonian basis for the American system of government. We cannot repeat often enough that in Newtonian dynamics and in the Newtonian system of the world, there is no balance, no equilibrium, but rather unbalanced forces producing accelerations. Adams did, however, on a quite different occasion, invoke Newtonian principles in the service of the form of government he preferred. This is our next subject.

Newton's Laws of Motion and the Structure of Congress

Some of Adams's political beliefs will surely seem odd to a reader at the end of the twentieth century. I have already mentioned that he was convinced that three contending political forces can often produce a stable system, but that two generally cannot. He believed strongly in a kind of "natural" aristocracy, though not in an aristocracy that would be, as in Europe, hereditary, giving special privileges based on conditions of birth.[63]

As a political conservative, Adams feared the results of vesting the legislative power in the hands of a directly elected legislature without any restraining body to check its possible excesses. On this point he differed from his political associate from Pennsylvania, Benjamin Franklin. Adams entered into public debate on this issue after reading a criticism of the constitutions of the American states by Turgot, a friend and associate of Benjamin Franklin, an economist of note, and the sometime finance minister under Louis XVI.[64] Turgot wrote out his opinions of these constitutions in a letter addressed in 1778 to Richard Price, in London. Price published Turgot's letter, first in the original French and then in English translation, as an appendix to his own *Observations on the Importance of the American Revolution* (1784, 1785).[65] One of Turgot's points of criticism was that the constitutions adopted by the American states generally had followed a British model, setting up different governing bodies (a legislature, a council, and a governor) instead of creating their own new system of government and "collecting all authority" into one center, that of the nation. Adams's somewhat faulty and overly narrow interpretation of Turgot's position is that Turgot, "by collecting all authority into one centre . . . meant a single assembly of representatives of the people, without a governor, and without a senate." Adams assumed, furthermore, that Turgot "meant [that] the representatives should be annually chosen."[66] Adams evidently paid no heed to the fact that Turgot, in that very same letter, had written of the need for "separating the objects of legislation from those of general administration and from those of particular and local administration" and had stressed the importance of establishing "local assemblies." Adams was apparently unable to make a distinction between Turgot's proposed centralism and unicameralism, a form of government in which there is only a single house or legislature. Adams abhorred any such expression of direct democracy and espoused the concept of a senate which could act as a brake on any immediate expression of popular will voted by the elected House of Representatives. For more than a century, according to the Constitution, members of the Senate were not elected by the direct vote of the citizens, but rather were chosen by the state legislatures.

Adams must have considered Turgot's letter as a kind of personal affront. Adams, after all, was not only the chief author of the Massachusetts constitution; he claimed that his ideas had influenced the constitutions of North Carolina and New York in 1776 and 1777. He also believed that his influence could be discerned in later alterations in the constitutions of South Carolina and Georgia, and even Pennsylvania. It has been remarked that there is "a fair amount of truth" in these claims, as in his belief that—by way of these state constitutions—some of his ideas had even influenced the structure of the federal government. His reply to Turgot eventually grew to three rather large volumes, of which the final one was published in 1788, under the title *Defence of the Constitutions of Government of the United States of America*. These learned tomes make hard reading today, especially because of the author's massive assembly of erudite references to authors and events from all periods of history.[67] The reader will sense at once the wide gulf that separates Adams's political principles from any of the normally accepted ideas of popular government. In the present study, we have no need to be concerned with this work as a whole, but only with a few pages in which Adams engages in a running debate with Franklin on the issue of whether a sound system of government requires one or two houses in the legislature.

Adams's disagreement with his political opponent occurs in the midst of the first volume, where Adams felt "obliged to consider the reputed opinion of another philosopher, I mean Dr. Franklin."[68] According to Adams, "Shortly before the date of M. Turgot's letter, Dr. Franklin had arrived in Paris with the American constitutions, and among the rest that of Pennsylvania, in which there was but one assembly." Adams added that it had been reported "that the doctor had presided in the convention when it was made, and there approved it." M. Turgot, "reading over the constitutions, and admiring that of Pennsylvania, was led to censure the rest, which were so different from it." Adams then recounted a "common anecdote," which—he said—"is known to everybody."

In 1776, according to Adams, during the Pennsylvania convention, while the "project of a form of government by one assembly was before them," a motion was made "to add another assembly, under the name of a senate or council." During the debate, the opinion of the president of Pennsylvania (Benjamin Franklin) was solicited. Franklin, ever a skilled political hand, was aware of the power of wit in political debate. On this occasion, he told a story which was so outrageously ridiculous that it made the bicameral argument seem silly. Let Adams be our raconteur.

Having "two assemblies," Franklin is reported to have said, "appeared to him like a practice he had somewhere seen, of certain wagoners, who, when about to descend a steep hill with a heavy load, if they had four cattle, took off one pair from before, and chaining them to the hinder part of the wagon drove them up hill; while the pair before and the weight of the load, overbalancing the strength of those behind, drew them slowly and moderately down the hill."

Adams's pompous reply to Franklin's funny story invoked an attempt to muster arguments of physics against the Philadelphia sage and scientist. "The president of Pennsylvania might, upon such an occasion," replied Adams, in a schoolmasterish style which may remind us of his first employment after leaving Harvard College, "have recollected one of Sir Isaac Newton's laws of motion, namely,— 'that reaction must always be equal and contrary to reaction,' or there can never be any *rest*."

This reply to Franklin is of real interest to the student of science and political thought because Adams based his argument on the premise that science provides the supreme authority in a political debate. He cites the physics of Newton's *Principia* as a rebuttal of Franklin's ideas from which there can be no appeal. I know of no other example which so dramatically shows the enormous esteem which Adams felt toward science in general and toward Newton's *Principia* in particular. Here we may see the extremity of the gulf that separates the eighteenth-century Age of Reason with its reverence for science from our own Age of Anxiety. Who could imagine a political debate today concerning fundamental principles of government being based on the laws of physics!

When Newton wrote about "action" and "reaction," he envisaged that whenever a body A acts on another B, that second body B will equally react on body A. An example was a horse pulling a stone (the action), which produces an equal and opposite reaction of the stone pulling on the horse in the other direction. Adams had a different kind of picture in mind, a case of equilibrium, of a state of balance produced by two equal and opposite forces, since he said that this was a condition without which there could "never be any *rest*." As we have seen, however, for a condition of equilibrium to occur when two forces are acting, the forces must not only be equal and opposite, but must also act on the same body. But in the case of Newton's third law, the law of action and reaction, the forces are acting on different bodies. Accordingly, the third law has nothing whatever to do with conditions of equilibrium. It must be noted, however, in Adams's defense that the meaning of Newton's third law is far from obvious. Adams's error in interpreting and applying the third law is a common one, based on a misunderstanding that has been widespread ever since Newton announced the law in 1687. What is important about this episode, therefore, is not the accuracy of Adams's memory of the laws of physics, but rather the significance of his having appealed to the fundamental laws of Newton's *Principia* as the highest possible authority in the course of a political debate on the ideal form of government.

A second Newtonian paragraph follows Adams's first one. Here Adams turns to principles of cosmology, by referring to "those attractions and repulsions by which the balance of nature is preserved," which he finds exemplified in "those centripetal and centrifugal forces by which the heavenly bodies are continued in their orbits, instead of rushing to the sun, or flying off in tangents among comets and fixed stars, impelled or drawn by different forces in different directions." The belief that the planets stay in their orbits because of a balance between centripetal forces drawing them in toward the Sun and centrifugal forces driving them outward is non-Newtonian, but Cartesian. Yet many writers of Adams's day used the old Cartesian language of balance as if the Newtonian inertial motion along the tangent could be considered a "force" and could be thought to act outward rather than tangentially.[69]

Electrical Theory in Defense of a Bicameral Legislature

Adams was not content to place the authority of the greatest scientist of the modern period against Dr. Franklin; he even tried to use

Franklin's own science against him. Franklin, he went on, "might have alluded to those angry assemblies in the heavens, which so often overspread the city of Philadelphia, fill the citizens with apprehension and terror, threatening to set the world on fire, merely because the powers within them are not sufficiently balanced." That is, Franklin "might have recollected, that a pointed rod, a machine as simple as a wagoner, or a monarch, or a governor, would be sufficient at any time, silently and innocently, to disarm those assemblies of all their terrors, by restoring between them the balance of the powerful fluid, and thus prevent the danger and destruction to the properties and lives of men, which often happen for the want of it."

In this case Adams was hoping to use Franklin's own discoveries to combat his political views. Unfortunately, he did not remember perfectly the electrical theory he had learned from Professor Winthrop. Franklin had shown that a lightning discharge occurs when a sufficiently charged cloud passes close enough to the earth for a discharge to take place. He had at first thought that such a discharge consists of a motion of a cloud's excess electric fluid to the earth and had concluded that a pointed, properly grounded, metal lightning rod would act to neutralize the cloud by quietly drawing off some of the excess charge and safely conducting it into the earth so as to prevent a stroke from occurring. Once lightning rods were put in place, he quickly learned that such a rod will also protect an edifice even when there is a stroke of lightning. In this event the rod will attract the stroke and once again safely conduct the electric fluid into the ground.

In Franklin's theory, as we saw in Chapter 3, a positive cloud (or any charged body) has an excess of electric fluid,[70] that is, has more than what Franklin called its "normal" quantity of such fluid. In a discharge, such a cloud will lose all (or some large part) of its excess electric fluid and so be restored to its condition of having only its "normal" quantity. Adams, however, referred to clouds as "assemblies [of which] the powers within them are not sufficiently balanced."

It would require a high degree of poetic license to legitimate Adams's characterization of the acquisition of a charge by a cloud as a condition of internal "powers" not being "sufficiently balanced." But Adams obviously did not appreciate that lightning rods act by conducting a stroke safely into the ground. He saw their purpose only as "restoring between them [i.e., between those assemblies or clouds] the balance of the powerful fluid," by which I suppose he meant an action to prevent any stroke from passing from one such charged cloud to another ("between them"). While some strokes can consist of a discharge from cloud to cloud, as Franklin knew full well, the ones that produce "danger and destruction to the properties and lives of men" are not of this sort but rather consist of discharges from clouds to the earth. Or, as Franklin later discovered, many such discharges proceed from the earth to clouds.

Adams concluded in a condescending manner by granting that "allusions and illustrations drawn from pastoral and rural life are never disagreeable," and suggesting that if Franklin had better understood his own story the bucolic analogy could have been as "apposite" as it would have been if it had been "taken from the sciences and the skies." There "is no objection to such allusions," he went on, "whether simple or sublime," so long as "they may amuse the fancy and illustrate an argument." He only found it surprising that Franklin should have "misunderstood" the "real force of the simile" which he had introduced and he argued that "if there is any similitude, or any argument in it, it is clearly in favor of two assemblies." Instead of appreciating that Franklin was poking fun at him and all who held his beliefs, Adams tried to interpret Franklin's account of the oxen hitched to the same cart in opposite directions as if it had been a description of actual practice.

"The weight of the load," Adams wrote, "would roll the wagon on the oxen and the cattle on one another, in one scene of destruction, if the forces were not divided and the balance formed; whereas, by checking one power by another, all descend the hill in safety, and avoid the danger." And then, not knowing where to stop, Adams continued, "It should be remembered, too, that it is only in descending uncommon declivities that this division of strength becomes necessary."[71]

Adams's Attitudes toward Some Other Aspects of Science

After serving his term as president, Adams returned to his farm in Quincy, Massachusetts, seeking solace in the hard manual labor of practical day-to-day farming. As he wrote to Samuel Dexter, he "had exchanged honor and virtue for manure."[72] He announced his prediction that the Presbyterians will find "a mathematical demonstration that my taste for agriculture is only a fruit of my arbitrary disposition and despotic principles."[73] Suffering from a palsy, he nevertheless kept active, mentally as well as physically. In these first years of retirement from active political life, he indulged in scientific speculations, two of which may attract our attention. One involved the reproduction of marine animals, the other the properties of the magnet. In a letter to his friend Francis Adrian Van der Kemp of Oldenbarnveldt, New York, in 1801, Adams voiced his opinion concerning the "little nautilus, or what sailors called the Portuguese man of war," which Adams believed was only "a shellfish in embryo." He thought that the eggs float on the surface of the water until they are hatched by the Sun's heat; he begged his correspondent to send him "whatever information" he might "possess on this subject."[74] While we take note of Adams's interest in natural history, we may be struck by his ignorance concerning the deadly sting of the Portuguese man-of-war. In those days, however, ocean bathing and swimming were not the universal habit that we know today.

Adams's speculations about the properties of magnets were a revival of the interest which he had taken in this subject while he was in France in association with Benjamin Franklin. The principle by which the magnet acts to draw bits of iron to itself was a mystery, he concluded, part of "the arcana of nature." This led him to the opinion that "nature herself is all arcanum." He believed "it will remain so." That is, he concluded, "It was not intended that men with their strong passions and weak principles should know much," since— without "a more decisive and magisterial moral discernment"— "much knowledge would make them too enterprising and impudent."[75]

Toward the end of his life, at the age of eighty-four, Adams conceived that life itself is a chemical process. "We are all Chymists from our cradles," he opined. "All mankind are Chymists from their cradles to their graves." Furthermore, he continued, "The Material Universe is a chymical experiment." The occasion for the expression of these sentiments was a letter written in 1817 to John Gorham, professor of chemistry at Harvard College. Adams's notions about life and chemistry reflect a shift in scientific interest following the great chemical revolution inaugurated by Lavoisier, which transformed that branch of science into the modern discipline as we largely know it today. Biographers of Adams have not paid any attention to his views on chemistry largely because the main focus of historians within the sciences is on the Newtonian natural philosophy, which had been at the center of intellectual concern during Adams's early youth. But by the early 1800s chemistry was emerging as the important path-breaking science of the new century, in token of which its name was being changed from the older "chemical philosophy" to "chemistry."

With regard to the limitations of this new science, and the nature of its subject matter, Adams considered his ideas to be so far-out that he concluded the letter to Gorham with these words: "I pray you consider this as confidential." He was concerned lest the opinions expressed in the letter might "get abroad," in which case he "should be thought a candidate for the new Hospital." What beliefs, the reader will surely wonder, did Adams believe to go so against the universally held convictions of the scientists of his day that he feared they might become generally known?

In his letter to Gorham, Adams recommended that chemists pursue their "experiments with indefatigable ardour and perseverance." By so doing, they will give us "the best possible Bread, Butter, and Cheese, Wine, Beer and Cider, Houses, Ships and Steamboats, Gardens, Orchards, Fields, not to mention Clothiers or Cooks." Should such "investigations lead accidentally to any deep discovery," he went on, let chemists "rejoice and cry 'Eureka!' " In other words, Adams saw chemistry as a limited subject that should be exclusively a pursuit leading to better products for use. Thus far, there was nothing so strange that it might cause his commitment to a mental hospital. But at that point, Adams inserted a minatory sentence which exposed the unorthodoxy of his beliefs. He did not believe in atoms! He recalled that, in "former times," when he looked into classics, and "very little indeed it was," he was "fascinated with the Numbers of Lucretius," but he "could not comprehend his Atoms." In "after times," he recollected, when he was "delighted with the eloquence of Buffon," he just "could not help laughing at his Molecules." And so he warned John Gorham and his fellow chemists "never [to] institute any experiment with a view or a hope of discovering the first and smallest particles of Matter."[76]

Fully to appreciate Adams's reactionary stance with respect to atoms, molecules, or a corpuscular theory of matter, we must take account that all the principal founders of modern physical science, from Galileo and Newton onward, had—to varying degrees—been atomists. Franklin's theory of electricity was based squarely on the concept of an electric fluid composed of mutually repelling particles. In the *Opticks,* published first in 1704 and issued in revised form in 1717 / 1718, Isaac Newton had stated the atomists' creed clearly and precisely. In the conclusion of the *Opticks,* Newton wrote: "All these things being considered, it seems probable to me, that God in the Beginning formed Matter in solid, massy, hard, impenetrable, moveable Particles, of such Sizes and Figures, and with such other Properties, and in such Proportion to Space, as most conduced to the End for which he formed them." He also set forth the position that "the Rays of Light [are] very small Bodies emitted from shining Substances." In rejecting notions of atomicity or corpuscularity of matter, Adams was ranging himself, no doubt unwittingly, on the side of William Blake and others who abhorred modern science and all forms of rationalism connected with science. Blake made fun of "the Atoms of Democritus and Newton's Particles of light." Adams not only rejected the concept of atoms, but warned chemists not to make any experiments that would disclose fundamental particles of matter. Such beliefs will seem to today's reader to be less odd than Adams's belief that chemists' "investigations" would only rarely and "accidentally" lead "to any deep discovery." This opinion about the limitations of chemistry as a science is puzzling. He had been in France during the time when Lavoisier and others were producing what is known as the Chemical Revolution, transforming the subject of chemistry into a true modern science. Was it his limited view of chemistry or his disbelief in atoms that Adams thought might warrant commitment to a mental hospital? He does not tell us.

The Importance of Science for the Future of the Nation

It is a fitting conclusion to this chapter to recall Adams's hierarchy of the sciences. In thinking about the present and future needs of America in 1780, Adams quite correctly observed that it is "not indeed the fine arts which our country requires," but rather "the useful, the mechanic arts," by which he meant the arts of road and canal building, ship and factory building, bridge building, engineering, and the techniques of commerce and finance. As for himself, as he wrote to his wife, Abigail, it "is my duty to study" the "science of government . . . more than all other sciences." As of this moment in history, he believed that "the arts of legislation and administration and negotiation ought to take the place of, indeed to exclude, in a manner, all other arts."

And then in a noble expression of sentiment which evokes our respect, he wrote, "I must study politics and war," so that "my sons may have liberty to study mathematics and philosophy," that is, natural philosophy or physical science. "My sons," he concluded, "ought to study mathematics and philosophy, geography, natural history and naval architecture, navigation, commerce and agriculture, in order to give their children a right to study painting, poetry, music, architecture, statuary, tapestry and porcelain."[77] In Adams's expression of his hopes for a future America in which there would be leisure for the creative arts, he was also giving voice to the Baconian role of science. Like Jefferson and Franklin, he envisioned practical consequences of abstract science and mathematics in producing material benefits for the nation.

5

Science and the Constitution

.....

Science in the Constitution

The Constitution is directly related to the central theme of science and the Founding Fathers in a number of different ways, even though only one of the individuals whom we have been studying was present at the Convention in Philadelphia in 1787. First of all, there is an explicit reference to "science" in the actual text of the Constitution. The framers of the Constitution were citizens of the Age of Reason and they introduced metaphors drawn from the sciences into their discussions and debates. Moreover, historians and political scientists have debated whether James Madison and his fellow delegates to the Constitutional Convention were influenced in their deliberations by scientific considerations. For more than half a century, some prominent American historians and political scientists have argued that the Constitution is a Newtonian document. In the time of the Founding Fathers, many individuals sought for links between principles of science and constitutional government. We have seen, in Chapter 4, an example of this endeavor in John Adams's discussion of the structure of the legislature in terms of Newton's third law of motion.

Jefferson and Adams were not directly active in the Constitutional Convention, since they were not in attendance either as delegates or as spectators. But Adams's ideas on government were of some influence through the state constitutions that bore his mark and through his book, *Defence of the Constitutions*. Franklin, as we have seen, was actually a delegate, but his role was minimal and his cherished plan for a unicameral legislature was not endorsed by his fellow delegates. His chief contribution was to offer the celebrated compromise plan that gave the larger states more power in the House of Representatives while giving all states equal power in the Senate. Franklin also gave the final oration, endorsing the Constitution, at the conclusion of the business of the Convention.

The only direct reference to science in the Constitution occurs in Paragraph 8 of Section 8 of Article I, setting forth the powers of Congress.[1] Here the power is explicitly assigned "To promote the Progress of Science and useful Arts, by securing for limited Time to Authors and Inventors the exclusive Right to their respective Writings and Discoveries." This sentence is a curious hodgepodge. First there is stated the need to "promote the Progress of Science and useful Arts." But then it is said that the means of achieving this bipartite goal is "by securing for limited Time" to "Inventors the exclusive Right to their . . . Discoveries" and to "Authors" a similar "exclusive Right" to their "Writings." A good case can perhaps be made that granting patents to inventors would promote "the Progress of . . . useful Arts," but it is difficult to see how such patents might equally promote "the Progress of Science." The greatest scientific advances— Newton's rational mechanics and gravitational cosmology, the Newtonian mathematics of the calculus, Linnaeus's system of classification of plants—had no aspect that would have been patentable, and hence the granting of patents would hardly have been a stimulus to their production. It is equally difficult to conceive how copyrights granted to authors could possibly "promote" either "the Progress of Science" or "the Progress of . . . useful Arts." That is, even if it is supposed that a single portion of Article I should be devoted to the advancement of science and also to patents and copyrights, the phrasing in the Constitution is confusing at best. A further difficulty arises for the reader of today because to us the word "science" without any qualifications suggests a subject like physics, chemistry, or biology, whereas to an eighteenth-century reader of the Constitution "science" without a qualifier could imply knowledge in general and even, by extension, the knowledge conveyed by original works in prose or in any literary form.

A historical narrative may help us to understand how the structure of this puzzling sentence came into being. During the Constitutional Convention, on Saturday, 18 August 1787, certain "additional powers" were proposed by James Madison "to be vested in the Legislature of the United States." Among the proposals for additional powers were a number that are relevant to the eventual science clause. First,

> To grant charters of incorporation in cases where the public good may require them, and the authority of a single State may be incompetent.

Among the specifically designated powers, three are of special concern to us here:

> To secure to literary authors their copy rights for a limited time,
> To establish an University,
> To encourage, by proper premiums and provisions, the advancement of useful knowledge and discoveries.[2]

These "propositions," Madison records, were "referred to the Committee of detail which had prepared the Report." This was done by unanimous approval, at the same time that "several additional powers" were proposed by C. C. Pinckney, the delegate from South Carolina.[3] Among the powers endorsed by Pinckney were:

> To establish seminaries for the promotion of literature and the arts & sciences,
> To grant charters of incorporation,
> To grant patents for useful inventions,
> To secure to Authors exclusive rights for a certain time,
> To establish public institutions, rewards and immunities for the promotion of agriculture, commerce, trades, and manufactures.

In Madison's original handwritten list of these proposed powers, there was one which did not appear in that form in the printed journal. According to Max Farrand this read:

> To secure to the inventors of useful machines and implements the benefits thereof for a limited time.

Pinckney was an ardent advocate of assigning patents to inventors. At that time, his native state (South Carolina) was a leader in the matter of patent legislation, although by 1787 all the states but one had passed copyright acts and many states were actually granting patents. We may reasonably assume that Pinckney would also have sponsored the granting of copyrights to authors, as well as the establishment of seminaries for the promotion of the arts and sciences and the proposed rewards for the "promotion of agriculture, commerce, trades and manufactures."

Eventually a committee was appointed to put in order the unfinished parts of the Constitution. The members were Nicholas Gilman, Rufus King, Roger Sherman, David Brearly, Gouverneur Morris, John Dickinson, Daniel Carroll, James Madison, Hugh Williamson, Pierce Butler, and Abraham Baldwin.[4] In September 1787 Brearly reported to the Convention their agreed-upon propositions. Gone were the university and the seminaries, as well as the rewards or premiums. For economy of space and expression, the previous statement of "secure to literary authors their copy rights for a limited time" was now coupled with what had previously been termed the powers to "encourage . . . the advancement of useful knowledge and discoveries," to "grant patents for useful inventions," and to "secure to the inventors of useful machines and implements the benefits thereof for a limited time." In the final version, however, the specific references to "patents" and "useful inventions" were eliminated in favor of granting to "inventors" the "exclusive right" to their "discoveries," while the dynamic function of encouraging "the advancement of useful knowledge and discoveries" was reduced to a mere promotion of "the progress of science and useful arts." There resulted the following version (as quoted by Madison):

> To promote the progress of Science and useful arts by securing for limited times to authors & inventors, the exclusive right to their respective writings and discoveries.[5]

This was "agreed to nem: con:" (that is, *nemine contradicente* or unanimously), and then went to the Committee of Style, which fiddled with punctuation, put the word "and" in place of an ampersand, and altered the capitalization of some nouns. No one on the Committee of Style—composed of William Johnson, Alexander Hamilton, Gouverneur Morris, James Madison, and Rufus King—evidently was concerned by the curious result of combining features of several early propositions, that is, the apparent statement that granting to literary authors an "exclusive Right to their . . . Writings" could, in some way, "promote" either the "Progress of Science" or "useful Arts."

Despite the open-sounding resonance of the introductory words, "To promote the Progress of Science," the rest of the sentence, as it stands in the final version, leaves no doubt that this promotion is strictly limited to the assigning of "exclusive Right[s]." These are restricted furthermore to "Authors and Inventors," who are said to have granted to them "exclusive Right[s]" to their respective "Writings and Discoveries." It must be assumed by the context, therefore, that "Discoveries" does not refer here to scientific discoveries in general, as we would use the word today, but rather to patentable inventions. This intent is made clear in the several preliminary statements that produced the final text.

We must not forget, however, that the framers of the Constitution were generally fluent in Latin and that the Latin word "inventor" means "one who finds out, a contriver, author, discoverer" (from the verb "invenio" meaning *sensu stricto,* "I come upon," "I find," "I discover"), and not just a mechanical artificer. In the *Principia,* for example, Newton often used this verb, always in the sense of find out or discover. So, for the Founding Fathers, "invention" and "discovery" were words that could be used interchangeably.

Let us observe, finally, that in the Constitution two very different ideas about science are being expressed. First, it said that the Congress (or "legislature") shall have the power (and presumably the responsibility or obligation) to "promote the Progress of Science and useful Arts." A second power is to provide patents and copyrights. In terms of real financial support, however, the federal government did not assume general responsibility for the promotion of the "Progress of Science" until after World War II. The framers of the Constitution certainly did not envisage that the federal government would become the primary source of financial support of scientific research. This position would have implied assigning to the federal government a far greater power than the framers were willing to assign. Although, at one time, consideration was being given to the power to "establish an University," to "establish seminaries for the promotion of literature and the arts & sciences," and even to "encourage, by proper premiums and provisions, the advancement of useful knowledge and discoveries," these prerogatives were quickly abandoned.

Whoever reads the record of the Constitutional Convention will be led to wonder whether Madison, Pinckney, and their collaborators who proposed Section 8 would have had any specific inventions or types of inventions in mind. Or were they merely echoing a general belief of the eighteenth century, deriving ultimately from Francis Bacon and from René Descartes, that all true science would ultimately prove to be useful? There are at least two major types of invention with which delegates might have been familiar. One is what may be called mechanical inventions: putting known elements together to produce a new device or a better machine. Sometimes this process requires no more than mere mechanical ingenuity coupled with a gift for innovation. Examples from that time were bifocal eyeglasses, a mechanical "long arm" that would serve to retrieve books from top shelves, and a rocking chair—all inventions of Benjamin Franklin. In a later age, this type of invention would be exemplified in the McCormack reaper and the sewing machine.

A complex machine of Convention days was the steamboat. This invention intruded into the deliberations of the Convention in a spectacular way. On 22 August 1787, when the propositions of Madison and Pinckney were presented, there was an opportunity to witness the first trial of one of John Fitch's steamboats, just launched. Some members of the Convention even sailed as passengers on this vessel in its maiden voyage along the Delaware River.[6]

A second type of invention results from the applications of scientific knowledge, often originally obtained without any motivation other than learning about natural phenomena. This process is often known today as putting "pure science" to work or finding practical uses for "basic research." Although Bacon and Descartes predicted that any basic scientific knowledge would lead to practical innovations of use to people everywhere, by 1787 there had been only one spectacular example of disinterested or basic research in science that had led to an important invention. I refer here to the lightning rod, invented by a member of the Convention, the venerable president of Pennsylvania, Benjamin Franklin. Franklin invented the lightning rod on the basis of his fundamental discoveries concerning grounding and insulation, the power of points to draw off charge from an electrified body, and the world-famous sentry-box experiment that had proved the lightning discharge to be an electrical phenomenon. When he started his research, the only possible practical application envisaged for the new subject of electricity was in giving electric shocks to the limbs of paralytics in the hope that this treatment would restore the power of motion. Franklin, as a matter of fact, did not believe in this supposed power of electric shocks. He began his research because he wanted to learn about the operations of nature, and he had no idea where that research would lead. Well into the nineteenth century the lightning rod was cited as the premier example of the way in which fundamental scientific advances may produce practical inventions.[7]

The sentence about copyrights and patents became Paragraph 8 of Section 8 of Article I of the Constitution on Wednesday 5 September 1787.[8] In the First Session of the First Congress it soon became clear that a general patent act was needed. Accordingly, on 23 June 1789, a bill for that purpose was introduced, but Congress ended its session without taking any action on it. When the Second Session convened in January of the following year (1790), President George Washington addressed the need for encouraging inventions. Noting that the Congress was committed to the advancement of agriculture, commerce, and manufacturing, Washington nevertheless expressed his concern about the need for "giving effectual encouragement" to both "the introduction of new and useful inventions from abroad" and "the exertions of skill and genius in producing them at home." He coupled his recommendation with a remark on the special needs of post office and post roads for "facilitating the intercourse between the distant parts of our Country." And he concluded: "Nor am I less persuaded . . . that there is nothing which can better deserve your patronage than the promotion of Science and Literature."[9]

Newtonian Science and the Structure of the Constitution

A large number of writers on American history have suggested that some of the primary concepts and principles and even the general architecture of the Constitution were conceived in relation to science. Discussions of this topic are not usually couched in terms of science in general but tend to refer specifically to the science of Isaac Newton. There is no point in seeking merely for references to Newton in the speeches and writings relative to the Constitution or its eventual ratification. After all, many of the delegates were lawyers and had been trained to seek, and to cite, authorities and precedents, thus setting a style which was widely used during and after the Convention. An example is a patriotic oration in Boston on 5 March 1781, commemorating the Boston Massacre, in the course of which

Thomas Dawes, Jr., referred to Isaac Newton. An examination of this work of oratory shows, however, that Newton was but one of a list of authorities that included also—among others—Marcus Aurelius, the *Spectator,* Ovid, Pope, Seneca, Blair, Juvenal, Addison's *Cato,* Blackstone, and even the Bible.[10]

A more serious Newtonian inquiry would be to seek for analogues of the *Principia* in the concepts and structure of the American constitutional system of government. Those who claim that they have found evidence for the influence of Newtonian science in the Constitution usually concentrate on three or four topics or themes. These include, first and foremost, the separation of powers and the allied notion of an equilibrium produced by the complexity of checks and balances. Second is the notion that the constitutional form of government resembles a Newtonian machine, in particular, what is often known as the Newtonian world-machine. The balance and equilibrium of constitutional government is thus seen as having been modeled on Newtonian physics and the Newtonian system of the world. In addition, there is the mystique of Newtonian mechanics and / or the Newtonian system of the world as the inspiration for a system of rational politics, as the example proving how countervailing forces can act to produce harmony, or even as proof of the benefits of democratic or republican government.

A discussion of the possible influences of the Newtonian philosophy must take us into several rather different kinds of inquiry. We must find out exactly what positions have been taken by those who believe in the Newtonian quality of the Constitution. We must also examine carefully the records of the Constitutional Convention in order to find out to what extent and in exactly what ways the Founding Fathers made use of scientific concepts, laws, and theories in their deliberations. These must be analyzed in order to specify what sciences were being used. It will be necessary, then, to make clear what are and are not instances of Newtonian natural philosophy. We will thus be able to decide whether—or in what sense or senses—the Constitution may be said to be Newtonian. Finally we will turn to the most fundamental question of all: the role of scientific analogy and metaphor in political thought and the significance of the scientific analogies and metaphors used by the Founding Fathers.

Woodrow Wilson on the Constitution as a Newtonian Document

So far as I have been able to determine, the first overt declaration that the Constitution is a Newtonian document or that the Constitution should be interpreted as an expression of Newtonian principles was made by Woodrow Wilson in a series of lectures given at Columbia University in 1907 and published in 1908 as a book entitled *Constitutional Government in the United States.*[11] Wilson was then president of Princeton University. Afterward, in 1912, Wilson explicitly claimed that the concept of a parallelism between Newtonian science and the Constitution was his own invention.[12] He made the same claim for his corresponding idea, which will be discussed later in this chapter, that a Darwinian view of the Constitution is more fitting, particularly in an age in which Darwinian science exerts the kind of influence exerted by Newtonian science in the eighteenth century.

Wilson's Newtonian interpretation of the Constitution was based on the notion that the "government of the United States was constructed upon the Whig theory of political dynamics" and that the latter "was a sort of unconscious copy of the Newtonian theory of the universe." Wilson explained what he meant in some detail. According to his conception of the Newtonian universe,

Some single law, like the law of gravitation, swung each system of thought and gave it its principle of unity. Every sun, every planet, every free body in the spaces of the heavens, the world itself, is kept in its place and reined to its course by the attraction of bodies that swing with equal order and precision about it, themselves governed by the nice poise and balance of forces which give the whole system of the universe its symmetry and perfect adjustment.

The Whigs, Wilson argued, "had tried to give England a similar constitution" by the plan "to surround and offset" the power of the king "with a system of constitutional checks and balances which should regulate his otherwise arbitrary course and make it at least always calculable." The Whig politicians, however, "made no clear analysis of the matter in their own thoughts," Wilson observed, noting that it had "not been the habit" of either "English politicians, or indeed of English-speaking politicians on either side of the water, to be clear theorists." In fact, Wilson remarked, there was needed a Frenchman,

Montesquieu, "to point out to the Whigs what they had done." Montesquieu recognized that "they had sought to balance executive, legislature, and judiciary off against one another by a series of checks and counterpoises, which Newton might readily have recognized as suggestive of the mechanism of the heavens."

In his enthusiasm over creating a gloss on Montesquieu's interpretation, Wilson not only asserted that the "makers of our federal Constitution followed the scheme as they found it expounded in Montesquieu," but he insisted that they did so "with genuine scientific enthusiasm." In particular, Montesquieu is said by Wilson to have been the source of the fervent exposition of the many varieties of checks and balances displayed in the *Federalist*. The "statesmen of the earlier generations quoted no one so often as Montesquieu, and they quoted him always as a scientific standard in the field of politics." Wilson concluded on the rhapsodic note that politics "is turned into mechanics" under Montesquieu's "touch," so that the "theory of gravitation is supreme."

In this presentation, we may see how Wilson's argument advances step by step by force of rhetoric and arbitrary assumptions. He does not cite a single fact, produce a single quotation, or offer a single bit of real evidence in support of his notion that the Constitution was influenced by science, much less that the Constitution shows traces of Newtonian science. He begins by some statements concerning what he believes to be the essentials of the Newtonian system, a congeries of gravitational forces in a "nice poise and balance" producing some "perfect adjustment," much like "a system of constitutional checks and balances" whose function is to "regulate." We are asked to believe that such "checks and counterpoises" would "readily" have been "recognized" by Newton himself "as suggestive of the mechanism of the heavens." Having supposed an analogy between what he considers to be the principles of the Newtonian "mechanism" and the political system of checks and balances, Wilson then declares that this system (as "expounded in Montesquieu") was followed with "scientific enthusiasm," although it is far from clear what he intended by "scientific." Then, carried away by the force of his own rhetoric, Wilson presents Montesquieu as a "scientific standard" for the Founding Fathers. Wilson concludes this rhapsody with a conversion of politics into "mechanics" and a declaration that the "theory of gravitation is supreme." A few lines later, Wilson supplies a gloss on his own paragraph by asserting that "our constitutional law" was "conceived in the

Newtonian spirit and upon the Newtonian principle." This display of rhetoric does not, of course, take any note of what meaning is to be assigned to either "the Newtonian spirit" or "the Newtonian principle."

In assessing this remarkable bit of exposition, we must take note that Montesquieu, however important he may have been in the thought of the framers of the Constitution, did not understand the simplest aspects of Newtonian science. He certainly did not in any way base his *Spirit of the Laws* on any "Newtonian principle." Furthermore, as we have seen, the notion of checks and balances, or of "poise and balance," or "checks and counterpoises," is not a feature of Newtonian dynamics but rather of the pre-Newtonian systems which it was the goal of the *Principia* to replace.[13] Nor is Wilson's political analogy of "a machine governed by mechanically automatic balances" in any sense Newtonian. The image of a machine, with the notion of "perfect adjustment" and the concept of being regulated, suggests the image of a clock, a metaphor which Newton himself did not use. In addition, as will be shown below, the Newtonian universe fails in one major aspect to qualify as a true machine and in fact was severely criticized for this very failing.

We know that Wilson's analysis of the American form of constitutional government was greatly influenced by the writings of Walter Bagehot. Wilson had earlier stated explicitly that in his first book, *Congressional Government*, he intended to do for the American Constitution what Bagehot had done for the English Constitution in his essays on that subject.[14] In *The English Constitution*, Bagehot made many comparisons and contrasts between the American and the British systems, often disparaging the American form of constitutional government, which he did not fully understand.[15] In particular, Bagehot made an unfavorable contrast of the extreme system of checks and balances in America with the relatively greater freedom of action possible to the British government. Wilson seems to have been deeply impressed by Bagehot's extensive analyses. Bagehot, however, did not develop the Newtonian theme espoused and probably invented by Wilson.

For the genesis of his idea about the Newtonian and Darwinian concepts of the Constitution we have Wilson's own account, presented in an address delivered to the Economic Club of New York on 23 May 1912, while he was still governor of New Jersey and about a month before he was nominated as presidential candidate by the

Democratic National Convention.[16] Wilson explained that when he was president of Princeton University he had enjoyed the opportunity of conversing with "a very interesting Scotchman who had been devoting himself to the philosophical thought of the seventeenth century" and that one of this scholar's theories was "that in every generation all sorts of speculation and thinking tend to fall under the formula of the dominant thought of the age that has preceded that." One example given by the historian of philosophy was a transition from the use of Newtonian analogies to the use of Darwinian analogies in thinking and in the expression of thought. At this time Wilson was looking for a way to pull together some of his "political ideas" and found that his interlocutor's observation was just what he "had been waiting for." As he listened to "this interesting man," Wilson declared, "it came to me . . . that the Constitution of the United States had been made under the dominion of the Newtonian theory." Indeed, he added, this is "written on every page" of the *Federalist* papers:

> They speak of the "checks and balances" of the Constitution and use to express their idea the simile of the organization of the universe, and particularly of the solar system—how by the attraction of gravitation the various parts are held in their orbits, and represent Congress, the judiciary, and the President as a sort of imitation of the solar system.[17]

Then Wilson presents what he thinks of as the current and more adequate Darwinian analogy.[18]

Wilson clearly regards as his own the characterization of the Constitution as Newtonian and Darwinian. But who is the "very interesting Scotchman" whose thought was a catalyst for Wilson's? He seems very likely to have been Norman Kemp Smith, who came to Princeton from Scotland after having been interviewed by Wilson in Edinburgh during the summer of 1906. At Princeton Norman Kemp Smith was Stuart Professor of Psychology from 1906 until 1913 and then McCosh Professor of Philosophy until 1919. From 1916 until 1919, however, he was on academic leave in Britain because he wanted to work for his country in time of war. In 1919, when he was about to return to Princeton, he accepted the chair of logic and metaphysics at Edinburgh University, having received a letter of recommendation from Wilson, then president of the United States.[19]

Norman Kemp Smith could well have been the person characterized by Wilson in 1912 as "a very interesting Scotchman who had been devoting himself to the philosophical thought of the seventeenth century" and as one of the "thoughtful men from all over the world" whom Wilson "had the pleasure of entertaining" when he was president of Princeton.[20] In 1902, under the name of Norman Smith,[21] the Scot had published his important *Studies in the Cartesian Philosophy.*[22] In September 1906, some time after Wilson had interviewed Smith in Edinburgh, he wrote to a friend about "our new Professor of Psychology, Norman Smith, a fellow I feel sure everybody will like."[23] In December of the same year, in his annual report for 1905–1906, Wilson wrote further:

> For ten years before coming to us, Mr. Smith had steadily grown in the confidence and admiration, not only of his colleagues at the University of Glasgow, but also of all those who observe the progress of philosophical thought in Great Britain. In him we once more bring from Scotland some of the most vigorous influences of our thought.[24]

Discussions with Kemp Smith's student and assistant George Davie, and correspondence with his daughter, Janet Ludlam, strengthen the plausibility of his having spoken with Wilson at Princeton about how

> after the Newtonian theory of the universe had been developed, almost all thinking tended to express itself upon the analogies of the Newtonian theory, and since the Darwinian theory has reigned amongst us everybody tries to express what he wishes to expound in the terms of development and accommodation to environment.[25]

It is almost certain that Norman Kemp Smith's conversation enabled Wilson to discover and develop his idea that the American government was originally intended to be "Newtonian" but must now be regarded as "Darwinian."

Later Views on the Constitution as a Newtonian Document

Although Wilson's presentation of Newton versus Darwin in the context of the Constitution was only a minor part of his general presentation, his stress on Newtonian thought has continued to be a source of discussion or comment. Some two decades later, in 1922, Carl Becker boldly declared the importance of "Newtonianism" in eighteenth-century American political thought in a different context, as part of the background of thought reflected in the Declaration of Independence. Becker, however, as we have seen in Chapter 2, used Newton's science primarily in relation to the eighteenth-century idea of nature. He did not draw on any concepts, laws, or principles of that science (which would have been beyond his powers of understanding), nor did he refer specifically to the influence of science on the Constitution. In fact, Becker's book does not even mention the Constitution, although there is some discussion of the influence of the philosophy of the Declaration of Independence on several of the state constitutions.

The theme of science and the Constitution, and more specifically Newtonian science and the Constitution, inaugurated by Woodrow Wilson, became quite popular in the 1920s and later decades.[26] This interpretation was boldly trumpeted in J. G. Crowther's book *Famous American Men of Science* (1937) and was adopted by a group of very respectable scholars concerned with American history.[27] Many of the writers on science and the Constitution drew heavily on Carl Becker's monograph on the Declaration, transferring to the Constitution the background of ideas developed by Becker in relation to the Declaration. Becker wrote with authority because he was a leading historian of American thought. His fellow historians were impressed by Becker's magnificent prose style, setting forth in resoundingly clear terms the general theme that the Newtonian philosophy of nature had created an intellectual environment that was in some undefined way conducive to a special sort of political thought. Crowther himself drew heavily on Becker in presenting his "scientific" interpretation of the Constitution and the system of government it engendered.

Crowther was not a historian of science or in any sense a trained historian, but rather a British science journalist with openly declared Marxist leanings. His ideas on science and the Constitution were put forth in a preliminary version in a series of lectures given at Harvard University, at the special invitation of President James B. Conant, during the academic year 1936–1937. I can still vividly recall Crowther's presentation, which I attended along with other members of George Sarton's graduate seminar in the history of science. I was astonished to hear Crowther spin an intellectual and social web of generalities, but with no hard evidence either to support his theme of Newtonian science and the Constitution or to validate his special Marxist interpretation of Newton's science. Like Becker, whom he cited with approval as an American ally, Crowther rarely got down to any specifics with which a critic could take issue and, on the few occasions when he did so, he was plainly wrong. As a professional journalist, he based much of his presentation on secondary works rather than on a study of primary sources. The published version does not differ much from Crowther's original oral presentation.[28] The part of his argument that is of concern to us here is that "the form of the Constitution is partly due to the influence of certain scientific ideas."

Crowther's study is entitled "Science and the American Constitution." It begins with the intellectual exchange between John Taylor and John Adams in 1814, which we have already examined in Chapter 4. This discussion of checks and balances and "wheels within wheels" in relation to a "complication of machinery" is cited at length without comment, other than to remark that Woodrow Wilson and W. A. Robson had used this passage to show the degree to which "the notions of 'checks and balances' entered into the structure of the Constitution." Crowther, claiming to follow Wilson, assumes that "under Newtonian influence any system, including that of government, is conceived as a system of bodies moving according to the laws of mechanics and gravitation, in which action and reaction are equal and opposite, and all bodies are nicely poised by a balance of the forces acting on them." This wholly erroneous concept of the Newtonian system of the world is said by Crowther, still allegedly paraphrasing Wilson, to be the counterpart of "a balanced constitution."

From Wilson and the Whig theory of politics, Crowther turns to "the mental conflicts, or 'checks and balances,' in puritan psychology," which are alleged to exhibit "parallels" with the " 'checks and balances' of the American Constitution." This leads Crowther to the rather odd opinion that the "psychological frustrations of an Adams are related to the principles of the Constitution." No gloss is provided to help the reader. Crowther then finds it "significant" that Franklin's freedom "from complexes and frustrations" accordingly made him "philosophically opposed" to the Constitution. Indeed!

With regard to the influence of Newton, Crowther leaves the reader in no doubt. He states explicitly that the "notions of checks and balances, and mechanical equilibrium," introduced into the Constitution, were "Newtonian."[29] Declaring that these notions "were introduced by philosophic statesmen and lawyers who were not scientists," Crowther then uses this "incident" as a cautionary "example of the dangers of the misapplication of scientific ideas by politicians who do not properly understand them."

In developing his theme of science—and particularly Newtonian science—in relation to the Constitution, Crowther also introduces what is said to be Madison's concept of "the Constitution as a machine." He concludes this part of his presentation with the comment that Adams's discussion of the themes of the Constitution was influenced by Montesquieu's *Spirit of the Laws* and then declares that "Montesquieu's book contains examples of Newtonian modes of thought," quoting in full the only paragraph in that book that even remotely might seem Newtonian, the discussion by Montesquieu of honor and the equilibrium of forces of gravitation and repulsion.[30] Crowther was unaware that the example he cited showed that Montesquieu did not understand one of the most basic concepts of Newtonian science.

In conclusion, Crowther asserts once again the importance for the Constitution, and for the philosophy of the Declaration of Independence, of "the notion of mechanical equilibrium developed by Newton."[31] He argues that Franklin was at odds with the "makers of the Constitution" because "their philosophy involved scientific ideas [which were] already getting out of date." Crowther would have us believe that "Franklin had little influence on the construction of the American Constitution" because the "makers of the Constitution" believed "they were including in their work the supposedly imperishable ideas of Newtonian mechanics, which they did not understand." The reason why "they distrusted Franklin" was partly because he was not Newtonian. As "the forerunner of modern experimental scientists," Crowther opines, Franklin "could not sympathize with lawyers in love with a scientific point of view already old-fashioned."

This final curious view of history is so much at odds with what we know about Franklin and his science that it leaves a critical reader gasping. Franklin, of course, had the highest veneration for Newton. He was himself a Newtonian scientist, a follower of the great scientific tradition that stemmed from Newton's great masterpiece the *Opticks,* which became the vade mecum of all of the scientists of the eighteenth century who saw that there were new sciences to be created by experiments, by the direct interrogation of nature. This group of scientists included Stephen Hales, the founder of experimental plant physiology; Joseph Black, who established the basic concepts of heat; and Antoine-Laurent Lavoisier and the other men who established chemistry on a wholly new basis. All of these scientists were convinced Newtonians, following the precepts and examples of Newton's *Opticks* rather than the *Principia*.

Crowther's analysis rests on the commonly held but erroneous belief that Newtonian physics is posited on the concepts of balanced forces, equilibrium, and a machine. A careful study of the *Principia* or of any of the first-rate commentaries on that work will show, however, that these three concepts are not part of the central doctrines of that work. The *Principia* is concerned almost exclusively with cases of forces acting in situations where there is no equilibrium, with unbalanced forces that produce accelerations, and with resistances to such forces.[32] The Newtonian system of the world, which embodies the application of these principles of force and motion to the planets and their satellites and to the motion of comets, is definitely not a machine or like a machine. In his own time, Newton was repeatedly criticized for having deserted mechanical principles, for having created a system of the world that was definitely not mechanical. Finally, as we have already had occasion to note in Chapter 1, Montesquieu was not a Newtonian and the passage quoted by Crowther shows specifically that Montesquieu did not understand the simplest and most basic principles of Newtonian science.

A more recent presentation of the thesis that "Newton's Laws Shaped the Constitution" was published in 1987, in an article boldly entitled "Science and the American Experiment."[33] Here it is argued that the Founding Fathers, rejecting "the political ideas of the past," turned for "models after which to pattern government" to "nature itself, whose laws were viewed as templates for social laws." The most valid "interpretation of nature" was that of Isaac Newton. Rather than argue for a direct influence of Newton on the Founding Fathers, the author concentrates on David Hume: "Taking Newton's empiricism as his model, Hume held that there can be no knowledge of anything beyond experience." Because of the influence of Hume, it is concluded, "It is no accident" that such documents as Adams's *Defence of the Constitutions* and the *Federalist* "are studded with the language of natural science." Arguing the similarity between Hume's positions and Adams's, the author concludes that "proof" for Adams's idea of a political balance "came from the classical mechanics of Newton (for every action, a reaction)." Wholly apart from the error in stating Newton's third law (the "reaction" must be specified to be equal and opposite to that "action"), this principle was used by Adams in rebuttal of Franklin but not—so far as any evidence would show—to support Adams's own primary belief in a bicameral legislature. Furthermore, as we have seen, this law does not support the notion of a balance or equilibrium.

The only other specific reference to Newton or Newtonian science made in this article is in relation to "two key features in the Constitution," a "separation of powers between the executive, legislative, and judicial branches of the federal government" and "a division of power between the federal and state governments." Although it is boldly declared that these "features were linked . . . intimately to Newtonian theories of motion and force," there is nothing in any way specifically Newtonian about them. Indeed, the author omits any argument that might show just how these concepts were, in his opinion, related to Newtonian physical principles.

In conclusion, the author would have us believe that the "Newtonian cosmology" somehow gave "the framers confidence that the Constitution should accord with the ways of nature." Asserting that "the design of government rests on the principles of physics and geometry," he concludes that "the American government would be one of *mechanisms* rather than of men." In evidence a quotation is submitted from the writings of the nineteenth-century American poet James Russell Lowell, who referred to "a machine that would go of itself." Wholly apart from the question of whether the system of the Constitution is or is not a machine, the Newtonian system of the world proposed in the *Principia*—as has been noted above—is definitely not a machine in any accepted sense in which that term is ordinarily used. What is perhaps the most extraordinary aspect of this presentation is the complete absence of any reference to the *Principia*. There is no citation of any Newtonian principle, law, or theorem other than a misstated version of the third law as a parenthetical observation in relation to Adams and not to the framers of the Constitution. The author stresses the factor of empiricism, which is hardly the characteristic feature of the *Principia*, but he does not so much as mention the primary aspects of the science of that work, the mathematical proofs and constructions by the use of ratios and proportions, geometry and trigonometry, algebra and infinite series, and even the calculus. There is not even so much as a reference to Newton's supreme achievement, the discovery of the law of gravity and its universal application. Nor is there so much as a hint that Newton's mathematically derived results could be tested by accurate predictions and retrodictions of terrestrial and celestial phenomena.

A Different View of Science and the Constitution

There is no need to make a critical analysis of all the historians who have expressed a belief that the Constitution may, in some sense, have been Newtonian.[34] But attention may profitably be paid to the conclusions of Clinton Rossiter, who was one of the most brilliant students of American government in the generation immediately after World War II. His early death at the age of fifty-three was a real loss to American historical scholarship. A historian with Rossiter's keen analytical mind would never be satisfied with the crude generalities of a Crowther. Nevertheless, he did conclude that a fundamental principle of American political philosophy was that "the political and social world is governed by laws as certain and universal as those which govern the physical world." This opinion, that there can be a science of government which is an image of the physical or biological sciences, was widely shared in the later eighteenth century and appears prominently in many works of the nineteenth. The belief in such a science appears in the many references to the "science of government," the "science of politics," and the "science of federal government."

Rossiter, however, was not concerned only with such general notions, which he found in the political literature of the eighteenth and nineteenth centuries; he also proposed three specific ways in which he believed that "Newtonian science quickened the advance toward free government." The first of these was the conquering of superstition and ignorance and the parallel exaltation of the power of human reason. This feature is a consequence of science that is easy to demonstrate in examples. It certainly does not characterize the science of Newton in any special way that would differ in any factor other than degree from the science of that age. We have seen, in Chapter 3, an example of this conquering of superstition by science in Franklin's explanation of the lightning discharge as an ordinary or regular phenomenon of nature and his subsequent invention of the lightning rod. A Newtonian example of such an effect may be the analysis of the motion of comets in the third book of the *Principia*, in

which Newton shows that comets "are a sort of planet." He demonstrated that many comets must move in elliptical orbits, returning
periodically from the far reaches of outer space to the neighborhood
of the Sun. Clearly, if comets move in determinate orbits, and if their
appearance is a regular and therefore predictable event, a comet is
not a providential sign and is not to be understood as a special warning or message from an angry God.

In Rossiter's view the second way in which Newtonian science
acted as a force toward free government was that scientific method
had a "kinship" with democratic procedure and that therefore "the
advance of science popularized other methods and assumptions that
were essential to [its] conduct." This thesis clearly has merit, although
it is not easily provable for science in general and in all ages and even
for science in the age of the Founding Fathers. It is certainly true
that in our own century we have witnessed the decline of science
under authoritarian regimes in Communist Russia and in Nazi Germany, but a century earlier science flourished under the rigid government of Prussia and Imperial Germany. In the eighteenth century,
Voltaire—like other French thinkers of his day—compared the state
of relative freedom and liberty in Britain with the restrictive authoritarian conditions in France. His argument stressed such issues as the
freedom of the press. Voltaire was particularly impressed by the fact
that nobles of the highest rank paid honor to Isaac Newton at his
funeral, even though Newton was born a commoner, the son of a
yeoman farmer. But in the second half of the century, autocratic
France—despite the restrictions of censorship—gained the ascen-
dancy, becoming the major scientific country of the world, the home
of d'Alembert, Buffon, Réaumur, Lagrange, Laplace, and Lavoisier.

"Finally," according to Rossiter, "the new science had a direct
influence on the development of American political and constitutional thought." According to Rossiter,

> Basic to the Newtonian system were the great generalizations of
> a universe governed by immutable natural laws and of harmony as
> the pattern and product of these laws. The first of these gave new
> sanction to the doctrine of natural law; the second had much to do
> with the growing popularity of the Whiggish principles of balanced
> government.[35]

Certainly there can be no valid grounds for disagreement with the
first proposal on the support for natural law, but the second, on the

"growing popularity of the Whiggish principles of balanced govern-ment," is open to question as a too facile generalization. It is to be observed, however, that Rossiter was careful not to argue that the concept of checks and balances or political equilibrium was related directly to some specific law or principle of Newtonian science.

Science in the Deliberations at the Constitutional Convention

The framers of the Constitution were primarily concerned with practical rather than philosophical issues. In particular, there were heated expressions of concern over specific powers assigned to the proposed federal government and others assigned to the individual states, and there was always the problem of some kind of balance between the larger states and the smaller ones. Accordingly, in the debates on the Constitution, both those that occurred during the Convention and those that took place during the time of ratification, the role of science was minimal. We shall see, however, that science in general and the Newtonian philosophy in particular served to pro-vide acceptable metaphors for discussion or argument. This is quite a different role for science than to be a fount of methods, principles, and concepts that guided or were a fundamental part of the thinking of those who produced the Constitution. To me the evidence seems overwhelming that there is no single aspect of the Constitution, or of the doctrines of government it sets forth, which was either con-sciously or unconsciously designed or formulated as a direct political analogue of a doctrine, principle, law, concept, or rule of Newton's or, for that matter, of the physical or biological sciences or mathe-matics.

The physicist J. Robert Oppenheimer once looked into this ques-tion in a rather straightforward way. Noting that the *Principia* embod-ied a system of mathematical deductions associated with the results of careful experiments and controlled observations, Oppenheimer asserted there is no trace of the Newtonian science in either Euro-pean or American politics. "What there is of direct borrowing from Newtonian physics," he wrote, whether "for chemistry, psychology, or politics is mostly crude and sterile." Furthermore, "What there is in eighteenth-century political and economic theory that derives from Newtonian methodology is hard for even an earnest reader to find." He concluded that the "absence of experiment and the inappli-cability of Newtonian methods of mathematical analysis make that inevitable. These were not what physical science meant to the Enlightenment."[36]

As we shall see, however, there are a number of references to science in both the debates and discussions at the Constitutional Convention and in the arguments concerning the Constitution in the *Federalist*. These leave no doubt that some of the framers of the Constitution sought to establish a harmony between their political tenets and proposed practices and the universe of nature as interpreted and understood by the science of their day. As we have seen in the case of John Adams's rebuttal of Franklin, an appeal to science—particularly to the *Principia* of Newton and its laws—was an invocation of the highest authority and would make the strongest case imaginable.

A close reading of Madison's minutes of the Constitutional Convention, however, does not disclose a single example in which the physical and the biological sciences provided an important concept, model, power, or restriction used in framing the principles of the new government. Yet, not surprisingly, these minutes display quite a number of rhetorical references to the sciences, just as we should expect to be the case in an age of reason.

Among the references to science, those dealing with the planetary system and forces that keep the system stable are among the most plentiful. As an example, consider Madison's own argument concerning the power of the federal government "of negativing" laws passed by state governments. This power "was absolutely necessary," he wrote, likening it to the "attractive principle which would retain [*probably should have been* restrain] the centrifugal force" in our solar system and noting that without it the "planets will fly from their orbits." Here Madison was not introducing the static notion of a balance between a centripetal and a centrifugal force, a condition of equilibrium which is absolutely non-Newtonian; had he done so he would have written about a balance or an opposing force exerted by the centrally directed attractive force rather than a restraining principle that retains planets in their orbits.

Madison was quite correctly applying the principles of Newtonian celestial physics which he had learned during his undergraduate days at Princeton. As we have seen, in the common language of Newtonian science, the inertial component of orbital motion, in the direction of a tangent to the orbit, was often referred to as centrifugal force. The reason is that if there were no restraining force (such as the Sun's gravity) that constantly makes the planets fall inward toward the Sun, the "force of inertia" of the planets would cause them to "fly from their orbits"—just as Madison said—outward along a tangent, as if there were some outward or "centrifugal" force.[37] Because these two forces (the so-called centrifugal force and the centripetal accelerating force of gravity) are of different kinds and act in different directions, there can be no question of balance between them, no equilibrium. This is, therefore, an example of an unbalanced central force of gravity and there is a resulting continual change in motion or acceleration of falling inward.

A somewhat similar use of science in a political context occurred during the Convention in an elaborate discussion of whether senators should be elected or appointed. John Dickinson "compared the proposed National System to the Solar System, in which the States were the planets, and ought to be left to move freely in their proper orbits." He alleged that James Wilson wished to "extinguish these planets." In rebuttal, Wilson declared that he "was not for extinguishing these planets as was supposed by Mr. D.," but "neither did he on the other hand, believe that they would warm or enlighten the Sun." His concern was to see that they were kept within "their proper orbits."[38] In another report, Dickinson referred to "the British house of lords and commons, whose powers flow from difference sources." By providing "mutual checks on each other," he said, two such differently constituted branches of the national legislature could act to "promote the real happiness and security of the country." This led him to the following astronomical example:

> . . . a government thus established would harmonize the whole, and like the planetary system, the national council like the sun, would illuminate the whole—the planets revolving round it in perfect order; or like the union of several small streams, would at last form a respectable river, gently flowing to the sea.[39]

Here Dickinson has mixed his metaphors, combining into one the "harmony" of the planets both circling and being illuminated by the Sun and the confluence of small streams into a river.

Dickinson's use of a scientific analogy has another feature that shifts the discussion from the general Copernican notion of a central Sun surrounded by a system of orbiting lesser bodies or planets to a consideration of the forces active in the system. In one report, he would have "the Genl. Govt. be the Sun and the States the Planets repelled yet attracted."[40] A more detailed version would have him say:

> This prerogative of the General Govt. is the great pervading principle that must control the centrifugal tendency of the States; which, without it, will continually fly out of their proper orbits and destroy the order & harmony of the political system.[41]

In this latter case Dickinson could have been invoking a Newtonian "centrifugal tendency," which must be controlled or restrained in order to prevent the circulating bodies from flying off along a tangent, from flying "out of their proper orbits." But this rhetorical flourish could equally imply the non-Newtonian position of a supposed equilibrium or balance of opposing forces, one a force of attraction, the other a force of repulsion. This is the sense of the previous extract. In the second version, he says that if the planets were not "controlled" by a centripetal force of gravity, they would fly "out of their proper orbits." He does not, however, specify whether they would then fly out along a tangent in accord with Newtonian principles or whether they would then shoot directly outward in a non-Newtonian manner, along a direction straight out from the Sun.

Dickinson's degree of understanding of the dynamics of the solar system may be ascertained by examining other examples of his use of the metaphor. In arguing for the powers of the several states, we have seen that Dickinson "compared the proposed National System to the Solar System, in which the States were the planets, and ought to be left to move freely in their proper orbits." But if the planets were simply to "move freely," that is, without any outside force, then according to the principle of Newtonian physics they would not remain in "their proper orbits" but would fly off along a tangent. He also referred to the "Congress" as "the Sun of our political World," which acts as "a check to the centrifugal Force which constantly operates in the several States to force them off from a common Centre, or a national point." It would seem, from the conclusion of this sentence, that Dickinson had in mind a centrifugal force that would, if not counterbalanced, force the planets to move straight out along a line directed outward from a central point, from the Sun at their "common Centre." In this case, the metaphor would be wholly non-Newtonian. It is to be observed, however, that Dickinson does not introduce either the term "balance" of forces or "equilibrium," which would have specified a non-Newtonian or erroneous conception of the physics of planetary motion.

In the first of the examples just cited, we have seen Dickinson refer to a planetary system and then leap to an image of the confluence of small streams to form "a respectable river." This odd juxtaposition of rivers and the solar system is typical of the rhetoric of poetical orators, piling metaphor on metaphor, example on example. The position taken in the argument is not based on an analogy between a political situation and a principle of the sciences, nor is the conclusion based on a theorem or law of the sciences. Rather, the sciences provide rhetorical authority to buttress a position or conclusion already adopted, just as a lawyer's argument is reinforced by citations of legal precedents or the names of great authorities. This was an aspect of political debate and oratory that was then well recognized by all. For instance, in the midst of a purely political analysis of the states' powers, a speaker compared the national government with the "Sun the Centre of planetary System."[42] No use was made of this reference to celestial physics in developing the argument. The authority of science, of nature itself, was simply being invoked to support and even to legitimate an adopted political position.

Other scientific metaphors in the discussions at the Convention were drawn from the science of matter, and even the phenomena of medicine and of vegetable and animal life. James Wilson wrote that "A Vice in Representation, like an error in the first concoction, must be followed by disease, convulsions, and finally death itself."[43] Wilson was using "concoction" here as a technical term for the medicine given at the early stage of a disease. In eighteenth-century therapeutics, it was believed that a concoction would act to separate the "morbid" matter by a "ripening preparatory to its discharge." We have seen (Chapter 1) that Wilson, in his lectures on law, frequently made use of the Newtonian natural philosophy.

These examples, illustrations, and metaphors drawn from the sciences—from the life sciences as well as from the physical sciences and notably the Newtonian natural philosophy—impress us today by their very presence. One could hardly imagine that such introductions of the sciences could occur as part of the political oratory or rhetoric of our own time. In this context there are two factors that must be kept in mind. The first is that many of the framers of the Constitution were well educated in science and so could readily introduce metaphors based on science into their political rhetoric. Equally important is the fact that their fellow delegates were also generally conversant with scientific principles and so could understand the references. This double feature of science as an element of the intellectual background is a splendid index of the high esteem accorded to the sciences in the Age of Reason.

James Madison's Scientific Education

James Madison's early education included the study of Latin and Greek, history, rhetoric, and some mathematics: arithmetic, algebra, and geometry.[44] At the age of eighteen he entered the College of New Jersey at Princeton, now grown into Princeton University, rather than following the traditional path of young Virginians to William and Mary. The college had been founded in 1746, twenty-three years before Madison's arrival as a freshman. In 1768, a year before Madison entered as a student, John Witherspoon came from Scotland to take over the presidency of the college. Witherspoon had been greatly influenced by the thinking of some of that group of intellectuals who spearheaded what is sometimes known as the Scottish Enlightenment. At Princeton, he devoted himself, among other duties, to improving the teaching of science.[45] It was during Witherspoon's presidency that the college obtained the celebrated Rittenhouse orrery, which exhibited to the students a model of the motions of the planets and their satellites. Madison concentrated his work so as to combine the normal four years into two and then stayed on at Princeton for some months after graduation in order to pursue additional studies with Witherspoon of history, moral philosophy, and divinity.

In the first year, Princeton students read classic works in Latin and Greek, including Horace, Cicero, Lucian, Xenophon, and the Greek Testament. Sophomores began work in "the sciences, geography, rhetoric, logic and mathematics." Juniors read additional Latin and Greek works and continued with mathematics and "natural philosophy" (physics and astronomy) along with "moral philosophy." Seniors spent their time in review and composition, covering "the most improving parts of the Latin and Greek classics, parts of the Hebrew Bible, and all the arts and sciences."[46]

There are no surviving notes concerning Madison's study of mathematics and the sciences as an undergraduate. A manuscript note-

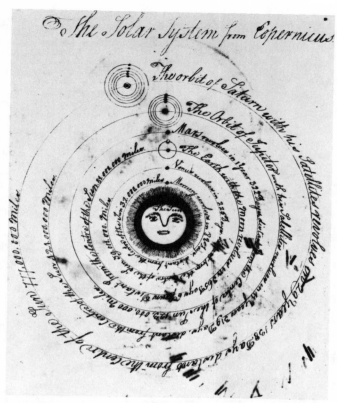

Fig. 38. James Madison's Drawing of the Copernican System. This undated drawing used to be considered an exercise from Madison's secondary-school days, but today this diagram (and Figs. 39 and 40) is believed to be part of a student notebook kept by Madison while an undergraduate in the College of New Jersey (Princeton).

book, devoted to literary or philosophical subjects and devoid of science or mathematics, does contain diagrams (mainly mathematical) without any accompanying text. One of these is a very elementary and greatly simplified diagram of the Copernican system of the universe (Fig. 38). Another (Fig. 39) displays Madison's command of trigonometry in the computations for a sundial, reckoned for the latitude of the part of Virginia where Madison lived. Another (Fig. 40) is intended to show the projection of a geometric solid, obviously an exercise in geometry of which the results are not very convincing. These diagrams are on so elementary a level, compared to the kind of work being done at that time by students at William and Mary or Harvard or Yale, that for a long time they were considered to have

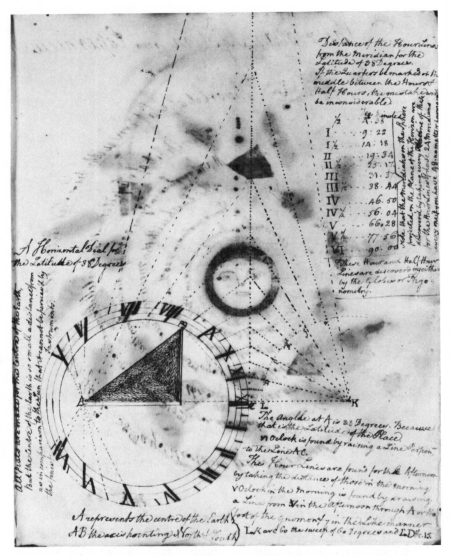

Fig. 39. Madison's Calculations for a Sundial. In this calculation, Madison has computed the angles of inclination and the "hour lines" for a sundial to be used at the latitude of 38°, approximately the latitude of his home in Virginia.

been made while Madison was in secondary school rather than in college.

There is considerable evidence that the teaching of science at Princeton during the years when Madison was a student there suffered because of a serious lack of adequate apparatus for demonstra-

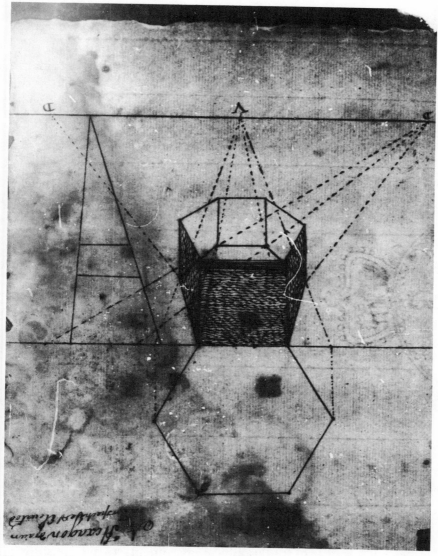

Fig. 40. An Exercise in Perspective. Part of the study of mathematics in Madison's student days was the science of perspective drawing. In this exercise, Madison has sketched out the projections in perspective of a hexagonal box.

tions and experiments. Students did, however, apparently receive sound instruction in mathematics and the elementary principles of physics (including some Newtonian science) and astronomy. In 1769, just before Madison entered college, the trustees recognized the need of paying more serious attention to mathematics and science; in that year, Mr. Jeremiah Halsey was "unanimously elected" to be the "Professor of Mathematics and Natural Philosophy."[47] Halsey did not accept the chair, however, and it remained vacant until 1771, when William Churchill Houston, senior tutor in the college, became the professor of science. In that same year, 1771, a committee was appointed to draw up a list of the types of scientific apparatus which they deemed "the most necessary and immediately wanted" in order to provide satisfactory education in science for the undergraduates.[48] These measures were taken too late to be of much use to Madison, who was not as fortunate in this regard as John Adams, who studied under a great master, Professor John Winthrop, and who used the rich resources of the Harvard collection of scientific apparatus, or Thomas Jefferson, who had the good luck to be taught by William Small and who could use the various scientific instruments available in Williamsburg.

Some sense of the poor state of science instruction in the days just before Madison entered college may be gained from a *Catalogue of Books in the Library of the College of New Jersey* compiled in 1760 by President Samuel Davies. Few "modern Authors," he complained, "adorn the Shelves" of the college library. In particular, he noted, "this Defect is most sensibly felt in the Study of Mathematics, and the Newtonian Philosophy, in which the Students have but very imperfect Helps, either from Books or Instruments." Nine years later, a report of the trustees still expressed deep concern about the "comparatively small assortment of modern authors" and the lack of "works on mathematics and the Newtonian philosophy, in which the students have but very imperfect helps, either from books or instruments."[49] The official historian of Princeton observed in sorrow that "the lack of [scientific] equipment in the College of New Jersey seems pathetic,"[50] as may be seen by the fact that the trustees "repeatedly" appealed "for gifts of 'philosophical instruments' to aid the study of astronomy, physics and geography." In 1760, it was reported that the college possessed "a few instruments and some natural curiosities."[51] In 1764, five years before Madison arrived in Princeton, the trustees once again lamented that the college was "but very indifferently furnished" with "mathematical instruments and an apparatus for experiments in natural philosophy."[52] We may well understand President Witherspoon's concern to purchase the orrery made by David Rittenhouse, which was duly installed in Nassau Hall in April 1771. While this complex instrument was a marvel of artisanal ingenuity and craftsmanship, it did not take the place of the needed apparatus for demonstrations and experiments in physics and chemistry as part of instruction in the Newtonian natural philosophy.

An advertisement for the college in 1772 contained the prediction that when the apparatus that "is commissioned and is now on its way" will have arrived in Princeton, the college's "apparatus for Mathematics and Natural Philosophy will be equal, if not superior, to any on the continent." In the following year "the Publick" was informed that "There has been lately brought from *London* a complete philosophical Apparatus, every part of which was examined and approved by Persons eminently skilled in Natural Philosophy, Mathematicks, and Astronomy, to which is added the Orrery constructed by *David Rittenhouse,* Esq., of *Pennsylvania.*"[53]

Despite the boasts made in the 1770s, the shortage of apparatus for instruction in the sciences was still severe in 1790, when Professor Walter Minto made an appeal in vain for funds to purchase new apparatus that would be adequate for teaching mathematics and the sciences. In 1796, James Madison, Aaron Burr, and other devoted alumni subscribed a fund of $122 for the purchase of "instruments and materials most necessary to exhibit the leading experiments in chemistry."[54]

Throughout his later life, Madison cultivated an interest in the sciences as a devoted amateur, as a cultivated citizen of the Age of Reason. He thus referred, in a letter of 1784, to "the philosophical books I had allotted for incidental reading."[55] Madison's adult education in the sciences was influenced and even to some degree directed by Thomas Jefferson, with whom he began to be intimate in the late 1770s. One of the advantages to Madison of this relationship was that it opened to him the vast resources of Jefferson's personal library. In 1784, when Jefferson went on his mission to France, he acted as Madison's agent and adviser, procuring for him books in French on the history and theory of government and on general philosophy, but also works on science.[56]

Jefferson especially stimulated Madison's interest in natural history. It was apparently a result of Jefferson's influence, if not his direct tutelage, that Madison was encouraged to study the works of such leading French naturalists as Buffon and Daubenton. Madison even made careful and detailed measurements of the organs of a weasel and a mole found on his plantation at Montpelier. These were sent on to Jefferson to be used as ammunition in the war against the theory espoused by Buffon and others that animals which were similar in the Old World and the New were necessarily smaller in the New World.[57]

There is a record of a discussion between Jefferson and Madison in late 1783, when they were together in Annapolis and in Philadelphia, concerning the central heat of the Earth. This was the time when Madison was leaving the Continental Congress and Jefferson was entering the Congress once again. Buffon had published his theory of the Earth's heat four years earlier in his *Epoques de la nature.* Jefferson, who tried to keep up with the latest science, had studied this work but Madison had not yet encountered it. In the course of their subsequent correspondence, they exchanged calculations and proposals for experimental tests. Jefferson, always the good teacher, corrected what Madison later described as "my misconception."[58] This exchange shows that Madison had become very adept at mathematical computations, even if the principles of physics were not his strong point.

During the course of their epistolary discussion of the Earth's heat, Madison informed Jefferson that he had been reading "the 45th volume of the London Magazine" and had found an interesting remark on "page 373" by "Doctor Hunter" (the anatomist William Hunter). According to Hunter, Madison reported, there are in the British Museum "grinders of the Incognitum which were found in Brazil and Lima." The significance of this information was that Jefferson had hypothesized that "no bones of that animal" had been encountered "so far to the south." Jefferson replied that if this was the only example ever found in the Southern Hemisphere of bones identical to those found in the Northern Hemisphere, he would rather suppose that they had been brought there long ago by migrating Indians or even later on by Spaniards.

Madison was aware that he was ignorant of chemistry and asked Jefferson to send him "two Boxes, called *La nécessaire chémique,*" together with "a good elementary treatise" on this subject.[59] We shall see below that in the *Federalist,* Madison drew on his study of chemistry for a political metaphor. His own observations of the world of nature and his reading in the scientific books sent to him by Jefferson so increased his competence in science that in 1785 he was elected a member of the American Philosophical Society, in a group that included the British chemist Joseph Priestley, who would later emigrate to America, the American naturalist Menasseh Cutler, and

Thomas Paine, a technologist and inventor as well as a political pamphleteer.

Madison and Science in the Federalist Papers

In the *Federalist,* as in the debates and discussions at the Constitutional Convention, science appears rather frequently as a source of metaphor. In particular, our attention may be drawn to the papers in this collection written by James Madison, rather than to those of his fellow authors, Alexander Hamilton and John Jay, primarily because metaphors based on science are more characteristic of Madison's contributions. In addition, Madison's contributions to the *Federalist* are of particular importance in the context of the present inquiry because of his role as a principal architect of the Constitution. Whatever he said in explanation or defense of any section or principle must command our attention. We have seen, however, that Hamilton did make use of the principles of mathematics in at least one of his contributions.

The *Federalist* is a collection of articles first published anonymously in a New York paper with the purpose of explaining and defending the principles of the Constitution and thus gathering support for ratification among the voters. Often reprinted then and now, the *Federalist* is highly regarded as a major document in the development of the theory of democratic government. Revered as the bible of American political thought, this work has been analyzed minutely by scholars of many kinds. The *Federalist Concordance* enables scholars to find the occurrence and context of every word in the collection. It is on this basis that we can assert categorically that "science" appears eleven times and "sciences" four times in the *Federalist.*[60] The *Concordance* also contains a statistical study of the frequency of words which authenticate the attribution of authorship to the papers in this collection.

In *Federalist* No. 10, Madison made use of a chemical metaphor. "Liberty is to faction," he wrote, "what air is to fire, an aliment, without which it instantly expires." By analogy, he argued, "It could not be a less folly to abolish liberty" ("which is essential to political life") on the grounds that liberty "nourishes faction," than "it would be to wish the annihilation of air, which is essential to animal life, because it imparts to fire its destructive agency."[61] This combination of the properties of air in supporting both combustion (or fire) and respiration has been a feature of science teaching ever since the days of Robert Boyle, who devised an air pump in the decades just before the *Principia*. A variety of experiments linked respiration and combustion. If a lighted candle is placed in a bell jar which is then evacuated, the flame goes out at once. So, too, a mouse placed in a bell jar will expire when the air is drawn out of the jar.

Madison used an analogy from the life sciences in discussing (No. 37) the "line of partition" between the authority of the federal government and that of the state governments. In this case, he pointed to the lack of accuracy on the part of "the most sagacious and laborious naturalists" in "tracing with certainty" the line of separation between "vegetable life" and "unorganized matter."

The *Federalist* papers make much use of such expressions as "legislative ballances and checks" (No. 9), "equilibrium" (Nos. 14, 15, 21, 31, 49, 66), "equilibrium of power" (No. 15), "balance" (Nos. 10, 34, 35, 45, 71, 74, 85), "the equilibrium of the national house of representatives" (No. 66), and "the constitutional equilibrium of the government" (No. 49). In not one of these, however, is there a reference to the dynamics of the solar system or to anything that might even vaguely suggest that the author had Newtonian science in mind.

The source of the greatest number of scientific metaphors in the *Federalist* is not physical science but the life sciences and particularly medicine. In *Federalist* No. 10, Madison refers twice to a "cure" and declares that "the extent and proper structure of the Union" discloses "a Republican remedy for the diseases most incident to Republican Government." In No. 14, he mentions that "almost every State will on one side or other, be a frontier," which he says implies that "the States which lie at the greatest distance from the heart of the union . . . of course may partake least of the ordinary circulation of its benefits." Here the concept of the heart and the circulation of the blood provides a metaphor for the political argument. Again, in No. 19, Madison discusses conditions which "render" an "empire a nerveless body." Expanding this physiological metaphor, he explains that such a nerveless empire is "incapable of regulating its own members" and is "insecure against external dangers"; it is, furthermore, "agitated with unceasing fermentations in its bowels." In No. 50, Madison discusses "the existence of the disease" in a political system and decries "the inefficacy of the remedy." In No. 38, there is a lengthy comparison of a sick person and an ailing government.

Alexander Hamilton, in his contributions to the *Federalist,* equally made use of metaphors from medicine and the life sciences. Among them are a "natural cure for an ill administration" (No. 21), the "two most mortal diseases of society" (No. 34), and his observation that "seditions and insurrections are unhappily maladies as inseparable from the body politic, as tumours and eruptions from the natural body" (No. 28). In No. 28, he referred to "the reveries of those political doctors, whose sagacity disdains the admonitions of experimental instruction." Hamilton also referred (No. 17) to "a known fact in human nature," that "affections are commonly weak in proportion to the distance or diffusiveness of the object." We may note here the proportion of the inverse of the distance, not the Newtonian inverse square, similar to Madison's reference to the factor of distance in relation to the circulation.

There are no direct references to Newtonian science in the *Federalist.*[62] Nor does the dynamics of the solar system occur in any prominent way. There is no mention of either the force of universal gravity or of centrifugal force as such. One of the few instances in which any dynamical aspect of the solar system is introduced occurs in a discussion by Hamilton (No. 9) of a "principle" which "has been made the foundation of an objection to the New Constitution." He had in mind what he called "the ENLARGEMENT of the ORBIT within which such systems are to revolve either in respect to a single State, or to the consolidation of several smaller States into one great confederacy." This metaphor hardly requires any knowledge of astronomy or of Newtonian physics. And it is much the same for another introduction of astronomy by Hamilton (No. 15), where he was concerned with the situation of "every political association which is formed upon the principle of uniting in a common interest a number of lesser sovereignties." Without exception, he wrote, there "will be found a kind of excentric tendency in the subordinate or inferior orbs, by the operation of which there will be a perpetual effort in each to fly off from the common center." I find this argument confusing at best. It is difficult to see, from the point of view of solar dynamics, why the inferior planets should have a "perpetual effort . . . to fly off" and not the superior planets as well. Nor am I at all sure what is meant by an "excentric tendency" unless possibly Hamilton had in mind the special features of the orbit of the innermost planet Mercury.

On a few occasions, Hamilton makes use of metaphors that relate to mathematics. He refers (No. 16) to "the ordinary rules of calcula-

tion" and he mentions (No. 21) "the saying . . . that 'in political arithmetic, two and two do not always make four.' " He also discusses the
possibility (No. 22) that "the frame of the government" may be "so
compounded, that the laws of the whole are in danger of being contravened by the laws of the parts." Even more interesting is his introduction (No. 41) of the "rational calculation of probabilities." We
have seen, in Chapter 2, that Hamilton also (Nos. 31, 80) referred to
axioms of geometry and of political science, even comparing geometry and politics.

Although there are a fair number of metaphors drawn from the
sciences in the *Federalist,* these are not a primary feature of this document. Indeed, unless readers are especially on the lookout for such
instances, they will probably not even be aware that there are such
metaphors in these papers. What is significant, therefore, is not that
science provided metaphors in a prominent way for the authors of
the *Federalist,* but rather the fact that there are any such metaphors
at all.

If not Newton, Darwin? Significance of Scientific Metaphors in Political Thought

In his discussion of science in relation to the Constitution, Woodrow Wilson declared that in "our own day, whenever we discuss the
structure or development of anything, whether in nature or in society, we consciously or unconsciously follow Mr. Darwin." Before "Mr.
Darwin," he explained, people "followed Newton."[63] We might carry
Wilson's metaphor further into our time by noting that we today, in
a similar fashion, make use of concepts originating in the science of
Sigmund Freud.

Wilson made two independent assertions. One was to distinguish
between the notion of government as an organism and as a machine,
the other to establish a specific Darwinian interpretation of the Constitution. On this score, we should take note that Wilson was not
arguing for an influence on the framers of the Constitution of any
pre-Darwinian ideas of evolution, such as those of Buffon, which
were current in the eighteenth century. In this feature, his argument
for a biological interpretation differs from his remarks about the
older model based on physics. Newtonian ideas, he held, had—perhaps unconsciously—influenced the founders of the Whig theory of

government, which was imitated by the framers of the Constitution. The reason for adopting Darwinian ideas, according to Wilson, was that a sound analysis of the actual workings of constitutional government must take cognizance that "government is not a machine, but a living thing."[64] In this extreme statement, Wilson has gone beyond metaphor or analogy, actually asserting identity, much as Herbert Spencer asserts that society "is" an organism. The conclusion which Wilson draws is that government "falls, not under the theory of the universe, but under the theory of organic life."

In advocating a biological basis for understanding government, Wilson argues that any government is "modified by its environment" and is "shaped to its functions by the sheer pressure of life." No animate being, he explains, can live if its organs are "offset against each other as checks." Rather, the life of an organism depends on the "quick coöperation" of its organs, "their ready response to the commands of instinct or intelligence," and "their amicable community of purpose." A government is "a body of men," he continues, whose cooperation is essential for survival. Governments cannot survive, he concludes, "without the intimate, almost instinctive, coördination of the organs of life and action." This, he maintains, "is not theory but fact." The consequence is that "living political constitutions must be Darwinian in structure and in practice."[65]

The discussion of the biological interpretation of government concludes with a restatement of an assertion, which we have explored earlier in this chapter, that "the definitions and prescriptions" of constitutional law were "conceived in the Newtonian spirit and upon the Newtonian principle." Fortunately, however, Wilson observes, these definitions and prescriptions have been "sufficiently broad and elastic to allow for the play of life and circumstance." In other words, despite the Newtonian design, the constitutional government of the United States did not emerge as "a machine governed by mechanically automatic balances" but has rather "had a vital and normal organic growth" which has made it adaptable to "the changing temper and purposes of the American people from age to age."[66]

Both a conception of government or of society as an organism and an interest in Darwinism can be traced to a stage in Wilson's thinking much earlier than that of 1907. Some examples from 1885 and 1889 are illustrative. In his first published book, *Congressional Government,* which appeared in 1885,[67] Wilson is already contrasting the original system of checks and balances with ideas of organic development. In particular he uses the "figure" of the Constitution as a root from which the present "delicate organism" of government has grown.[68] In "The Modern Democratic State," written in December 1885,[69] he describes the modern democracy as

> a vast state with autonomous life in every part and yet a common life and purpose, with as many centres of self-directive energy as units of organization and yet with voluntary ardour and unhampered facility of coöperation—not a head organism with huge dependent parts, but itself a single giant organism with a stalwart, common strength and an abounding life in individual limb and sinew.[70]

This vivid image is reiterated in May 1889, in an address on the "Nature of Democracy in the United States," largely extracted from "The Modern Democratic State."[71] Of the United States Wilson says that a "huge stalwart organism like our own nation, with quick life in every individual limb and sinew, is apt, too, to have the strength of variety of judgment."[72] More abstractly, in *The State*, published in that same year of 1889, Wilson writes that society "is as truly natural and organic as the individual man himself" and that, "like other organisms," it "can be changed only by evolution." One of the sections in which Wilson makes these points is entitled "Society an Organism, Government an Organ."[73] A review of *The State* by David MacGregor Means, printed in *The Nation* for 26 December of the same year,[74] criticizes the assumptions and methodology lying behind such attitudes:

> It is to be feared that a tendency exists among our rising school of historical writers to adopt rather a scientific vocabulary than a scientific method. We hear much talk of society as an organism; of growth and development, of crystallization, and integration, and solidarity. These are convenient expressions, but they may sometimes serve to conceal a lack of thought. It is quite possible to mistake a metaphor for a demonstration, and to deal so nimbly with abstractions as to make them pass for the concrete.[75]

But in July of that very year, if the tentative dating given in the *Papers* is correct,[76] Wilson had himself addressed the same or at least a similar problem in a manuscript memorandum:

The figure and idea of evolution, it would seem, is approaching the misfortune of being overdone. Everybody knows that govt. was never invented and can never be safely altered by invention; but everybody is now grown weary of hearing . . . that govts. are not made, but grow. . . . The formula of evolution is easy, we know; but it is not wholly safe, we suspect. True Mr. Bagehot was a great thinker upon such topics and an incomparable critic of politics, whether old or new; and Mr. Bagehot, in his quite incomparable *Physics and Politics* deliberately applies the laws of heredity and natural selection to political society. For my part, I acknowledge, I religiously believe most that the book contains. If it be not true, it is a sin to have presented it so charmingly. It is made to appear so irresistibly probable that no man of historic imagination can long hold out against believing it. But of course Mr. B. was speaking by analogy. He knew that Politics was *not* Physics: that they proceeded not by laws of nature but laws of character and mind. And that is just the fact that makes the analogy risky. We need a fresh formulation of the principles [of] political change, and a . . . somewhat shifted point of view.[77]

Wilson, of course, had particular reasons in the 1880s for paying attention to theories of evolution. For example, his uncle, James Woodrow, who was both a scientist and a minister, had been the subject of great controversy because he had declared himself a supporter of Darwinian evolution while holding the Perkins Professorship of Natural Science in Connection with Revelation at Columbia Theological Seminary in South Carolina.[78] References to the situation are scattered through Wilson's correspondence. On 26 June 1884, for instance, he wrote to his future wife, Ellen Louise Axson:

The attack on uncle James Woodrow which I predicted has begun. . . . If uncle J. is to be read out of the Seminary, Dr. McCosh [president of Princeton since 1868 and known in particular as a supporter of Darwinism[79]] ought to be driven out of the church, and all private members like myself ought to withdraw without waiting for the expulsion which should follow belief in evolution.[80]

Nevertheless, Wilson's evolutionism as expressed in *Constitutional Government* must be termed Spencerian rather than Darwinian. Moreover, it still seems to show the influence not only of Herbert Spencer but also of Walter Bagehot in its presentation of an organismic model of government.

Spencer was one of the most influential thinkers at the end of the nineteenth century. Although an evolutionist, he did not adhere to Darwinian principles. Instead of maintaining Darwin's central doctrine of "natural selection," according to which random variation occurs in individuals and only innate characteristics can be passed on to descendants, Spencer argued for a natural selection in which evolution could occur in entire societies and in which acquired characters could be inherited. In this way, he believed, it was possible for organisms—including social organisms—to determine the future course of their own evolution. Wilson's Spencerism is shown by his stress on adaptation, on the modification produced by the "environment," the way in which government is "shaped to its functions by the sheer pressure of life."[81] This is in no way Darwinian, since in Darwinian evolution the continual random variation among individuals brings it about that those individuals who survive are the ones best adapted to their particular environment and since, for Darwin, evolution occurs only if those characteristics that cause certain individuals to succeed in the "struggle for existence" are heritable.

Wilson's ideas about an evolutionary interpretation of government were greatly stimulated by his study of the works of Walter Bagehot. Bagehot's *English Constitution,* which was very important for Wilson, does not make any use of biological evolution, and a paragraph quoted near the end of the book from Darwin's *Origin of Species* does not discuss concepts of evolution but rather the importance of "facts."[82] However, Bagehot's *Physics and Politics* gives prominence to evolutionary ideas.

Although the title of *Physics and Politics* appears to suggest a theme of our own time, such as the control and use of atomic energy, at that time the current meaning of "physics" was only just coming into general use. When Wilson was a professor at Princeton, for example, the books on physics in the university library were still classified as "natural philosophy." The historian of physics, Henry Crew, used to tell the story of his coming to Princeton as a freshman member of the class of 1882 and being disappointed that the university library seemed to have no section of books on "physics," the subject in which he planned to major. In his puzzlement, he asked the librarian why there were no books on physics. The reply was, "Oh, we classify them as 'natural philosophy.'" Bagehot was using the word in the older sense of the study of "natural things," derived from the Greek word "physis," that is, "nature." The biological orientation is revealed by the book's subtitle: *Thoughts on the Application of the Principles of "Natural Selection" and "Inheritance" to Political Society.* Bagehot's purpose was to explore how Britain had become a superior nation, and his chief authority was not Darwin but Herbert Spencer, together with the legal historian Sir Herbert Maine. The evolutionism that he expounded was clearly Spencerian and not Darwinian, as may be seen in his emphasis on a contest for survival among races or peoples rather than among single individuals.

A brief glance at some of Bagehot's discussions of "natural selection" is illuminating. In a chapter on "Nation-Making" he appears to be presenting a Darwinian concept, since he invokes "natural selection":

> Certainly . . . nations did not originate by simple natural selection, as wild varieties of animals (I do not speak now of species) no doubt arise in nature. Natural selection means the preservation of those individuals which struggle best with the forces that oppose their race.[83]

But a close examination shows this is not at all Darwinian, since Darwin's natural selection refers specifically to individuals in competition and not races or species or varieties. Here, as in other sections of the essay, Bagehot's focus of interest is the evolution of nations; his concept of natural selection relates to "polities" or "nations" rather than individuals. In the chapter on "The Preliminary Age," for example, he proposes that

> when once polities were begun, there is no difficulty in explaining why they lasted. Whatever may be said against the principle of "natural selection" in other departments, there is no doubt of its predominance in early human history. The strongest killed out the weakest, as they could.[84]

In a chapter on "The Use of Conflict," Bagehot lays down the three "principles" or "laws, or approximate laws" which must be acknowledged if one is to understand progress:

> First. In every particular state of the world, those nations which are strongest tend to prevail over the others; and in certain marked peculiarities the strongest tend to be the best.

Secondly. Within every particular nation the type or types of character then and there most attractive tend to prevail; and the most attractive, though with exceptions, is what we call the best character.

Thirdly. Neither of these competitions is in most historic conditions intensified by extrinsic forces, but in some conditions, such as those now prevailing in the most influential part of the world, both are so intensified.

These are the sort of doctrines with which, under the name of "natural selection" in physical science, we have become familiar; and as every great scientific conception tends to advance its boundaries and to be of use in solving problems not thought of when it was started, so here, what was put forward for mere animal history may, with a change of form, but an identical essence, be applied to human history.[85]

Here there is no doubt that Bagehot conceived of evolution as a contest among nations or races and believed that in biological science (which he calls "physical science"), natural selection applies to a contest among races or species rather than among individuals.

As has been indicated above, Wilson was able to distance himself somewhat from his greatly admired Bagehot, but his own attitude to evolution is obviously the Spencerian view with which Bagehot's influence had imbued him. Despite Wilson's invocation of Darwin, a close examination of his book on the Constitution and of his other writings shows that his thought does not display any of the specific features of Darwinian evolution. On the basis of his references to evolution in relation to the Constitution, we would have to conclude that he had no more understanding of Darwin's science than he had of Newton's. He showed no understanding of any of the fundamental principles of Darwinian evolution: random variation, common descent, struggle for survival among individuals, heritability.

The kernel of Wilson's analysis is that the framers of the Constitution were influenced by the belief that government is a kind of machine, whereas he believed that government should be viewed as an organism. He used the names of Newton and Darwin primarily as symbols for the scientific subjects with which each was concerned. Wilson, however, also allowed his rhetoric to carry him beyond the normal limits of analogy or metaphor or values into the dangerous realm of identification, arguing—as he put it—that "government is . . . a living thing."[86]

Conclusion: The Lessons of Metaphor

Wilson's position is extreme, both in his conception of the thinking of the framers of the Constitution and in the expression of his own views. That is, unlike the delegates to the Constitutional Convention, unlike Madison and Hamilton in their defense of the Constitution in the *Federalist,* Wilson wanted to make an essayistic, historical, almost philosophical generalization about the nature of the Constitution. He was like the Founding Fathers, however, in his mastery of rhetoric and in his concomitant use of metaphor in the service of his ideas about practical politics. The delegates to the Convention, as also those who commented on or discussed the Constitution, were primarily concerned with the powers being assigned to the federal government, not with the more philosophical question about whether the government or the Constitution is an animate being or a machine. Nor were they interested in exploring whether the features of the Constitution and of the system of government outlined in that document were like those of a machine or those of an organism. The discussions at the Constitutional Convention and in the *Federalist* show that for these politicians and thinkers the use of science was quite different, invoking metaphor and analogy in the service of rhetoric.

The Founding Fathers used science as a source of metaphors because they believed science to be a supreme expression of human reason. Science, furthermore, represented knowledge that was certain, removed from metaphysical suppositions and from ungrounded hypotheses and speculations. Scientific knowledge was based on sound method, the use of induction and of mathematical derivations. The certainty of science depended on its ultimate empirical foundation, on its firm base in experience, on the evidence of experiment and of controlled observation. Science not only could explain the phenomena of the present time but also could retrodict past events and accurately predict events to come. One of the most powerful arguments in favor of the Newtonian system was that Newton and Halley accurately predicted an event decades ahead, the return of a comet in 1759, the comet we now know as Halley's. Both the physics of Newton and the system of nature of Linnaeus embodied principles of tested truths, conveying values that every builder of a political system wanted to emulate.

Many thinkers of the Age of Reason, including some framers of the Constitution and other American political leaders or theorists, had a deep conviction that there is a parallelism between the world of nature and the world of human existence. They believed, accordingly, that sound systems of government and of the organization of society should display some analogy, some set of similarities in both values and actual forms, with the systems of nature. All of the framers of the Constitution would have agreed that no system of government or of society could be sound and stable if it contravened any of the fundamental principles of nature revealed by science.

The Founding Fathers were engaged in practical tasks, setting forth specific grievances against the king or working out the ways in which a central government would be powerful enough to assure the conduct of national affairs and yet not so strong that its powers would become tyrannical or impose on the rights of the individual states. Their insistence on checks and balances was not so much an expression of some abstract philosophical position about government being a machine rather than an organism as it was an expression of concern about limiting the power of the central government by a counterpoise of the power of the states and equally a balancing of the influence of the larger and more populous states at the expense of the smaller ones. The discussions at the Convention are remarkable in the degree to which practical problems of the day-to-day operations of government predominate over expressions of metaphysical principles and philosophical convictions about humankind and nature.

It is for these reasons that the use of metaphors, and notably the choice of metaphors, provides an important key to value systems, to inner convictions, and to most deeply held beliefs. The study of science in relation to political thought, especially in relation to the delegates who produced the Constitution and to their contemporaries, is thus a useful guide to supplement the declared principles that usually occupy the researches of political scientists and historians. The metaphors drawn from the sciences give us an index of the high esteem then accorded to the study and understanding of nature. Even more important, the analysis of these metaphors indicates an implied set of values and cherished beliefs that enlarges both our view of history and our understanding of the inner forces that motivated some of the primary founders of our country.

Supplements

· · · · ·

1. The Meaning of "Science" and "Experiment"

In interpreting documents of the age of the Founding Fathers, and in assigning meaning to terms or expressions used in those days, great care is needed to avoid reading those texts anachronistically, as if key words had then the same denotation that they have today. Two words in particular are traps for the unwary reader: "science" and "experiment." In the eighteenth century, the word "science" was used to denote any branch of organized or demonstrated knowledge and had not acquired the present more restricted sense of the physical or biological or earth sciences. Even today the word "science" retains its older sense in the name "political science." (See Supplement 11.)

The word "experiment" comes from the Latin "experimentum," which (like the Latin "experientia") is a noun formed from the verb "experior," to make trial of, to have experience of, to learn or find by experience. In classical Latin, "experientia" had an abstract signification, including the testing of possibilities, the experiencing of events, and even skills gained by practice (by experience), while "experimentum" had a more non-abstract sense, signifying method or means of testing, trial, demonstration, proof, or example. In Spanish, in Italian, and especially in French, the word derived from "experientia" ("experiencia," "esperienza" or "sperienza," and "expérience") is used for experiment and for experience. In the restricted context of science, however, from the seventeenth century onward, the word "experiment" has tended to signify what we would mean today by a scientific experiment in science. Thus, when Isaac Newton introduced a "Proof by Experiments" following each proposition in his *Opticks,* he had in mind the same kind of experimental evidence that we would understand by those words. He even made use of an expression, which he learned from Robert Hooke, of an "experimentum crucis" or "crucial experiment." Joseph Priestley, writing in the late eighteenth century, made a distinction which many—but by no means all—scientists have used. He said that an experiment was a means of testing a hypothesis or theory; he did not approve of using the word for empirical explorations or for trial and error. Franklin knew what an experiment was. His own theory of electrical action was based on extensive experimentation and, accordingly, his book on the subject bore the title *Experiments and Observations on Electricity.* Thus, when Franklin wrote that the government of the new nation was an experiment, he had in mind something more than mere trial and error. In this aspect he may have differed from those nonscientific contemporaries who wrote of the "experiment" they were making merely in the sense of trial. Today we often use the word in this latter sense when talking about an experimental course or even an experimental program, not meaning a test of a specific hypothesis or theory, but rather a trial.

This old notion of an experiment as a trial is still in common use. Thus we talk of an experiment as successful if the result is positive or favorable. But if the result is negative or unfavorable, we conclude that the experiment (as a test) was a failure. In a scientific experiment, however, the result is positive in both cases. An unsuccessful experiment is one that is inconclusive, that proved nothing.

Thomas Jefferson used the word "experiment" in this loose sense of trial in his opinion on the bill apportioning representatives (see Chapter 2). He explained that the Congress had not set forth the method adopted. Accordingly, it had been necessary to work backward from the results in order to determine how the numbers had been assigned. As he explained, he had "to find by experiment what has been the principle of the bill."[1] A somewhat similar use of "experiment" appears in Boswell's journal under the date 13 July 1788. Boswell wrote that "he went to make an experiment whether it was possible to get the least hospitality from Burgh." Here the word "experiment" was being used to mean making a trial of something.[2] This same sense of testing or seeing if something works appears in Washington's use of "experiment" in his Farewell Address. In discussing whether the Union can function, Washington referred to "experience" as the test, but almost immediately after spoke of a hoped-for "happy issue to the experiment," repeating that " 'Tis well worth a fair and full experiment."[3]

We may see the combination of the older sense of "experiment" and "science" in David Hume's *An Enquiry Concerning the Human Understanding,* Section VIII, "Of Liberty and Necessity." Here Hume discusses historical "records of wars" as "so many collections of experiments, by which the politician. . . . fixes the principles of his science." Hume's essay "That Politics may be Reduced to a Science" (as we have seen in Chapter 1) does not use the word "science" in the sense of the physical and biological sciences, but rather for organized knowledge. Even earlier, in Jonathan Swift's *Gulliver's Travels* (1726), there is a reference to the ignorant Brobdingnagians who had "not hitherto reduced Politicks into a *Science*." Swift meant only that this subject was not treated in a systematic or ordered manner.

The Founding Fathers, as we have had occasion to note elsewhere in this volume, spoke and wrote of "political science," a "science of government," and even a "science of federal government." In June 1782, Adams had declared that "politics are the divine science." In 1784, in a letter to A. M. Cérisier, Adams applauded the way in which French savants (Cérisier among them) had turned their attention to the subject of government; he voiced the judgment that "the science of society is much behind other arts and sciences, trades and manufactures." A year later, on 10 September 1785, in a letter to John Jebb written from London, Adams referred specifically to "the social science."[4] There is no reason to suppose that Adams had invented this name.

2. *No Balance of Forces in Newtonian Science*

Many writers about American government have supposed that the Newtonian natural philosophy and the system of the world developed in the *Principia* exhibit a balance of forces, a condition of equilibrium, what Woodrow Wilson (see Supplement 12) called an equipoise. Sometimes this alleged balance is associated with the belief that the Newtonian world system is a kind of machine, a world-machine, a gigantic system of clockwork, one that is so perfect that it can run perpetually in a self-regulating manner. From some or all of these assumptions, the analogy is made with a system of government, frequently the Whig theory of government, which is even held to have been an "unconscious" copy of Newtonian physics. Primarily, the conclusion is drawn that the equilibrium or equipoise of the Newtonian system of the world provides a model for a governmental system of checks and balances. As we have seen in Chapter 5, this line of argument has led to a Newtonian interpretation of the Constitution. A critical examination reveals, however, that the Newtonian natural philosophy of the *Principia* and the Newtonian system of the world do not exhibit either a counterpoise of forces or a perfect self-regulated world-machine that can "run of itself." Newton wrote clearly and unambiguously that the world was not a perfect machine that can run perpetually without some tinkering or adjustment by the divine artificer.[5]

The fundamental working principle in Newtonian rational mechanics, as developed in the *Principia,* is that orbital motion is the result of an *unbalanced* force and is in no sense a case of equilibrium or balanced forces. The Newtonian natural philosophy is concerned almost entirely with problems of dynamics, with the science of unbalanced forces and motions or—more exactly—forces (that is, unbalanced forces) and the motions they generate, the changes in motions they produce. Only in one very small section, a sort of appendix to the "Laws of Motion," does Newton even introduce the subject of statics or balanced forces, that is, forces acting on bodies that are and remain at rest. It is thus simply wrong to claim, as Woodrow Wilson and other scholars have done, that the science of the *Principia* is centered on the notion of an equipoise or balance of forces.

Much of the *Principia* is concerned with problems of orbital motions and the unbalanced force or forces that produce such orbital motion in a planet, a planetary satellite, or a comet, or indeed any body that follows a curved path. In the case of planetary orbits, or the orbits of planetary satellites or of comets, this unbalanced force is directed toward the central body, that is, toward the Sun for the planets and comets, toward the central planet for satellites. Newton invented the name "vis centripeta" for this kind of force, because it is centrally directed, a "centripetal force" pulling the planet or comet in the direction of the Sun and a satellite in the direction of the central planet. In the physics of the *Principia,* the necessity for having an unbalanced force in orbital motion (or in any curved motion) is a consequence of the first law of motion. That is, unless there is an unbalanced force, a body will either continue in its state of rest or continue in its state of moving uniformly straight forward, that is, along a straight line at constant speed. Accordingly, a condition of balanced forces or equilibrium can never produce a curved motion.

The reason why the planets do not actually spiral inward toward the center is that, although they are constantly falling toward the center, they each have also a forward motion. Thus, as the planet falls inward, it also moves forward and it does so—according to Newton's first law—in the direction of the straight line tangent to the orbit. And it is the same for the planetary satellites. The combination of the forward motion and the falling motion keeps the planet or satellite moving along its curved path or orbit.

At every instant, then, according to the Newtonian analysis, an orbiting body has two independent components of motion. One is an inertial tangential component of uniform motion, the other a centrally directed accelerated motion of falling—inward toward the center. The orbiting body seems to experience a tendency to move straight outward, to get farther and farther from the center, but this is the result of the tendency to move out along the tangent, a motion which would in fact take the body to a more outward orbit.

In the eighteenth century, some writers on orbital dynamics used the term "centrifugal force" in a way that can easily cause confusion to a modern reader unfamiliar with the older language. They referred to the inertial component of orbital motion in terms of what they called a "projectile force" and sometimes a "centrifugal force." A careful reading of the texts by good Newtonians who used these terms shows that they tended to be fully aware that the "force" in question is of a very different kind from the centripetal force. They knew that this "projectile force" can not produce an acceleration, that it is never exerted outward along a line of direction through the center of motion. Accordingly, they understood that there is no possible way in which a "projectile force" can produce a balance or an equilibrium with a centripetal force such as gravity. Not only are the projectile force and the centripetal force of wholly different kinds, but they act in different directions. One is directed inward, toward the center of orbital motion, the other along the tangent. These two directions are at right angles to each other. Under this condition of orientation, there is no way that two forces can ever produce an equilibrium or be in equipoise.

Of course, even though there is no "balance of forces," no equipoise, no equilibrium, there is in Newtonian science a kind of restraint. Without the centripetal force of gravity of the Sun, planets and comets would fly off along tangents and move out of the solar system. This restraining force, as Newton's disciple J. T. Desaguliers explained (see Supplement 3), "always checks the Projectile Force." That is, the Sun's gravitational force restrains the planets from their inertial tendency "to fly away from the Sun" along a tangent.

3. *Desaguliers's Poem on Newtonian Science and Government*

John Theophilus Desaguliers was a member of Newton's London circle, an able physicist of Huguenot descent. A gifted experimenter, he was the author of a popular two-volume text of Newtonian physics that demonstrated the principles by experiments rather than developing them mathematically. His original experiments in various areas of physics, among them optics and electricity, were published in the *Philosophical Transactions* of the Royal Society, of which he was a Fellow. He was the translator of W. J. 'sGravesande's two-volume textbook of Newtonian science. Desaguliers was also known for his active role in Freemasonry. He wrote a manual for Freemasons that was published in America by Benjamin Franklin, who had studied Desaguliers's own two-volume textbook of Newtonian science.

Many scholars, including Carl Becker, have cited the impressive title of Desaguliers's poem without giving any sign that they had ever read a line of it. One reason may be that the poem was published in a very limited edition and is a bibliographical rarity. Speaking with the real authority of a master of Newtonian physics, Desaguliers gave his poem a title which may easily arrest our attention: *The Newtonian System of the World, the Best Model of Government, an Allegorical Poem* (Westminster: printed by A. Campbell for J. Roberts in Warwick Lane, 1728). The subtitle goes on to declare that there is also "a plain and intelligible Account of the System of the World, by Way of *Annotations*." These "annotations" are a series of extensive notes which, with a set of "copper plates," provide an admirable and easily understandable presentation of the major features of the different systems of the world: Ptolemaic, Tychonic, and Copernican. A second poem, published as a kind of supplement to the primary offering, is entitled "Cambria's Complaint." This set of verses displays a rather silly argument against inserting "the *Intercalary* Day in the Leap-Year." The quality of the poetry in both offerings is low. One may guess that the only reason why Desaguliers's doggerel did not achieve a primary place in Alexander Pope's *Dunciad* must be that the edition was so small and the work so rare that Pope never read it.

The occasion for the poem on the calendar was "His present MAJESTY's Accession to the Throne." Desaguliers's own birthday was the first of March and he was annoyed that every four years the calendar required the insertion of an "intercalary day" (29 February) which advanced his birthday by a day. When he discovered that Queen Caroline (the wife of George I) was also born on the first of March, he was goaded into expressing his complaint about the "intruding Day" in verse.

The little booklet containing the two poems contains a dedication that begins with Desaguliers's assertion: "I never made *Politicks* my Study." He always had thought it his "Duty" to "take Care to be obedient," rather than "to look into the Management" of affairs of state by his "Superiors." He had, however, long regarded "Government as a Phaenomenon" and had deemed "that Form of it most perfect, which did most nearly resemble the Natural Government of our *System*." He found the essence of the Newtonian system to be "the most regular Attraction of universal *Gravity*, (or Attraction) whose Power is diffus'd from the Sun to the very Centers of all the Planets and Comets." This force "wonderfully" brings "back the Comets from their immensely distant *Aphelion*, in their very long Ellipses, by the same Laws that it keeps the nearest Planet *Mercury* in its Orbit." The force of gravity, he notes, "always checks the Projectile Force, (whereby the Bodies tend to fly from the Sun) in Proportion to the Quantity of that Force." As has been mentioned (in Supplement 2), this force of restraint does not produce a balance of forces, an equipoise, or a condition of equilibrium.

Desaguliers found a model of happy relationship in the way in which the falling inward toward the Sun, the effect of the Sun's gravity, draws the planets and comets away from their tangential motion. He thus asserted that "The *limited Monarchy*, whereby our Liberties, Rights and Privileges are so well secured," is "a lively Image of our System" of the world. He concluded that "the Happiness that we enjoy under *His* present MAJESTY's Government" makes us aware "that A T T R A C T I O N is now as universal in the Political, as the Philosophical World."

A few verses will suffice to indicate the level of poetry in Desaguliers's allegory:

> When Kings were not ambitious yet to gain
> Other's Dominions, but their own maintain.

Here is another:

> This *Ptolemaick Scheme,* his Scholars saw,
> No way agreed with the *Phaenomena.*

Finally, an example of schoolboy doggerel:

> In *Plato*'s School none cou'd admitted be,
> Unless instructed in *Geometry.*

The general message seems to be that harmony based on universal love, not fear, is the best source of governing power, that a limited monarchy, such as existed in Britain, provides the best of all possible worlds and the closest political counterpart of the Newtonian system of the world. The poem ends with a bathetic couplet:

> A T T R A C T I O N now in all the Realm is seen,
> To bless the Reign of G E O R G E and C A R O L I N E.

It is obvious why no political thinker seems ever to have cited this poem.

4. How Practical Was Jefferson's Science?

One aspect of Jefferson's science is often misunderstood, his concern for practicality and his often expressed belief that the only science that is valuable is science which is useful. This stress on practicality has led many historians to link together his activities in relation to science and his role as inventor and even to stress his activities in agriculture at the expense of his interest in physics. In interpreting Jefferson's statements, however, we must keep in mind that they were expressions of a philosophy then widely held and usually associated with Francis Bacon, one of Jefferson's "trinity" of great men whose portraits he had hung in his gallery of notables. Bacon, however, would never have limited science to what would seem immediately practical.

Rather, he taught that true science would always end up by producing practical applications of use to humankind in controlling the environment, protecting health, and easing the burdens of life.

In our efforts to understand Jefferson's expressed views concerning practical or useful science, we must go beyond his words and examine his deeds. We have seen that two of his favorite authors, among the group of four whose works he planned to devote himself to in his leisure, were Euclid and Newton. While, to be sure, a very minuscule part of Euclid's geometry may be useful in surveying and carpentry, the greater part by far is of purely intellectual interest and bears no relation whatever to any considerations of practicality. When John Adams, late in life, recalled the many hours he had spent in the study of mathematics, he found in retrospect that the only practical benefit had been to teach him patience and perseverance.

As for Newton's *Principia,* that work did yield Jefferson a few results which he used for practical purposes. We have seen that these included Newton's rule for the variation of the length of a seconds pendulum with latitude. As in the case of Euclid's geometry, the useful part consists of a minuscule portion of the whole, a single proposition out of the 150 or more that comprise the *Principia.* And it is much the same for Newton's *Opticks,* most of which is concerned with theoretical considerations about color production and the phenomena of diffraction.

Invoking the examples of Euclid and Newton brings to mind an anecdote about Newton and Euclid. According to the recollections of one of Newton's acquaintances, Newton had loaned someone a copy of Euclid. When the book was returned, the borrower asked Newton what use geometry might have, what might be the "good of it." According to the report, this was the only occasion on which Newton was known to have laughed.

Anyone who has studied the life of Jefferson is aware that a subject of major scientific interest and concern for him was paleontology, a science which had no practical application whatsoever. This fact suggests that Jefferson's expressions about useful science must be tempered by taking account of his actual scientific activity. He wanted to see science put to use, but he himself was interested in all kinds of scientific studies simply because he was curious about the operations of nature. Jefferson believed in fostering science on all levels, from the abstract or pure to the concrete or applied. As he wrote in 1799, he was "for the encouraging the progress of science in all its branches."[6] At the same time, however, he believed that we should be constantly on the alert for any useful or practical applications of scientific knowledge.

An enlightening approach to the problem of Jefferson's science and practicality is given by Gerald Holton, to whom reference has been made in Chapter 2. Holton finds that there are three major ways of doing science. He calls one the "Newtonian research program," an attempt to bring all of science into a single unified theoretical structure. A second is the "Baconian research program," which aims to enlarge human control of the environment and the many forms of the conduct of life. The first may exemplify "omniscience," the second "omnipotence." According to Holton the third way of doing science is to plan, envision, sponsor, or conduct research in basic science with the goal of solving the problems of society. A basic premise of this new "practicality" is the recognition that many of the "large-scale functional difficulties" faced today are not caused so much by the "progress of basic science" as by "the *absence* of some specific fundamental scientific knowledge." It is this "style of research . . . still struggling to come to prominence" which Holton sees as typified to a high degree by Jefferson and exemplified by the scientific aspects of his plans for the Lewis and Clark expedition.

Jefferson's philosophy of use may be illuminated by a comparison with the career of Benjamin Franklin. Like Jefferson, Franklin often expressed a Baconian point of view toward useful science. One of his experiments demonstrated that there is a different degree of absorption of solar heat in cloths of different colors. In describing these experiments, he wrote: "What signifies Philosophy [i.e., natural philosophy or science] that does not apply to some Use?" The use he foresaw for his discovery was that it could teach us that white or light-colored cloth should be used for garments in tropical regions, since they tend to reflect the solar heat and absorb less of it than materials of a darker hue.[7] On another occasion, Franklin wrote that it is important to know Newton's laws of motion, but we can explain why our china breaks when it falls without them. Such sentiments are much like those expressed by Thomas Jefferson, but they must be read in the light of Franklin's actual scientific activity.

The student of Franklin's scientific career will be aware that his comment about useful science in relation to the experiment with colored cloths was not made at the time of the discovery. Rather, it was written by Franklin some twenty years later. The actual experiments were originally made as part of the exploration of the phenomena of light and heat, with no practical end in view.[8] Even more significant is the fact that the scientific area in which Franklin achieved his worldwide reputation was electricity. Today we think of electricity as a major source for powering our machines, lighting our homes, and providing communication, a practical subject par excellence. But when Franklin began his experiments in this area, this was in no way a practical subject. The only use that anyone could then imagine was the therapeutic application of electric shocks to help sufferers from various kinds of paralyses. Franklin, however, doubted that the electric shock itself had any therapeutic value, attributing the beneficial effects to the desire of the patient to get well and to the actual exercise in going to Franklin's house for treatment.

In those days, electricity was primarily a subject for the entertainment of the curious, a kind of "toy physics," which did not even seem to be a significant part of science. The only reason for studying electrical phenomena was curiosity about the operations of nature. In 1747, as we have seen in Chapter 3, when Franklin discovered that a first version of his theory did not fit all the phenomena, he wrote that if "there is no other Use discover'd of Electricity, this, however, is something considerable, that it may *help to make a vain Man humble.*"[9] Hardly the words of someone working in a "practical" area of science!

Later on in the course of his research, Franklin was able put his discoveries to use in wholly unexpected and unanticipated ways: first, in proving that the lightning discharge is an electrical phenomenon and, then, inventing the lightning rod to prevent the destructive effects of this awful force of nature. Franklin's invention of the lightning rod, in fact, seemed all the more remarkable to his contemporaries because it was the first major and large-scale vindication of the Baconian thesis that advances in true knowledge would lead to practical innovations, to humankind's control of the force of his environment. Throughout the eighteenth century, scientists kept insisting that the fruits of their research would be new truths that not only would increase knowledge, but would also be practical applications based on those truths. But until well into the nineteenth century, there was only one major or large-scale example of such an application of fundamental scientific research, Franklin's lightning rod. As late as the 1830s, Franklin's invention of the lightning rod was still being cited in France as the major example that disinterested or pure or fundamental scientific research was to be valued because true scientific knowledge would eventually produce practical innovations.

5. *Jefferson and the Megalonyx or Megatherium*

Throughout his mature life, paleontology always remained for Jefferson a principal scientific interest. He was an avid collector of fossil bones and he even believed that some of the giant mammoths were still in existence somewhere in the wilds of America.[10] He not only was an active collector and student of fossils; he also encouraged others in the pursuit of paleontology and he sent many specimens to Paris where they are still preserved as an important collection.[11] One incident in his activity as paleontologist is worth noting. In 1796, Jefferson received some giant bones dug up in limestone caves in western Virginia. He was thrilled to discover that these belonged to an animal that had not as yet been described in the annals of science. Deciding that the bones came from a giant lion-like creature with a huge claw, he assigned this animal to the "family of the lion, tiger, panther, etc." and gave

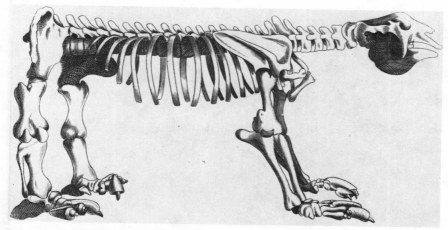

Fig. 41. Skeleton of a Megatherium. A print made from a drawing of a specimen from Paraguay, published in the Monthly Magazine and British Register *for September 1796. It was this print which drew Jefferson's attention to the fact that a fossil much like the one he had been studying had already become known in Europe.*

it the name of Megalonyx or giant claw. Since the skeleton was incomplete, with the thigh bones missing, it was not possible to estimate the exact size of the original animal.

It was the fossil bones of the Megalonyx that Jefferson brought with him to Philadelphia in 1797, when he was to be inaugurated as vice-president of the United States and president of the American Philosophical Society. Before he could deliver his paper on the Megalonyx to the American Philosophical Society, however, he encountered a copy of the London *Monthly Magazine,* containing an engraving of the skeleton of a gigantic animal that greatly resembled his own Megalonyx (see Fig. 41). This fossil skeleton had been discovered in Paraguay and sent on to Madrid to the Royal Cabinet of Natural History, where it arrived in 1788; the skeleton was assembled and mounted by the Spanish naturalist Juan Bautista Bru y Ramón.[12] The article in English describing this skeleton was an abbreviated translation of a longer publication by the young Georges Cuvier, destined to become the world's leading paleontologist. Cuvier gave this creature the name Megatherium and classified it among the "edentates" (that is, animals having few teeth or no teeth at all) and deemed it a distant relative of the sloth.[13]

Jefferson must have been bitterly disappointed on learning that Cuvier had already published an account of this animal. He became aware that his own classification as a member of the cat family, having great claws, might be wrong. Accordingly, he revised his own paper for the American Philosophical Society, deleting some parts and adding others, and noting the

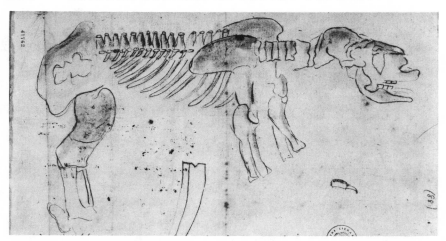

Fig. 42. Drawing of the Megatherium. A drawing of the same fossil shown in the Monthly Magazine and British Register *was sent to Jefferson by William Carmichael in 1789, just as Jefferson was leaving Paris for America.*

prior publications by Cuvier, which—he said—had not come to his attention until after he had completed his own report. He could not doubt the similarity between the two skeletons, but left open the final question of whether both actually represented the very same animal.[14]

It is one of the curiosities of history that almost a decade earlier, in 1789, the American chargé d'affaires in Madrid, William Carmichael, no doubt aware of Jefferson's interest in natural curiosities and in paleontology, had sent Jefferson (who was in Paris) a drawing and description of the skeleton made by Bru, a drawing still preserved among Jefferson's papers in the Library of Congress (see Fig. 42). Carmichael told Jefferson that the Spanish Academy of Natural History "will soon publish an account of this Animal" and that until then the description "should not be made public." Jefferson sent Carmichael an acknowledgment of the drawing and notes.[15] Jefferson was just then packing up his belongings for his return to America and he obviously forgot all about the drawing when he came home. Had this not been so, it has been remarked, he "might have had the honor of preceding the Spanish naturalists and Cuvier in identifying and naming the animal."

Some time later, Jefferson gained access to a copy of a publication by another Spanish naturalist, José Garriga, which included an engraving of the skeleton and antedated Cuvier's report. In 1804, Cuvier published a more complete account of the Megalonyx and the Megatherium, giving Jefferson full credit for the Megalonyx.[16] Eighteen years later, the French naturalist Anselm Desmarest named the Virginia specimen *Megalonyx jeffersonii* in recognition of Jefferson's pioneering efforts and this is the name by which it is still known today.

When Jefferson died, Cuvier made a formal recognition of the contributions of Jefferson, whom he described as a person "with an enlightened love for the sciences and a broad knowledge of scientific subjects to which he has made notable contributions."[17] In an official *éloge* for the Société Linnéanne de Paris, of which Jefferson had been an honorary member, Jefferson was particularly praised for the important collection of fossils which he had presented to France.

6. *The Mathematics of Plow Design: Newtonian Fluxions and the Shape of a Solid of Least Resistance*

A plow has several parts: a coulter which cuts the soil penetrated by the plowshare and a moldboard which turns the soil over. The function of the moldboard was described succinctly by Jefferson as "to receive the sod after the share has cut under it, to raise it gradually and reverse it."[18] Recognizing that this function should determine the form of the moldboard, he wrote that the "fore end of it should be horizontal to enter under the sod, and the hind end perpendicular to throw it over, the intermediate surface changing gradually from the horizontal to the perpendicular."[19]

One of the principal features of Jefferson's improvement of the plow was the design of a better moldboard, a problem which quickly took him from the level of simple trial and error into the higher reaches of Newtonian mathematics. Jefferson recognized early on that the most efficient kind of moldboard was one that would be drawn through the soil with the least possible resistance; obviously, some kind of wedge shape would be required. Since the soil would also have to be lifted and turned over, he wrote, the moldboard also "operates as a transverse or rising wedge." Hence the moldboard would essentially consist of a combination of two wedges, placed at right angles to each other. The first of these was set "in the direct line of the furrow" and its function was "to raise the turf gradually"; the second one was set "across the furrow," so as "to turn it over gradually." For both of these purposes, Jefferson wrote, the "wedge" must be "the instrument of the least resistance."

During a period of active work on the design of the moldboard, Jefferson sought technical advice from the astronomer Benjamin Rittenhouse, a skilled amateur mathematician, and also from Robert Patterson, a professor of mathematics at the University of Pennsylvania. Patterson corrected Jefferson's error in having assumed that "a *plane* sided wedge" was the "best form" to be adopted. He reminded Jefferson that finding the shape of a wedge of least resistance was a standard problem in the calculus and he referred Jefferson to the textbook of fluxions written by William Emerson. "Emerson," he observed, "in his doctrine of fluxions, makes it a solid of a *curvilineal surface.*"[20] Jefferson replied that "it had escaped me" that Emerson "had treated the question of the best form of a body for removing an obstacle in a single direction."

The book in question was William Emerson's *Doctrine of Fluxions* (London, 1747, and later editions). Emerson was an important Newtonian who wrote "A Short Comment on, and Defence of, the *Principia*" which was published in London in 1770 and later reprinted as the third volume of the 1819 edition of the English translation of the *Principia*. Jefferson owned a copy of Emerson's *Fluxions,* which was, as he wrote to Patterson, "the book I used at college."[21] It had not occurred to Jefferson, while designing his plow, to consult a mathematics textbook. The shape he had chosen seemed so obviously the best that he "did not think of questioning it." Jefferson was writing from Philadelphia and did not have his library available. Accordingly, he asked Patterson whether he could borrow the book for "a day or two." Patterson obliged and so Jefferson was able to study the problem from a new point of view, as an exercise in applied Newtonian mathematics.

When returning the book, he told Patterson that Emerson's treatment of the problem had "suggested a qualification" of his original supposition concerning "the form offering least resistance to the rising sod."[22] What Jefferson learned from Emerson was not only that the surface of the wedge should be curved and not merely have the shape of a twisted plane, as he had at first supposed, but that the curve should be of a special sort.[23] It was a rare event for Jefferson to find such a practical use for his study of Newtonian mathematics.

This episode is more significant than may at first appear. One of the most important applications of the Newtonian calculus has always been in problems of a "minimax" kind, that is, finding curves or solids that fulfill some kind of maximum or minimum condition. One of the public challenges made to Isaac Newton during the controversy over the invention of the calculus was whether he could use his method of fluxions to find the curve of least descent,[24] that is, the curve along which a frictionless bead will slide from a given starting point to a given terminus in a minimum time. Even today, students learning the calculus are given minimax problems to demonstrate the extraordinary power of the calculus.

Furthermore, the problem of finding the shape of the "solid of least resistance" had a position of great prominence in Newton's *Principia* (Book Two, prop. 34, scholium). It was this problem that gave rise to a comment by Newton, one of the very few of this kind in the *Principia,* that there might possibly be a practical application of the mathematical principles in fields other than astronomy. In this case, Newton suggested that the determination of the solid of least resistance might be of use in the design of the hulls of ships.[25] It was in the context of the shape of such a solid that Newtonians referred to a "solid of least resistance," an expression used by Emerson and by others who introduced this problem into their treatises on the Newtonian calculus. Furthermore, in the English translation of the *Principia* made by Andrew Motte, of which a copy was in Jefferson's library, there was a special editorial supplement dealing with this very problem, based on Newton's manuscripts. Jefferson's use of the Newtonian phrase "mould-board of least resistance" indicates a direct link between his design of an efficient mold-board and the science of Newton's *Principia*.

Wholly apart from the question of whether Jefferson's design of an improved plow was or was not a useful invention, this episode is important in the present context for a number of different reasons. It establishes as fact that Jefferson had studied the Newtonian calculus while an undergradu-ate at William and Mary, learning the calculus from William Emerson's text-book. It also shows that Jefferson was a master of this branch of Newtonian mathematics. So far I know, this aspect of Jefferson's ability to use Newton-ian mathematics has not heretofore been noted. Emerson is apt to be referred to, if at all, as an "English mathematician," who provided Jefferson with "a formula that would be useful to him," without even mentioning the calculus.

Even more important, perhaps, is the fact that the problem of the "solid of least resistance" is a Newtonian problem, one that goes back to the *Prin-cipia*. Thus the actual history of Jefferson's design of the plow provides firm evidence of Jefferson's knowledge of Newtonian mathematics and the sci-ence of the *Principia*.

7. *Jefferson Corrects Rittenhouse's Gloss on the* Principia

Jefferson's mastery of technical Newtonian science is exhibited in an episode that occurred during the preparation of his report on weights and measures (discussed in Chapter 2). In a letter of 25 June 1790, Jefferson's friend, the Philadelphia astronomer David Rittenhouse, gave him a mathematical dis-cussion of some aspects of the physics of pendulums. Essentially, Ritten-house's letter is a gloss on Newton's treatment of this subject in the *Principia*, Book Three, prop. 20. It was in relation to this letter that Jefferson is said (his *Papers*, vol. 16, p. 570, n. 2) to have written out the "quotation from Newton" that turns out to be inaccurate and that contains an error in Latin. As we have seen in Chapter 2, Jefferson neither misquoted Newton nor did he make the error in Latin attributed to him by his editors.

Rittenhouse's discussion of prop. 20 begins with the statement that "Sir Isaac Newton determines the Earth to be an Oblate Spheroid," that is, a sphere squashed down at the poles and bulging at the equator. The polar axis is to an equatorial diameter as 229 to 230. If we turn to the *Principia*, we find that Newton proves that for such a spheroid, the weights of bodies with equal masses at different places on the surface of the Earth will be "inversely as the distances of those places from the center." Therefore, "the lengths of pendulums oscillating [or vibrating] with equal periods are as the gravities."

These results are paraphrased by Rittenhouse. He writes, first, quite correctly that "the weights of equal Bodies [that is, bodies with equal masses] on the Earths surface will be inversely as their distances from the Center." But he then wrongly concludes that "consequently . . . the lengths of pendulums vibrating in equal times will be directly as the distance from the Center."

Jefferson was quite properly puzzled by the second of Rittenhouse's statements. He was aware that in the *Principia* Newton says that the weight is inversely proportional to the distance from the center and that the length (for pendulums with equal periods) is proportional to the weight. Together, these two statements imply that the lengths of pendulums with equal periods must be inversely proportional to the distance from the center. Accordingly Jefferson recognized that Rittenhouse had made an error. He had begun with a correct statement of Newton's result, that the weights are inversely proportional to the distances from the center, but he had then incorrectly concluded that the lengths of pendulums with equal periods are directly (rather than inversely) proportional to those distances.

As the manuscript shows clearly, Jefferson bracketed Rittenhouse's phrase "the distance from the Center," indicating that it was to be deleted, since it is wrong. He would substitute the phrase that he interlined, reading "these weights or the gravity of the place." In this way Jefferson converted Rittenhouse's incorrect statement into a correct one, declaring that the lengths of pendulums with equal periods are as the weights or the gravity of the place. Then, in order to provide the authority for his correction, Jefferson quoted an extract from Newton's *Principia*, "longitudines pendularum aequalibus temporibus oscillantium, sunt ut gravitates," which we have quoted in translation as "the lengths of pendulums oscillating [or vibrating] with equal periods are as the gravities." Jefferson's extract from the *Principia* differs from Newton's text only in having the indicative "sunt" for Newton's subjunctive "sint." As has been explained (in Chapter 2), Newton's statement occurs in a causal "cum" clause requiring the subjunctive, which Jefferson has transformed into a simple declarative sentence.

This episode shows that Jefferson knew enough of the physics of the *Principia* to spot Rittenhouse's error and to correct it. We may take note that Rittenhouse's letter runs to several pages of text and calculations which are not vitiated by the error we have been discussing. Rittenhouse wanted Jefferson to get the information he was sending as soon as possible and so sent his first draft, apologizing for the fact that he had "intended to have copied the above Scrawl." No doubt, he would have found his error if he had done so.[26]

8. *Jefferson's Changing Views concerning the Abilities of Black People*

One of the most remarkable features of the *Notes on the State of Virginia* is Jefferson's outspoken denunciation of slavery on almost every ground, from moral issues to practical economics. So outspoken was he on this issue that for a time he hesitated to put his book into general circulation. He feared, as he wrote to James Monroe, "that the terms in which I speak of slavery and of our constitution may produce an irritation which will revolt the minds of our countrymen against the reform of these two articles, and thus do more harm than good."[27] Charles Thomson, secretary of Congress, wrote Jefferson that he regretted "that there should be such grounds for your apprehension respecting the irritation that will be produced in the southern states by what you have said of slavery."[28] He told Jefferson not to be "discouraged," that slavery is "a cancer that we must get rid of."

Jefferson's message was unambiguous. He detested slavery and argued in strong language for emancipation. John Adams took special account of this feature of the *Notes* in a letter to Jefferson, acknowledging the receipt of a copy that Jefferson had sent him. While expressing his belief that the book "will do its Author and his Country great Honour," Adams especially praised Jefferson's "Passages upon Slavery," which he declared to be "worth Diamonds." He predicted that "they will have more effect than Volumes written by mere Philosophers."[29] As Jefferson's biographer Merrill Peterson has remarked, "No abolitionist of later time ever cried out more prophetically against slavery than Jefferson."[30] He believed that slavery was degrading to masters as well as to slaves, destroying the "morals" and "industry" of slave-owners.

In a letter to the Marquis de Chastellux, Jefferson explained that he hoped that what he had written would not "indispose the people toward the two great objects I have in view, that is, the emancipation of their [that is, his countrymen's] slaves and the settlement of their constitution on a firmer and more permanent basis."[31] In a letter to Bishop James Madison[32] about the idea of distributing the book to all the students at the College of William and Mary, he wrote that if the book should not produce his desired effect, he had "printed and reserved just copies enough to be able to give one to every young man at the College." Madison conferred with George Wythe, who suggested putting copies in the college library. He feared that a general distribution of the passages against slavery "might offend some narrow minded parents."[33] Jefferson was no doubt correct when he concluded that it would be "the rising generation," and not "the one now in power," who would achieve "these great reformations."[34]

In his autobiography, Jefferson wrote that the "public mind would not yet bear the proposition" of emancipating the slaves. He recalled how, in 1769, he had sought unsuccessfully, as a young member of the Virginia

House of Burgesses, "for the permission of the emancipation of slaves." In 1774 he accused George III of a "shameful abuse of power" in thwarting efforts to prohibit the importation of slaves. Jefferson repeated this charge in his draft of the Declaration of Independence, but the Congress struck it out in the final version. Again, in 1789, his "Plan of Government for the Western Territory" proposed that after 1800 there should be no slavery in those lands.[35]

Jefferson apparently never wavered in his strong belief in emancipation. In his *Notes on the State of Virginia*, he introduced this subject briefly in Chapter 8, "On Population," followed by a full discussion in Chapter 14, "Laws." He concluded that the liberated blacks should form a separate and independent country of their own to which "we should extend our alliance and protection." He gave a number of reasons why such a separation was preferable to a single nation of free whites and blacks. In explaining his position, he was very careful to explain how certain traits which he attributes to black slaves and former slaves are the results of their economic, social, and cultural conditions. At the same time, however, he writes of racial differences in a manner that is all the more astonishing in the context of an argument for emancipation. Jefferson's belief in racial differences in no way lessened his absolute belief in emancipation.

The anomaly is especially striking because in the same *Notes* Jefferson demolished Buffon's theory of degeneration by showing that the learned naturalist had relied on hearsay and travelers' stories rather than on the direct evidence required by science. In particular, as we have seen in Chapter 2, Jefferson showed that Buffon's statements concerning Native Americans were false. Many historians have been puzzled by the contrast in the *Notes* between Jefferson's praise of the talents of Native Americans and his disparagement of the abilities of black Americans. Jefferson's "vindication of the American Indian" has been described by Merrill Peterson as "a vindication of the American environment." Jefferson could not similarly envisage "a natural place in that environment" for African-Americans and so "his solution for them was not amalgamation but expulsion."[36]

In the *Notes on the State of Virginia*, Jefferson made reference to a proposed amendment to a bill that would "emancipate all slaves born after the passing of the act." This led him to a discussion of the differences between blacks and whites in the context of whether free blacks could live in harmony with their former masters. Such harmony would not be possible, he believed, because of "deep rooted prejudices entertained by the whites," the "ten thousand recollections by the blacks, of the injuries they have sustained," and "the real distinction which nature has made." Among the latter, Jefferson noted that black people "have less hair on the face and body." They "secrete less by the kidneys and more by the glands of the skin, which gives them a very strong and disagreeable odor." They "seem to require less sleep." While "they are more ardent after their female," their love seems "more an eager desire, than a tender delicate mixture of sentiment and sensation." They have a "disposition to sleep when abstracted from their diversions, and unemployed in labour." Finally, although "in memory they are equal to the whites," in "reason [they are] much inferior." Jefferson said, "I think one could scarcely be found capable of tracing and comprehending the investigations of Euclid." We may easily agree with Merrill Peterson that Jefferson's statement that he had never met or heard of a black person who had true eloquence, who could write a musical composition, paint a picture, or discover a truth, "came with ill grace from the man who demolished Raynal for employing the same line of argument against the American whites."[37]

Although Jefferson described what he sensed as differences between blacks and whites, he was also aware of the need "to make great allowances for the difference of condition, of education, of conversation." With regard to his catalog of differences, Jefferson the scientist could not help but appreciate that he was on unsure grounds. "The opinion that they are inferior to whites in the faculties of reason and imagination," he wrote, "must be hazarded with great diffidence." Fully aware that there was no basis in science for his statements of differences, he admitted that the opinion he was advancing was no more than "a suspicion," that black people "are inferior to the whites in the endowments both of body and mind." It must be repeated that his "suspicion" about the inferiority of black people did not lessen Jefferson's ardor for emancipation and that, in fact, his opinions about racial differences appear in the very book in which he makes an impassioned and logical argument for the abolition of slavery.

Jefferson's "suspicion" has often embarrassed his biographers and has made him the subject of direct attack.[38] Clearly, the very fact that he was capable of such racialism illustrates the force of prejudice. It is, therefore, all the more significant that he eventually changed his mind about the intellectual abilities of black people.

In 1791, only five years after the publication of the *Notes*, Jefferson received a manuscript ephemerides for the year 1792, computed by Benjamin Banneker, a free black planter living in Maryland.[39] This work was a table of the positions of the planets at various times during the year, together with information about coming eclipses and other celestial phenomena. Banneker was a self-taught mathematician and astronomer who was an "astronomical assistant" during the survey of the Federal Territory.

Banneker said, in a letter introducing himself, that he had sent this specimen of his work to Jefferson to show that black people are capable of doing mathematics and science. He was aware that Jefferson was "measurably friendly and well disposed" to black people. In acknowledging receipt of the ephemerides, Jefferson wrote: "No body wishes more than I do to see such proofs as you exhibit, that nature has given to our black brethren talents equal to those of the other colors of men" and that "the appearance of a want of . . . [those talents] is merely owing to the degraded condition of their existence in Africa & America." He added "with truth" that "no body wishes more ardently to see a good system commenced for raising the condition both of their body & mind to what it ought to be."[40]

Jefferson at once sent Banneker's calculations to the Marquis de Condorcet, the permanent secretary of the Paris Academy of Sciences, declaring that he was "happy to be able to inform you that we have now in the United States a negro . . . who is a very respectable mathematician." He mentioned to Condorcet that he had seen "very elegant solutions of Geometrical problems made by him." So much for the earlier expression of opinion that black people could not understand Euclid! Jefferson also told Condorcet that he would be "delighted" to see the example of Banneker "so multiplied as to prove that the want of talents observed" in black people "is merely the effect of their degraded condition" and does not proceed "from any difference in the structure of the parts on which the intellect depends."[41]

Banneker regularly published an almanac containing his calculated ephemerides. In the almanac for 1793, he printed his correspondence with Jefferson, which was also issued as a separate pamphlet. This publication gave fuel to Jefferson's political enemies, who drew attention to the great difference in sentiments between the letter to Banneker and the *Notes on the State of Virginia*. Among those who used these letters for political purposes was William Laughton Smith, a congressman from South Carolina. When Jefferson was a candidate for president, Smith asked: "What shall we think of a *secretary of state* thus fraternizing with negroes," a man going to the extreme of "writing them complimentary epistles"! He could not forgive Jefferson for "stiling them *his black brethren*." Smith excoriated Jefferson for "congratulating them on the evidence of their *genius*" and for "assuring them of his good wishes for their speedy emancipation." Jefferson's opponents also unfairly criticized him for having considered Banneker's accomplishments as evidence of "moral eminence" and not merely intellectual ability.[42]

In a letter to Henri Grégoire, the French statesman and Bishop of Blois, Jefferson wrote that "no person living wishes more sincerely than I do, to see a complete refutation of the doubts I myself entertained and expressed on the grade of understanding allotted to . . . [black people] by nature, and to find that in this respect they are on a par with ourselves."[43] His "doubts," he explained, had been "the results of personal observation" within "the limited sphere of my own State." There, "the opportunities for the development of their genius were not favorable, and those of exercising it still less so." He rejoiced that black people were "gaining daily in the opinions of nations" and that there were hopes that they would become reestablished "on an equal footing with the other colors of the human family."

9. The Laws of Nature and of Nature's God: Jefferson, Franklin, and Polly Baker

Scholars have sought in vain for a source of Jefferson's expression "Laws of Nature and of Nature's God." There were current in Jefferson's day a number of expressions similar to Jefferson's, among them Alexander Pope's lines:

> Slave to no sect, who takes no private road.
> But looks through Nature up to Nature's God.

Here we have a close association of nature and nature's God, but without the "laws." The "laws" do appear, however, in another couplet of Pope's, the famous one (cited in Chapter 2) about Newton, in which Pope invokes "Nature and Nature's Laws" which "lay hid in night." In a "Letter to Mr. Pope," Bolingbroke wrote of how a person who "follows Nature and Nature's God" is following "God in his works and in his word." This same invocation of nature and nature's God appears in a couplet used by Franklin's friend John Bartram, the Philadelphia botanist. Over the entrance to his greenhouse, Bartram placed the motto:

> Slave to no sect, who takes no private road
> But looks through nature, up to nature's God.

Jefferson's phrase, "the laws of nature and of nature's god," has become a familiar and concise expression of some fundamental principles of Enlightenment belief. Was this grouping of words original with Jefferson? Or did Jefferson encounter these memorable words in the course of his reading and merely quote them? Or, possibly, did Jefferson create this phrase by transforming some other and possibly less appropriate expression, such as Alexander Pope's? As of now, these questions remain unanswered. The scientific literature abounds in references to "the laws of nature" and Pope's couplet gave wide currency to "nature and nature's laws." But these differ from Jefferson's phrase.

There is, however, a statement that was current in Jefferson's day and that is very close to the words of the Declaration. It occurs in a famous speech allegedly delivered by one Polly Baker at her trial in Boston in the mid-eighteenth century. Polly had been brought to trial on a charge of fornication after having produced her fifth child out of wedlock. In a noble speech, defending her actions according to the highest philosophy, and wholly out of character with one who had borne five illegitimate children to different sires, Polly Baker defended herself by saying that she had merely been obeying the "great Command of Nature, and of Nature's God." She had been acting in accord with the laws which taught that one should

"Encrease and Multiply." In at least one printing, this story was given a happy ending. One of the judges was so impressed by her speech and her general demeanor that he married her forthwith and they lived happily ever after.[44]

Widely printed in English and American newspapers in the 1740s, some time after Jefferson was born, the story was still in circulation a half-century later, in the days just before the Revolution. We have evidence of the wide currency given to Polly's speech in the fact that it was printed and reprinted in various forms in England and America in the years before the Revolution[45] and it appeared in a retelling of the story by the Abbé Raynal in his book on the New World, one of the books advocating the theory of "degeneration" which Jefferson attacked in his *Notes on the State of Virginia*. First published in 1770 in a six-volume set, Raynal's *Philosophical and Political History of . . . the East and West Indies* was published in some twenty authorized editions, and more than fifty pirated editions, in both the original French and in various translations.[46] Raynal reported the trial of Polly as if it were established fact, quoting her speech at length. Her case was cited as proof of the extreme and unreasonable severity of the laws of New England. In this way Polly's ordeal was given worldwide circulation. By 1777, when an English translation of the Abbé's book appeared, the translator omitted this speech and story on the grounds, as he said, that it was too "well known."

Considering the wide circulation given to the story of Polly's trial, there is a strong possibility that Jefferson would have encountered Polly's address to the court with the phrase about the "Command of Nature, and of Nature's God." It is even possible that when Jefferson was composing the Declaration, he drew on an unconscious memory of this felicitous phrasing from Polly Baker and incorporated it into his text, merely transforming Polly's "Command of Nature" into "laws of nature." Some three years before Jefferson wrote the Declaration, on 16 February 1773, Polly's speech was printed in Salem, Massachusetts, in the *Essex Gazette;* this was reprinted a month and a half later (with acknowledgment to the Salem paper) in Jefferson's own Williamsburg, in the 1 April 1773 issue of the *Virginia Gazette*. There was, in other words, plenty of opportunity for Jefferson to have read about "the Command of Nature, and of Nature's God," which he could have either misremembered or transformed so as to produce the memorable phrase of the Declaration.

I have found no evidence to prove that, as early as 1776, Jefferson actually did know the story of Polly Baker, that he actually had encountered her famous speech. But we know for certain that not too long afterward, Jefferson was familiar with the Polly Baker episode, as the following anecdote will show. Our source for this anecdote is Jefferson's own "Anecdotes of Benjamin Franklin." One day, probably toward the end of 1777 or near the beginning of 1778, while Benjamin Franklin was still minister to the French court, he received a visit from the Abbé Raynal. Franklin's colleague, Cyrus Deane, was present. Franklin and Deane had been discussing the Abbé's book and Deane remarked to the Abbé that the book showed the Abbé to have been overly credulous and not careful enough about his facts. The Abbé at once took umbrage and defended the veracity of his book, demanding that Deane cite a single contrary example if he could. Deane gave as his evidence the episode of Polly Baker and her speech, which the Abbé had cited as a true story.

At this point they both became aware that Franklin could hardly contain his laughter. Franklin then admitted that the story was a hoax, that he had invented it out of whole cloth. He explained that back in the days when he had been a printer and publisher, "we were sometimes slack of news." In order "to amuse our customers," he went on, "I used to fill up our vacant columns with anecdotes and fables, and fancies of my own." There never was such a person as Polly Baker. Her supposed trial was a hoax invented by Franklin, a story made up to fill space. Her famous speech was a composition by Benjamin Franklin. Accordingly, if Polly Baker was Jefferson's source for "the laws of nature and of nature's god," then the famous phrase is a slight transformation of one that was written by Benjamin Franklin.[47]

Jefferson's own memoranda concerning the life of Benjamin Franklin not only provide information that Jefferson knew the Polly Baker story, but also prove that he was aware that the story was written as a hoax by Franklin. We have no way of telling when he learned of Franklin's conversation with the Abbé Raynal and Deane, with the revelation about Franklin's authorship, but a likely time would have been when he and Franklin were in Paris together, at the end of Franklin's mission to France, that is, during the years 1784–1785, almost a decade after Jefferson had written the Declaration of Independence.

There is still another possible source that I have found for Jefferson's phrase. Two phrases very like "the laws of nature and of nature's god" appear in a work by Franklin's associate Thomas Pownall, whose use of scientific metaphors has been introduced in Chapter 1. In a work entitled *Principles of Polity* (London, 1752), published just before Pownall came to America, there is a reference (p. 102) to "the Law of Nature, and of God" and another (p. 103) to "the eternal Laws of Nature and God," both of which are very close to "laws of nature and of nature's god." Jefferson was familiar with Pownall's writings and had a number of his works (although not this one) in his personal library.

The fact that both Polly Baker's words and Pownall's are close to Jefferson's may suggest that the ultimate source of the famous phrase was neither one. Rather, both Franklin (through the mouth of Polly) and Pownall—like Jefferson—may simply have been using an expression of a commonly accepted higher authority that was widely shared during the eighteenth century. In this case the problem is not so much to discover who was Jefferson's ultimate source as to recognize that the most forceful argument in the Age of Reason was to invoke both "nature and nature's laws" and "nature's god."[48]

10.   *Adams's Speculations about Respiration and Magnetism*

While in the midst of negotiations for a treaty of trade and friendship, Adams found time for scientific speculation. Juxtaposed to a long set of reflections about European politics he interposed ("without explanation or transition") a query concerning certain new discoveries in the chemistry of combustion, a topic then being elucidated by the studies of Joseph Priestley and Antoine-Laurent Lavoisier, both good friends of Franklin's. "What is it in the air which burns?" Adams asked.[49] "When we blow a spark with the bellows, it spreads. We force a current of air to the fire, by this machine, and in this air are inflammable particles. Can it be in the same manner that life is continued by the breath? Are there particles conveyed into the blood of animals through the lungs which increase heat of it or is the pulse caused by rarefying the blood or any part of it into vapor?" In these speculations, Adams was in fact pursing the lines of thought then being explored by the chemical researches of Priestley and Lavoisier, especially linking the analysis of combustion with the parallel analysis of respiration. Priestley adhered to the theory that both combustion and respiration are processes whereby a special substance (he and others of his belief called it by the name "phlogiston") is emitted into the air. When the air becomes saturated with this substance, and can no longer absorb any more of it, the air will not support either combustion or respiration. If a burning candle is inserted into a closed vessel filled with this saturated air (called "phlogisticated" air) the flame will become extinguished; if a mouse or other living creature is placed in such saturated air, it will be unable to breathe and die. The rival theory, developed by Lavoisier, held that both processes—combustion and respiration— were acts of chemical combination with oxygen, a component of ordinary air. The older notion of "saturated air" was thus to be replaced by a new concept that respiration and combustion would no longer be possible when these processes had used up the available oxygen.

Adams seems to have been leaning to the Lavoisier theory. "The external air," he wrote, "drawn into the lungs in breathing through the mouth or nostrils, either leaves some particles behind, in the lungs, or in the blood, or carries some particles off with it. It may do both, i.e., carry in some particles that are salubrious and carry out others which are noxious." In Lavoisier's theory the salubrious particles brought into the lungs would be oxygen, the noxious ones emitted through the lungs would be carbon dioxide and other waste products. In any event, as Adams noted, "The air once breathed is certainly altered. It is unfit to be breathed again." In writing down his interpretations of the state of the chemical theory of respiration, Adams couched his ideas in terms of "particles," that is, atoms or molecules of matter. Some three decades later, when writing to Harvard's Professor John Gorham, Adams (as explained in Chapter 4) eschewed the notion of atoms and warned chemists not to try experiments that might reveal the existence of such fundamental particles of matter.

In the essay, Adams suddenly shifts his intellectual gears, turning abruptly from the chemistry of respiration to the physics of magnetism. "The loadstone [or natural magnet] is," he wrote, "in possession of the most remarkable, wonderful and mysterious property in nature." Magnetism, he declared, "is in the secret of the whole globe," and "must have sympathy with the whole globe." It is, furthermore, "governed by a law and influenced by some active principle that pervades and operates from pole to pole, and from the surface to the centre and the antipodes." Break "the stone to pieces," he noted, "and each morsel retains two poles, a north and a south pole, and does not lose its virtue." He then turned to the topic of the "magnetic effluvia," which he said "are too subtle to be seen by a microscope" and yet "have great activity and strength." Here he was invoking a theory of magnetic "effluvia" similar but not identical to the theory of electricity invented by Benjamin Franklin, which was based on a subtle electric fluid composed of particles that could not been by a microscope.

Adams then set forth a research program to determine how magnetism works and to explore specifically its similarity (Adams called it "sympathy") with electricity. "Has it been tried," he asked, "whether the magnet loses any of its force in a vacuum?[50] While he expressed the belief that the "springs of nature may be too subtle for all our senses and faculties," he also held that "no subject deserved more the attention of [natural] philosophers or was more proper for experiments" than what he called "the sympathy between iron and the magnetical and electrical fluid." In particular, he suggested grinding up the loadstone to powder in order to see if the magnetic "virtue" is retained and whether the "virtue" is "affected, increased or diminished" by steeping "the stone or the dust in wine, spirits, oil and other fluids." He suggested that experiments be instituted to discern whether "boiling or burning the stone destroys or diminishes the virtue," whether "earth, air, water, or fire any wise applied" affects the virtue of the stone "and how." In fine, he wanted to see whether there is some "chemical process that can be formed upon the stone or the dust to discover what it is that the magnetic virtue resides in." These speculations reveal only the current state of knowledge, or of ignorance, concerning magnetism. Advances in this subject were not made through chemical analysis, as Adams believed, but by physical experiments that revealed the properties rather than the ultimate seat of the magnetic force. Adams's remarks are of interest, therefore, primarily as an

index of the chemical focus of his scientific thinking. But we may be astonished to find that he was still thinking of the four Aristotelian elements (air,
earth, fire, and water) rather than invoking the chemical concepts of his
own times.

11.   "Science" and "Useful Arts"

In the eighteenth century, the term "science" was used in association with
"arts" in a manner somewhat unlike our general understanding of these
words today. In fact, there are several rather different traditions in which
"science" and "art" appear in company. As a result, the lexicographers of the
eighteenth century were puzzled about the exact sense to be given to this
pair of words. We may see the problem clearly by looking at Ephraim Chambers's *Cyclopaedia, or an Universal Dictionary of Arts and Sciences* (seventh edition: London, 1752), the foremost scientific dictionary in English at the time
of the Constitution. According to Chambers, "science" is "a clear and certain
knowledge of any thing, founded on self evident principles, or demonstrations." In the commentary it is noted that science "is more particularly used
for a formed system of any branch of knowledge comprehending the doctrine, reason, and theory, of the thing, without any immediate application
thereof to any uses or offices of life." In "this sense," Chambers remarks,
"the word is used in opposition to *art*." He observes, however, that "the precise notion of an art, and *science*, and their just adequate distinction, do not
seem to be yet well fixed." For Chambers, the terms "philosophy" and "natural philosophy" would be closer to our "science."

In his preface, Chambers discusses at some length the differences
between sciences and arts. He confesses, however, that although "Philosophers have long laboured to explain and ascertain" the "notion and difference" in these two terms, all "their exploration amounts to little more, than
the substituting [of] one obscure notion for another." The result has been
that their "attempts have usually terminated in some abstract definition,
which rather casts obscurity, than light, on the subject." Chambers himself
ends by making the distinction that an "*art* and a *science* . . . only seem to
differ as less and more pure." His not particularly enlightening conclusion
is that an "art, in this light, appears to be a portion of science, or general
knowledge, considered, not in itself, as science, but with relation to its circumstances, or appendages."

Samuel Johnson's *Dictionary* (London, 1755) notes that there are six different meanings for the word "art." Only the first three are of concern in
the present context, the others being "cunning," "speculation," and "art-
fulness, skill, dexterity." The usage example given by Johnson is "the power

of doing something not taught by nature and instinct." For example, "to *walk* is natural, to *dance* is an *art*." The second meaning is a "science, as the liberal *arts*," while the third is a "trade," as "the *art* of making sugar." In reading this presentation, we must remember that for Johnson (as for his contemporaries) "science" was defined first of all as "knowledge" and second as "certainty grounded on demonstration." His third definition of "science" was "art attained by precepts, or built on principles," as in an extract quoted from Dryden that "*Science* perfects genius, and moderates that fury of the fancy which cannot contain itself within the bounds of reason." A fourth definition was "any art of species of knowledge," while a fifth was any "one of the seven liberal arts, grammar, rhetorick, logick, arithmetick, music, geometry, astronomy." In the latter case, Johnson was invoking the traditional meaning of the "arts" as the seven subjects of the ancient curriculum— the trivium (grammar, logic, and rhetoric) and the quadrivium ((arithmetic, geometry, music, and astronomy)—known also as the "seven arts" or the "liberal arts."[51]

In the eighteenth century, the subjects we call science tended to be known as "philosophy," often with the qualification of "natural philosophy" or "experimental philosophy" or—later in the eighteenth century—"chemical philosophy." In the *Federalist,* the term "science" generally appears coupled with a qualifying noun or adjective, as when Alexander Hamilton uses this word (*Federalist* No. 9) in relation to a "science of finance" and a "science of politics," which—he observes—has, "like most other sciences, . . . received great improvement." In a somewhat similar context (*Federalist* No. 37), Madison wrote of "political science" and "science of government," and also (*Federalist* No. 8) of the "science of federal government." These expressions involve the same sense of "science" that appears in our present designation of "political science." It will be obvious, also, that there is no unambiguous reference to "science" as physical or biological science.

The only time that the word "science" occurs in the *Federalist* without any reference to a science of something (politics, government) is in No. 43, in the statement of Article I, Section 8, Paragraph 8, of the Constitution, where it is coupled with "useful arts." It is much the same in the records of the Federal Convention, where we find references to "political science" and the "Science of Government." Here, in addition, there is a mention of "science" by itself, that is, without a reference to politics or government, in the various versions and statements of Article I, Section 8, Paragraph 8. We may note that at this time in history, "science" appeared with a similar usage in the writings of the British jurist Blackstone about the "science of the law." In a pamphlet on the Constitution, published in October 1787, Noah Webster wrote of "the science of legislation."

The framers of the Constitution were familiar with yet another sense of "science," one that often is found in association with "art" or "arts" and especially to set a distinction between the two. In this context "art" has nothing to do with the trivium or quadrivium, but rather implies a skill "in doing anything" that is acquired by knowledge and practice or implies human skill "as an agent" or "human workmanship" as opposed to what occurs or is found in nature. This sense of the word appears in such derivative terms as "artifice" and "artificial." In this context "science" was usually conceived to be a set of general rules and to exhibit a concern with "theoretic truth," in contrast to an "art," which was a set of practical methods "for effecting certain results."[52] This stress on practice rather than theory explains why the framers of the Constitution introduced "practical arts" rather than simply "arts."

Sometimes, however, the term "science" was used to denote a "department of practical work," but only if it "depends on the knowledge and conscious application of principles," in distinction to "art," which generally (as in "practical art") was "understood to require merely knowledge of traditional rules and skill acquired by hand." Thus, in 1724 Isaac Watts (in his *Logic*) referred to the "remarkable distinction between an art and a science, viz. the one refers chiefly to practice, the other to speculation." In 1796, Richard Kirwan wrote that "Previous to the year 1780, mineralogy, though tolerably understood by many as an art, could scarce be deemed a science." This distinction continues to the present in at least one area, the health sciences, where we make a distinction between theory and practice by referring to the "science" and "art" of medicine. A traditional expression of this difference was given by Robert Southey in 1862, when he wrote that "The medical profession . . . was an art, in the worst sense of the word, before it became a science; and long after it pretended to be a science, was little better than a craft." In 1870, the economist and philosopher of science William Jevons explained that "a science teaches us to know and an art to do."[53]

The conclusion, therefore, would seem to be that in Paragraph 8 of Section 8 of Article I, the juxtaposition of "Science" and "useful Arts" tells us that what the framers sought to promote was not the progress of science at large, as we would understand the word "science" to imply, but more narrowly and specifically those theoretical or general principles of practice that are associated directly with useful inventions or that lead to economic benefits or financial rewards.

12.   *Woodrow Wilson on the Constitution, Newtonian Philosophy, and the Whig Theory of Government*

Recent scholarship indicates a possible reason for Wilson's invention and advocacy of a Newtonian interpretation of the Constitution. According to

Harvey Mansfield, this was not a mere excursion into historical analysis, but rather had a political purpose in setting up a basis for criticizing the Constitution. That is, Wilson would appear to have blamed the Newtonian mechanistic origins of the Constitution for the system of checks and balances, which he saw as a barrier to any real progress. "For Wilson," according to Mansfield, "the mechanical Constitution with its checks and balances prevented progress." Accordingly, in order to "expose the mechanistic error, Wilson went behind the arguments of political science to uncover an osmosis of Newton's mechanics from the scientific into the political portion of the Founders' brains of which they were unconscious." In this way, he blamed the older Newtonian natural philosophy for having produced what he saw as the constricting force of an outdated political framework and he proposed in its place a new and radical scientific basis for our political system, namely, Darwinian evolution.[54]

The following autobiographical statement, describing how Wilson came to consider the Whig theory of government as a copy of the Newtonian system of the world, is taken from an address given at the annual banquet of the Economic Club of New York on 23 May 1912; it is reprinted from Arthur S. Link (ed.): *The Papers of Woodrow Wilson*, vol. 24 (Princeton: Princeton University Press, 1977), pp. 415–417.

One of the chief benefits I used to derive from being president of a university was that I had the pleasure of entertaining thoughtful men from all over the world. I can not tell you how much dropped into my granary by their presence. I had been casting around in my thought for something by which to draw several parts of my political ideas together when it was my good fortune to entertain a very interesting Scotchman who had been devoting himself to the philosophical thought of the seventeenth century. His talk was so engaging that it was delightful to hear him speak of anything, and presently there came out of the unexpected region of his thought the thing I had been waiting for. He called my attention to the fact that in every generation all sorts of speculation and thinking tend to fall under the formula of the dominant thought of the age that has preceded that.

For example, after the Newtonian theory of the universe had been developed, almost all thinking tended to express itself upon the analogies of the Newtonian theory, and since the Darwinian theory has reigned amongst us everybody tries to express what he wishes to expound in the terms of development and accommodation to environment. Now, it came to me as this interesting man talked, that the Constitution of the United States had been made under the dominion of the Newtonian theory. You have only to read the papers of the Federalist to see it written on every page. They speak of the "checks and balances" of the Constitution and use to express their idea the simile of the organization of the universe, and particularly of the solar system—how by the attraction of gravitation the various parts are held in their orbits, and represent Congress, the judiciary, and the President as a sort of imitation of the solar system.

No Government, of course, is a mechanism; no mechanical theory will fit any Government in the world, because Governments are made up of human beings, and all the calculations of mechanical theory are thrown out of adjustment by the intervention of the human will. Society is an organism, and every Government must develop according to its organic forces and instincts. I do not wish to make the analysis tedious; I will merely ask you, after you go home, to think over this proposition: that what we have been witnessing for the past hundred years is the transformation of a Newtonian constitution into a Darwinian constitution. The place where the strongest will is present will be the seat of sovereignty. If the strongest will is present in Congress, then Congress will dominate the Government; if the strongest guiding will is in the Presidency, the President will dominate the Government; if a leading and conceiving mind like Marshall's presides over the Supreme Court of the United States, he will frame the Government, as he did. There are no checks and balances in the mechanical sense in the Constitution; historical circumstances have determined the character of our Government.

Wilson made a formal presentation of this theme in a series of lectures he gave at Columbia University, published under the title *Constitutional Government in the United States* (New York: Columbia University Press, 1908). The following extract is taken from pp. 54–57. Wilson's book has been reprinted in *The Papers of Woodrow Wilson*, vol. 18 (1974), pp. 105–107.

The government of the United States was constructed upon the Whig theory of political dynamics, which was a sort of unconscious copy of the Newtonian theory of the universe. In our own day, whenever we discuss the structure or development of anything, whether in nature or in society, we consciously or unconsciously follow Mr. Darwin; but before Mr. Darwin they followed Newton. Some single law, like the law of gravitation, swung each system of thought and gave it its principle of unity. Every sun, every planet, every free body in the spaces of the heavens, the world itself, is kept in its place and reined to its course by the attraction of bodies that swing with equal order and precision about it, themselves governed by the nice poise and balance of forces which give the whole system of the universe its symmetry and perfect adjustment. The Whigs had tried to give England a similar constitution. They had no wish to destroy the throne, no conscious desire to reduce the king to a mere figurehead, but had intended only to surround and offset him with a system of constitutional checks and balances which should regulate his otherwise arbitrary course and make it at least always calculable.

They had made no clear analysis of the matter in their own thoughts; it has not been the habit of English politicians, or indeed of English-speaking politicians on either side of the water, to be clear theorists. It was left to a Frenchman to point out to the Whigs what they had done. They had striven to make Parliament so influential in the making of laws and so authoritative in the criticism of the king's policy that the king could in no matter have his own way without their coöperation and assent, though they left him free, the while, if he chose, to interpose an absolute veto upon the acts of Parliament. They had striven to secure for the courts of law as great an independence as possible, so that they might be neither overawed by parliament nor coerced by the king. In brief, as Montesquieu pointed out to them in his lucid way, they had sought to balance executive, legislature, and judiciary off against one another by a series of checks and counterpoises, which Newton might readily have recognized as suggestive of the mechanism of the heavens.

The makers of our federal Constitution followed the scheme as they found it expounded in Montesquieu, followed it with genuine scientific enthusiasm. The admirable expositions of the *Federalist* read like thoughtful applications of Montesquieu to the political needs and circumstances of America. They are full of the theory of checks and balances. The President is balanced off against Congress, Congress against the President, and each against the courts. Our statesmen of the earlier generations quoted no one so often as Montesquieu, and they quoted him always as a scientific standard in the field of politics. Politics is turned into mechanics under his touch. The theory of gravitation is supreme.

The trouble with the theory is that government is not a machine, but a living thing. It falls, not under the theory of the universe, but under the theory of organic life. It is accountable to Darwin, not to Newton. It is modified by its environment, necessitated by its tasks, shaped to its functions by the sheer pressure of life. No living thing can have its organs offset against each other as checks, and live. On the contrary, its life is dependent upon their quick coöperation, their ready response to the commands of instinct or intelligence, their amicable community of purpose. Government is not a body of blind forces; it is a body of men, with highly differentiated functions, no doubt, in our modern day of specialization, but with a common task and purpose. Their coöperation is indispensable, their warfare fatal. There can be no successful government without leadership or without the intimate, almost instinctive, coördination of the organs of life and action. This is not theory, but fact, and displays its force as fact, whatever theories may be thrown across its track. Living political constitutions must be Darwinian in structure and in practice.

Fortunately, the definitions and prescriptions of our constitutional law, though conceived in the Newtonian spirit and upon the Newtonian principle, are sufficiently broad and elastic to allow for the play of life and circumstance. Though they were Whig theorists, the men who framed the federal Constitution were also practical statesmen with an experienced eye for affairs and a quick practical sagacity in respect of the actual structure of government, and they have given us a thoroughly workable model. If it had in fact been a machine governed by mechanically automatic balances, it would have had no history; but it was not, and its history has been rich with the influences and personalities of the men who have conducted it and made it a living reality. The government of the United States has had a vital and normal organic growth and has proved itself eminently adapted to express the changing temper and purposes of the American people from age to age.

Wilson made an informal presentation of his views in an address in Baltimore on behalf of a commission on civic reform. The following extract is taken from a report in the *Baltimore American*, 6 December 1911. The report has been reprinted in *The Papers of Woodrow Wilson*, vol. 23 (1977), pp. 576–577.

In order to get control you must get unity. I would, if I could, speak with entire reverence of our system of checks and balances in American government, but with due apologies I cannot. Because, leaving aside the greater sphere of government, the government of the States and the government of the nation, nothing has so played into the hands of corruption in our State governments as the attempt to establish in them a set of checks and balances, offsetting authorities against one another.

That means that you set up a series of co-equal departments intending to check one another, intended to foil one another's purposes, intended to see to it that nothing can be done which all of them do not agree upon.

After you have set up this series of independent authorities, what have you done? Why, if you will pardon me for what will seem a pedantic digression and which may prove a little interesting, I will say that you have constructed a government on the Newtonian theory of the universe.

Now, the universe, as I understand it—which is very imperfectly—is a mechanical contrivance, and each part of it is intended to stay where it is and mind its own business. I understand that there would be deep mischief if it did not. (Laughter.)

And the writers in the *Federalist* and all of the expounders of the checks and balances of the American system think in the terms of the Newtonian universe. They speak of the centripetal and centrifugal forces and of the powers of gravitation which hold the various bodies in their proper orbits and spheres.

Now we have no orbit and spheres; we talk about woman's sphere and man's sphere, but we are talking nonsense most of the time. (Applause.)

What we have to realize is that we and our Government are organisms, and there is only one law by which an organism can live, and that is the law pointed out not by Newton, on a mechanical analogy, but by Darwin on the analogies of living beings.

Our governments do not work that way. The parts do cooperate; if they did not, there would not be any government.

Notes

· · · · ·

1 Science and American History

1. Jefferson, as Julian Boyd has pointed out, left "no doubt which he valued most." In proof, Boyd cited Jefferson's letter accepting the presidency of the American Philosophical Society, declaring that "the suffrage of a body which comprehends whatever the American world has of distinction in philosophy and science in general is the most flattering incident of my life, and that to which I am the most sensible."

Julian P. Boyd: "The Megalonyx, the Megatherium, and Thomas Jefferson's Lapse of Memory," *Proceedings of the American Philosophical Society* 102(1958): 420–435, esp. p. 424.

2. Examples of such "organismic" sociologists are Herbert Spencer, Paul von Lilienfeld, and A. E. F. Schäffle. On this topic see my *The Natural Sciences and the Social Sciences: Some Critical and Historical Perspectives* (Dordrecht / Boston / London: Kluwer Academic Publishers, 1994), ch. 1.

3. The term "analogy" is commonly used in the sciences to denote some similarity in function, in which case the companion term "homology" is applied to the restricted case of a similarity in form.

4. Charles Darwin: *The Origin of Species* (London: John Murray, 1859; reprint, Cambridge: Harvard University Press, 1964), p. 484.

5. These examples, and others, are explored at some length in my *The Natural Sciences and the Social Sciences;* see also my *Interactions: Some Contacts between the Natural Sciences and the Social Sciences* (Cambridge / London: MIT Press, 1994).

6. For details see my *The Natural Sciences and the Social Sciences;* also Margaret Schabas: *A World Ruled by Number: William Stanley Jevons and the Rise of Mathematical Economics* (Princeton: Princeton University Press, 1990).

7. Berkeley's critique of the foundations of Newton's theory of fluxions, that is, Newton's version of the calculus, was a real challenge to mathematicians. His *Siris* was an attempt "to assimilate Newtonian concepts to the more complex phenomena of chemistry and animal physiology." In his *De Motu* he analyzed "Newtonian concepts of gravitational attraction, action and reaction, and motion in general." See Gerd Buchdahl: "Berkeley, George," *Dictionary of Scientific Biography* (New York: Charles Scribner's Sons, 1970), vol. 2, pp. 16–18.

8. Berkeley fully understood Newton's explanation. He gave the correct Newtonian reason why the planets do not actually fall inward so as to join together at the center. They "are kept from joining together at the common centre of gravity," he wrote, "by the rectilinear motions the Author of nature hath impressed on each of them." This tangential or linear component, he continued, "concurring with the attractive principle," produces "their respective orbits round the sun." He concluded that if this linear component of motion should cease, "the general law of gravitation that is now thwarted would shew itself by drawing them all into one mass" (George Berkeley: "The Bond of Society," *Works*, ed. A. A. Luce and T. E. Jessop, vol. 7 [London / Edinburgh: Thomas Nelson and Sons, 1955], pp. 226–227).

9. Ibid., pp. 225–228; cf. George Berkeley: "Moral Attraction," *Works*, ed. Alexander Campbell Fraser, vol. 4 (Oxford: Clarendon Press, 1901), pp. 186–190. See also Supplement 2.

10. For additional materials concerning Berkeley's Newtonian sociology, see my "Newton and the Social Sciences, with Special Reference to Economics: The Case of the Missing Paradigm," to appear in Philip Mirowski (ed.): *Natural Images in Economics: Markets Read in Tooth and Claw* (Cambridge / New York: Cambridge University Press, 1994 [in press])—Proceedings of a Symposium at Notre Dame on "Natural Images in Economics," October 1991.

11. The eminent sociologist Pitirim A. Sorokin translated Berkeley's correct Newtonian physics into a hodgepodge of incorrect pre-Newtonian explanations. Sorokin not only would have Berkeley make use of the misleading notion of a balance of centrifugal and centripetal forces, but continued his travesty by saying that Berkeley concluded that "Society is stable when the centripetal forces are greater than the centrifugal." This is plainly nonsense even in pre-Newtonian physics; if the centripetal forces should be greater than the centrifugal forces, then obviously there would be no stability but an instability, a lack of balance or equilibrium, and a resultant motion inward, as Berkeley clearly stated would be the case under such circumstances. See Pitirim A. Sorokin: *Contemporary Sociological Theories* (New York / London: Harper & Brothers, 1928), p. 11.

12. On this subject see the suggestive essay by Douglass Adair: " 'That Politics may be Reduced to a Science': David Hume, James Madison, and the Tenth *Federalist*," *The Huntington Library Quarterly* 20(1957): 343–360.

13. See the writings of Duncan Forbes, notably his introduction to the reprint of Hume's *History of Great Britain* (Harmondsworth: Penguin Books, 1970); "Sceptical Whiggism, Commerce and Liberty," pp. 179–201 of A. S. Skinner and T. Wilson (eds.): *Essays on Adam Smith* (Oxford: Oxford University Press, 1976); "Hume's Science of Politics," pp. 39–50 of G. P. Morice (ed.): *David Hume: Bicentenary Papers* (Edinburgh: Edinburgh University Press, 1977). Also James E. Force: "Hume's Inter-

est in Newton and Science," ch. 10 of James E. Force and Richard Popkin: *Essays on the Context, Nature, and Influence of Isaac Newton's Theology* (Dordrecht / Boston / London: Kluwer Academic Publishers, 1990).

14. David Hume: *A Treatise of Human Nature,* ed. L. A. Selby-Bigge (Oxford: Clarendon Press, 1888), pp. 12–13.

15. If, as Hume believed, human behavior and social action are regulated by social laws, there is implied the possibility of a social science, one in which—as Hume wrote—"consequences almost as general and certain may sometimes be deduced . . . as any which the mathematical sciences afford us." Seeking to establish a kind of psychology of individual action, Hume seems to have envisioned the construction of a new theoretical science that would ultimately find expression in practice. On the certainty of social laws compared to mathematics, see David Hume: "That Politics may be Reduced to a Science," *Essays: Moral, Political, and Literary,* ed. T. H. Green and T. H. Grose (London: Longmans, Green, and Co., 1882; reprint, Aalen [Germany]: Scientia Verlag, 1964), vol. 1, p. 99.

16. Fourier's social physics was based on a system of twelve human passions and a fundamental law of "passional attraction" or "passionate attraction." This led him to conclude that only a carefully determined number of individuals could live together in "harmony" in what he called a "phalanx," which was the basis of the Fourierist communities that were later established.

17. Cf. *Design for Utopia: Selected Writings of Charles Fourier,* intro. Charles Gide, new foreword by Frank E. Manuel, trans. Julia Franklin (New York: Shocken Books, 1971 [orig. *Selections from the Works of Fourier* (London: Swan Sonnenschein & Co., 1901)]), esp. p. 18; *The Utopian Vision of Charles Fourier: Selected Texts on Work, Love, and Passionate Attraction,* trans., ed., intro. Jonathan Beecher and Richard Bienvenu (Boston: Beacon Press, 1971), esp. pp. 1, 8, 10, 81, 84; *Harmonian Man: Selected Writings of Charles Fourier,* ed. Mark Poster, trans. Susan Hanson (Garden City: Doubleday & Company—Anchor Books, 1971). On Fourier, see Nicholas Y. Riasanovsky: *The Teachings of Charles Fourier* (Berkeley / Los Angeles: University of California Press, 1969) and Frank E. Manuel: *The Prophets of Paris* (Cambridge: Harvard University Press, 1962).

18. Adam Smith: *An Inquiry into the Nature and Causes of the Wealth of Nations* (Oxford: Oxford University Press, 1976—The Glasgow Edition of the Works and Correspondence of Adam Smith, II), bk. 1, ch. 7, p. 15 (§15). The "Cannan edition"— Adam Smith: *An Inquiry into the Causes of the Wealth of Nations,* ed. Edwin Cannan (London: Methuen & Co., 1904; reprint, Chicago: University of Chicago Press, 1976; reprint, New York: Modern Library, 1985)—is easier to read and has the advantage of useful postils. A postil (1976 ed., p. 65; 1985 ed., p. 59) repeats the message: "Natural price is the central price to which actual prices gravitate."

19. Adam Smith: *Essays on Philosophical Subjects,* ed. W. P. D. Wightman and J. C. Bryce (Oxford: Oxford University Press, 1980—The Glasgow Edition of the Works and Correspondence of Adam Smith), vol. 3, pp. 33–105, "The History of Astronomy."

20. Claude Ménard: "La machine et le coeur: essai sur les analogies dans le raisonnement économique," in *Analogie et Connaissance,* vol. 2, *De la poésie à la science* (Paris: Maloine Editeur, 1981—Séminaires Interdisciplinaires du Collège de France), pp.

137–165; trans. Pamela Cook and Philip Mirowski as "The Machine and the Heart: An Essay on Analogies in Economic Reasoning," *Social Concept* 5(1988): 81–95.

21. *The Spirit of the Laws*, trans. Thomas Nugent (revised ed., London: George Bell and Sons, 1878; reprint, New York: Hafner Press, 1949), bk. 3, §7, "The Principle of Monarchy."

22. See Supplement 2.

23. On this score see Henry Guerlac: "Three Eighteenth-Century Social Philosophers: Scientific Influences on Their Thought," *Daedalus* 87(1958): 6–24; reprinted in Henry Guerlac: *Essays and Papers in the History of Modern Science* (Baltimore: Johns Hopkins University Press, 1977), pp. 451–464.

24. James Wilson: *The Works*, ed. Robert Green McClosky, 2 vols. (Cambridge: Belknap Press of Harvard University Press, 1967), p. 305.

25. See further Supplement 2.

26. Wilson, p. 192. See also Wilson's references to the circulation of the blood in his discussion of analogy (pp. 390–391).

27. Ibid., p. 560. Cf., e.g., pp. 183 and 388–390.

28. John A. Schutz: *Thomas Pownall, British Defender of American Liberty* (Glendale, Calif.: Arthur H. Clark Company, 1951).

29. Thomas Pownall: *Principles of Polity, being the Grounds and Reason of Civil Empire* (London: printed by Edward Owen, 1752), pp. 57–58.

30. Thomas Pownall: *The Administration of the Colonies* (London: printed for J. Wilkie, 1764), pp. 32–33.

31. Thomas Pownall: *The Administration of the British Colonies*, 5th ed., vol. 1 (London: printed for J. Walter, 1774), p. 45.

32. On this book and its editions see Schutz, pp. 256–262.

33. Thomas Pownall: *A Memorial, Most Humbly Addressed to the Sovereigns of Europe, on the Present State of Affairs between the Old and New World*, 2nd ed. (London: printed for J. Almon, 1780), pp. 2–3.

34. Ibid., p. 4.

35. T. Pownall: *A Memorial Addressed to the Sovereigns of America* (London: printed for J. Debrett, 1783), p. [i].

36. Thomas Pownall: *Intellectual Physicks: An Essay Concerning the Nature of Being, and the Progression of Existence* (Bath: printed by R. Cruttwell, 1795), e.g., pp. 25–28, 219–220. This book was printed in 1795 but not really made public until 1801; see Charles A. W. Pownall: *Thomas Pownall, M.P., F.R.S., Governor of Massachusetts Bay, Author of The Letters of Junius* (London: Henry Stevens, Son & Stiles, 1908), p. 442.

37. On the two varieties of Newtonian science see my *Franklin and Newton: An Inquiry into Speculative Newtonian Experimental Science and Franklin's Work in Electricity as an Example Thereof* (Philadelphia: American Philosophical Society, 1956; Cambridge: Harvard University Press, 1966).

38. In the preface to the first edition of the *Principia*.

39. See Supplement 3.

40. In addition, certain presentations of Newtonian science made use of a form of mechanical planetarium known as an "orrery," which showed some of the principal features of the motions of the planets and even of planetary satellites. (See Figs. 12–18.)

41. Carolus Linnaeus: *Genera Plantarum*, 5th ed. (Stockholm: Impensis Laurentii

Salvii, 1754), no. 536 (p. 208); facsimile reprint, intro. William T. Stearn (Weinheim; Codicote, Herts.: H. R. Engelmann [J. Cramer] and Wheldon & Wesley; New York: Hafner Publishing, 1960).

42. Carolus Linnaeus: *Species Plantarum,* vol. 1 (Stockholm: Impensis Laurentii Salvii, 1753), pp. 450–463; facsimile reprint (Berlin: W. Junk, 1907).

43. Ibid, p. 457.

44. For whatever microscopic observations required, Trembley generally used a "simple" microscope, a powerful magnifier consisting of a single lens. In those days, before the introduction of "corrected" lenses, the simple microscope gave far clearer images than the compound microscopes (with two lenses) then available, which suffered from extreme chromatic and spherical aberration. Trembley performed most of his microscopic observations in a darkened room, mounting his lens in any desired position above the specimens, which were illuminated by a candle. But many observations could be made in natural light with the aid of a magnifying glass.

45. See especially A. Trembley: *Mémoires pour servir à l'histoire d'un genre de polypes d'eau douce, à bras en forme de cornes* (Leiden: Chez Jean & Herman Verbeek, 1744). Trembley's studies of the polyp have been translated, with an extensive commentary, by Sylvia G. Lenhoff and Howard M. Lenhoff: *Hydra and the Birth of Experimental Biology—1744: Abraham Trembley's Mémoires concerning the Polyps* (Pacific Grove, Calif.: Boxwood Press, 1986).

46. Translated from the French in Virginia P. Dawson: *Nature's Enigma: The Problem of the Polyp in the Letters of Bonnet, Trembley, and Réaumur* (Philadelphia: American Philosophical Society, 1987), p. 140.

47. We shall see in Chapter 3 that Franklin based an important political metaphor on a "joynt" snake that could be cut in parts and then joined together again and survive.

48. Translated from the French in Dawson, p. 7.

49. Dawson, pp. 143–144.

50. Pownall, *Administration of the British Colonies,* 5th ed., vol. 2, pp. 10–11, quoted in Schutz, pp. 238–239.

51. Peter Collinson to Cadwallader Colden, 30 March 1745, *The Letters and Papers of Cadwallader Colden,* vol. 3, 1743–1747 (New York: New-York Historical Society, 1919—Collections of the New-York Historical Society), p. 110.

52. *Gentleman's Magazine* 15(1745): 193–197; *American Magazine and Historical Chronicle* 2(1745): 530–537.

53. I. B. Cohen: *Benjamin Franklin's Science* (Cambridge / London: Harvard University Press, 1990), ch. 4, 5.

54. J. A. Leo Lemay: *Ebenezer Kinnersley: Franklin's Friend* (Philadelphia: University of Pennsylvania Press, 1964).

55. *The Papers of Benjamin Franklin,* vol. 5, ed. Leonard W. Labaree (New Haven / London: Yale University Press, 1962), pp. 521–522; cf. vol. 4 (1961), p. 480, n. 6. (See Note 1 to Chapter 3 for further information on this edition.)

56. Quotation in Garry Wills: *Explaining America: The Federalist* (New York: Doubleday & Company, 1981), p. 23, this comment appears in a letter published in the *New York Journal* on 7 January 1788. The complete communication as reported by the journal (volume 42, number 6, page [3]) deserves to be quoted in the context of the present study: "A correspondent, having observed the attempt of the present

voluminous writer upon the new constitution, to explain the meaning of its several abstruce parts, by MATHEMATICAL demonstration, and his endeavours to prove its *right angular* construction, begs leave to propose, since he has failed in this mode, that he next have recourse to CONC SECTIONS, by which he will be enabled, with greater facility, to discover the *many windings* of his favorite system."

57. G. Washington's "Circular to the States," Head Quarters, Newburgh, 8 June 1783, John C. Fitzpatrick (ed.): *The Writings of George Washington from the Original Manuscript Sources*, vol. 26 (Washington, D.C.: Government Printing Office, 1938), p. 485. On this subject see Douglass Adair's article cited in Note 12.

58. According to Newton's disciple J. T. Desaguliers, Locke "asked Mr. *Huygens*, whether all the mathematical *Propositions* in Sir *Isaac*'s *Principia* were true" and was told "he might depend upon their Certainty." See Desaguliers's *Course of Experimental Philosophy*, 3rd ed. (London: A. Millar, 1763), preface, p. viii.

2   Science and the Political Thought of Thomas Jefferson: The Declaration of Independence

1. The primary source for Jefferson's correspondence and other documents is the edition in progress of *The Papers of Thomas Jefferson*, of which the founding editor was Julian P. Boyd; vol. 1 was published by Princeton University Press in 1950. This edition is cited as *Papers*. For the later period of Jefferson's life, not yet covered by *Papers*, there are several older collections, notably what is sometimes called the Memorial Edition, edited by Andrew A. Lipscomb and Albert Ellery Berg, which appeared in 20 volumes as *The Writings of Thomas Jefferson* (Washington, D.C.: Thomas Jefferson Memorial Association, 1903–1905). It is cited here as *Writings*. A convenient source is Merrill D. Peterson's one-volume anthology (of 1600 pages) in the "Library of America" series, 1984.

The life of Jefferson has been portrayed in meticulous detail by Dumas Malone in six volumes under the general title *Jefferson and His Time: Jefferson the Virginian, Jefferson and the Ordeal of Liberty, Jefferson and the Rights of Man, Jefferson the President 1801–1805, Jefferson the President 1805–1809,* and *The Sage of Monticello* (Boston: Little, Brown and Company, 1948–1981), cited here as Malone, vols. 1–6. A splendid one-volume biography has been written by Merrill D. Peterson, *Thomas Jefferson and the New Nation: A Biography* (New York: Oxford University Press, 1970), cited here as Peterson. Another work by Peterson, *The Jefferson Image in the American Mind* (New York: Oxford University Press, 1960), is also very useful. There are many other biographies of Jefferson and numerous scholarly studies of various aspects of his career, far too many to be listed here. I would, however, acknowledge a special indebtedness to Daniel J. Boorstin's *The Lost World of Thomas Jefferson* (New York: Henry Holt and Company, 1948) and Adrienne Koch's *The Philosophy of Thomas Jefferson* (New York: Columbia University Press, 1943) and *Jefferson and Madison: The Great Collaboration* (New York: Alfred A. Knopf, 1950).

Jefferson's activities as a scientist have been the subject of four volumes (by Silvio Bedini, Charles A. Browne, I. B. Cohen, and Edwin T. Martin), listed in Note 5. There are also a number of studies of Jefferson and the Declaration of Independence (notably by Carl Becker, Morton White, and Garry Wills), listed in Note 91.

2. TJ to Pierre-Samuel Dupont de Nemours, 2 March 1809, *Writings,* vol. 12, p. 260, toward the end of his second term as president.

3. Others have had training as engineers, such as Herbert Hoover and Jimmy Carter, who was a graduate of the U.S. Naval Academy.

4. *Washington Post,* 30 April 1962, p. B-5; quoted in Noble E. Cunningham, Jr.: *In Pursuit of Reason: The Life of Thomas Jefferson* (Baton Rouge / London: Louisiana State University Press, 1987), p. xiv. See also Theodore C. Sorenson: *Kennedy* (New York: Harper & Row, 1965), p. 384.

5. Silvio A. Bedini: *Thomas Jefferson, Statesman of Science* (New York: Macmillan Publishing Company, 1990); Charles A. Browne: *Thomas Jefferson and the Scientific Trends of His Time* (Waltham: Chronica Botanica Company, 1944); I. B. Cohen (ed.): *Thomas Jefferson and the Sciences* (New York: Arno Press, 1980); Edwin T. Martin: *Thomas Jefferson: Scientist* (New York: Henry Schuman, 1952). Another work of major importance is John C. Greene: *American Science in the Age of Jefferson* (Ames: Iowa State University Press, 1984).

6. Gerald Holton: *The Advancement of Science and Its Burdens: The Jefferson Lecture and Other Essays* (Cambridge: Cambridge University Press, 1986).

7. Gerald Holton: *Science and Anti-Science* (Cambridge / London: Harvard University Press, 1993), ch. 1.

8. TJ to Pierre-Samuel Dupont de Nemours, 2 March 1809, *Writings,* vol. 12, pp. 259–260.

9. TJ to Harry Innes, 7 March 1791, *Writings,* vol. 8, p. 135.

10. TJ to Giovanni Fabbroni, 8 June 1778, *Papers,* vol. 2, p. 195.

11. TJ to Lafayette, 4 August 1781, *Papers,* vol. 6, p. 112.

12. TJ to Thomas Mann Randolph, Jr., 1 May 1791, *Papers,* vol. 20, p. 341.

13. Further information concerning this fossil is given below.

14. Sarah N. Randolph: *The Domestic Life of Thomas Jefferson* (New York: Harper & Brothers, 1871), pp. 245, 249, 262–263, as quoted in Martin, p. 5.

15. Ibid.

16. TJ to Dr. Caspar Wistar, 3 February 1801, *Writings,* vol. 10, pp. 196–197.

17. Martin, p. 6.

18. Ibid., pp. 6–7.

19. Ibid., p. 7.

20. Charles Francis Adams (ed.): *Memoirs of John Quincy Adams, Comprising Portions of His Diary from 1795 to 1848* (Philadelphia: J. B. Lippincott & Co., 1874–1877), vol. 1, p. 317.

21. Ibid., pp. 472–473.

22. Bernard Mayo: "A Peppercorn for Mr. Jefferson," *Virginia Quarterly Review* 19(1943): 222–235; Charles C. Sellers: *Charles Willson Peale* (Philadelphia: American Philosophical Society, 1947), vol. 2, p. 184.

23. Martin, pp. 11–12.

24. Ibid., p. 29.

25. An admirable account of Jefferson's inventions is given by Bedini in his *Thomas Jefferson, Statesman of Science.*

26. See the articles by Julian P. Boyd, Howard C. Rice, Jr., and George Gaylord Simpson cited in Notes 14, 11, and 10 to the supplements.

27. Sir Mortimer Wheeler: *Archaeology from the Earth* (Oxford: Clarendon Press,

1954), pp. 41–42; Wheeler based his information on Alexander F. Chamberlain: "Thomas Jefferson's Ethnological Opinions and Activities," *American Anthropologist* 9(1907): 499–509.

28. Bedini, p. 341.

29. Ibid., p. 342.

30. *Writings*, vol. 18, p. 160. For Jefferson's instructions to Meriwether Lewis, see Reuben Gold Thwaites (ed.): *Original Journals of the Lewis and Clark Expedition*, vol. 7 (New York: Dodd, Mead & Company, 1905), pp. 247–297. Cf. Paul Russell Cutright: "Jefferson's Instructions to Lewis and Clark," *Bulletin of the Missouri Historical Society* 22(1965–1966): 302–321; Ralph B. Guinness: "The Purpose of the Lewis and Clark Expedition," *Missouri Valley Historical Review* 20(1933): 90–100. See also Donald Jackson (ed.): *The Letters of the Lewis and Clark Expedition, with Related Documents, 1783–1854* (Urbana: University of Illinois Press, 1962). An admirable account of Jefferson and the Lewis and Clark expedition, written from the point of view of Jefferson and science, may be found in Holton, *Science and Anti-Science*, pp. 116–120.

31. Malone, vol. 1, p. 52.

32. Robert E. Schofield: *The Lunar Society of Birmingham: A Social History of Provincial Science and Industry in Eighteenth-Century England* (Oxford: Clarendon Press, 1963).

33. TJ to the Reverend James Madison, 29 December 1811, *Writings*, vol. 19, p. 183.

34. Malone, vol. 1, p. 55.

35. Ibid., pp. 75–78.

36. Autobiography, *Writings*, vol. 1, pp. 1–164.

37. There is a modern edition of the *Notes* by William Peden (Chapel Hill: University of North Carolina Press, 1955).

38. Malone, vol. 2, pp. 93–94.

39. Peden, p. xi.

40. Ibid.

41. For an analysis of the scientific content of the *Notes*, see Martin, ch. 6–8.

42. Jefferson himself summed up the first four of these propositions as follows: "The opinion advanced by the Count de Buffon, is 1. That the animals common both to the old and new world are smaller in the latter. 2. That those peculiar to the new are on a smaller scale. 3. That those which have been domesticated in both have degenerated in America; and 4. That on the whole it exhibits fewer species. And the reason he thinks is, that the heats of America are less; that more waters are spread over its surface by nature, and fewer of these drained off by the hand of man. In other words, that *heat* is friendly, and moisture adverse to the production and development of large quadrupeds." Martin, pp. 163–164.

43. John Robert Moore: "Goldsmith's Degenerate Song-Birds, an Eighteenth-Century Fallacy in Ornithology," *Isis* 34(1943): 324–327.

44. The history of the theory of degeneration has been explored in rich detail in Antonello Gerbi: *The Dispute of the New World: The History of a Polemic, 1750–1900*, trans. Jeremy Moyle (Pittsburgh: University of Pittsburgh Press, 1973). On Buffon, see Jacques Roger: *Buffon: Un Philosophe au Jardin du Roi* (Paris: Fayard, 1989).

45. Peden, p. 56.

46. "Regulae Philosophandi" can be translated literally as "Rules of Philosophiz-

ing," but Motte did better by writing "The Rules of Reasoning in Philosophy," implying rules of natural philosophy.

47. These quotations from Buffon are taken from Jefferson's *Notes*, ch. 6.

48. Ibid.

49. Jefferson responded to the Abbé's charge that there had never been a "good" American poet by predicting that this situation will no doubt have changed when "we shall have existed as a people as long as the Greeks did before they produced a Homer, the Romans a Virgil, the French a Racine and Voltaire, the English a Shakespeare and Milton." Jefferson added, in a footnote, that in any event there were only two poets, Homer and Virgil, who "have been the rapture of every age and nation: they are read with enthusiasm in their originals by those who can read the originals, and in translations by those who cannot."

50. Brook Hindle: *David Rittenhouse* (Princeton: Princeton University Press, 1964).

51. Howard C. Rice, Jr.: *The Rittenhouse Orrery: Princeton's Eighteenth-Century Planetarium, 1767–1954. A Commentary on an Exhibition Held in the Princeton University Library* (Princeton: Princeton University Library, 1954).

52. Quoted in Rice, p. 30.

53. From a manuscript description of the orrery, ibid., p. 84.

54. This description is taken from am account of the Pennsylvania orrery, probably written by William Smith in 1771 and printed in *The Pennsylvania Gazette*, reprinted in Rice, pp. 86–88.

55. Ibid.

56. Ibid.

57. Anna Clark Jones: "Antlers for Jefferson," *New England Quarterly* 12(1939): 333–348. See also Ruth Henline: "A Study of *Notes on the State of Virginia* as an Evidence of Jefferson's Reaction against the Theories of the French Naturalists," *Virginia Magazine of History and Biography* 55(1947): 233–246; also Gilbert Chinard: "Eighteenth Century Theories on America as a Human Habitat," *Proceedings of the American Philosophical Society* 91(1947): 27–57.

58. James Madison to TJ, 19 June 1786, *Papers*, vol. 9, pp. 659–665.

59. Malone, vol. 2, p. 100.

60. William Carmichael to TJ, 15 October 1787, *Papers*, vol. 12, p. 241.

61. TJ to John Harvie, 14 January 1760, *Papers*, vol. 1, p. 3.

62. TJ to Abigail Adams, 22 August 1813, *Writings*, vol. 19, p. 194.

63. TJ to James Madison, 20 December 1787, *Papers*, vol. 12, p. 442.

64. On the subject of apportionment, see Michel L. Balinski and H. Peyton Young: *Fair Representation: Meeting the Ideal of One Man, One Vote* (New Haven / London: Yale University Press, 1982); also Paul Hoffman: *Archimedes' Revenge: The Joys and Perils of Mathematics* (New York / London: W. W. Norton & Company, 1988), part 4.

65. According to Balinski and Young.

66. Ibid.

67. Jefferson was concerned—as he made abundantly clear in his memorandum to Washington—that the "bill does not say that it has given the residuary representatives *to the greatest fractions; though in fact it has done so.*"

68. TJ: Opinion on Apportionment Bill, 4 April 1792, *Papers*, vol. 23, p. 375.

69. An argument for keeping rather than discarding fractions can be made quite

simply on the basis of the numbers given in the first apportionment bill voted by the House. In this enumeration, it can be seen that Delaware's ideal apportionment would be 1.851, while Massachusetts' would be 15.844. If fractional parts were ignored, then Delaware's loss of 0.851 represents a net loss of 46 percent of its total, but Massachusetts' loss of an almost equal 0.844 is a net loss of only about 5 percent of its total. Thus, as a contemporary analyst remarked, "The injury arising from the unrepresented fraction of population is more severely felt by the smaller than by the larger States, as in the case of Delaware." In real numbers, with the discarding of fractions, Delaware would end up with one seat for 55,540 persons, whereas Massachusetts would have fifteen seats for a population of 475,327, or one seat for every 31,688 persons. These figures leave no doubt that to ignore fractional parts would be to favor the larger states at the expense of the smaller ones.

70. See Balinski and Young, Note 64.

71. The only exception to this rule permitted by the Constitution would be in cases where the population was so small that the "common ratio" would not otherwise give that state the minimum single representative.

72. According to Malone, vol. 2, p. 211, Jefferson's "instructions to Trumbull for the procurement of busts and pictures were so numerous that no one can easily keep them straight after a century and a half." In a letter to Trumbull on 15 February 1788, *Papers*, vol. 14, p. 561, Jefferson expressed his high regard for Bacon, Locke, and Newton "as the three greatest men that have ever lived, without any exception." They "laid the foundation of these superstructures which have been raised in the Physical and Moral sciences." See, further, *Papers*, vol. 14, pp. 467–468, 524–525, 561, 634–635, 663; vol. 15, pp. 38, 152, 157. On Hamilton and these portraits, see *Writings*, vol. 13, p. 4. The three portraits (*Papers*, vol. 15, p. 152) were copies made of originals that hung "in the Apartments of the Royal Society." The portraits of Bacon and Locke may still be seen at Monticello, but the Newton portrait has disappeared.

73. TJ to John Adams, 21 January 1812, in Lester J. Cappon (ed.): *The Adams-Jefferson Letters: The Complete Correspondence between Thomas Jefferson and Abigail and John Adams* (Chapel Hill: University of North Carolina Press, 1959), vol. 2, p. 291.

74. E. Millicent Sowerby (ed.): *Catalogue of the Library of Thomas Jefferson*, 5 vols. (Washington, D.C.: Library of Congress, 1952–1959). The evidence that Jefferson used this textbook as a college student is given in Supplement 6.

75. TJ to the Rev. James Madison, 19 July 1788, *Papers*, vol. 13, pp. 379–381. In this letter Jefferson also referred to Ingenhousz's "opinion that *light* promotes vegetation."

76. TJ to Joseph Willard, 24 March 1789, *Papers*, vol. 14, pp. 697–699.

77. In Jefferson's proposal, time was the astronomer's "mean time," based—as he wrote in a preliminary draft of his report—on the "motion of the earth round it's axis," which "is uniform and invariable." Rittenhouse suggested to Jefferson that he delete his assertion that the rotation of the Earth is in any sense absolutely equable. This motion is, he noted, "sufficiently uniform for every human purpose," even though there are good reasons to suppose that there are long-term and short-term reasons why this rotation "is not perfectly Equable." The final version incorporated Rittenhouse's correction and used his very words: "The motion of the earth round

it's axis, tho' not absolutely uniform and invariable, may be considered as such for every human purpose."

78. In the first draft of his report, Jefferson's first sentence says that "Sir Isaac Newton has determined the length of a pendulum vibrating Seconds in latitude   to be 39.2 inches = 3.2666 &c feet measuring from it's point of suspension to it's center of oscillation." Jefferson later said that he had based this value on his memory of Newton's text, since at the time of composition he was in New York, far from his own library, and unable to lay his hands on a copy of Newton's *Principia*. This fact explains why, in the quoted statement, there is a blank space for the exact latitude of London, which he could not remember. By the time of the second draft, Jefferson had been able to borrow a copy of the *Principia* from Professor Robert Patterson.

79. In Book Three, prop. 20, of the *Principia*, Newton gave a table from which the length of a seconds pendulum can be found (by extrapolation) for any latitude, but he did not give any such length for the specific latitude of London, 38° north. Furthermore, in Newton's presentation all lengths are given in French measure, that is, in French feet ("pieds") and "lines" (twelfth parts of a "pied"). Jefferson's statement would have been more accurate if he had written that according to calculations based on Newton's tables, the length of a seconds pendulum for latitude 38° north comes out to be 39.1285 inches.

Since all of Newton's numerical examples are given in French measures rather than in English feet, at best Jefferson's statement can mean that (1) by using Newton's tables, one can compute the length of a seconds pendulum for the latitude of London and then (2) convert the answer from French measures into English inches.

80. At the time when Jefferson was drawing up his report, similar proposals had been made in America, England, and France. There was even a tract on this subject, written by John Whitehurst, a copy of which was in Jefferson's library. A proposal for this kind of standard, written in manuscript by William Waring, was waiting in the papers of the secretary of state when Jefferson undertook the assignment. James Madison also had thought of using the seconds pendulum in setting up a standard of length. In fact, Jefferson adopted an important alteration in his original plan after encountering the discussion of this subject in the National Assembly in Paris by the Bishop of Autun, the celebrated Marquis de Talleyrand. Jefferson had at first thought of setting the latitude for the standard pendulum at 38° north, a central point of the continental United States that also happens to be the latitude of Monticello. He was convinced by Talleyrand's argument, however, that an international neutral latitude should be adopted, a point midway between the poles and the equator, the latitude of 45°, which happens to be within the boundaries of the United States, but far in the north.

81. The documents relative to Jefferson's *Report on Weights and Measures* are collected in *Papers*, vol. 16.

82. For a variety of mathematical and technical reasons, Jefferson chose to have his pendulum be a swinging metal cylinder or rod. He showed how to calculate its center of oscillation and so determine the length of the equivalent ideal pendulum, the basis of the unit of length in a system of weights and measures.

83. Rittenhouse to TJ, 21 June 1790, *Papers*, vol. 16, p. 546.

84. Jefferson had at first made an arithmetical error in presenting Newton's dis-

cussion of the change of length of an iron rod with temperature. This was corrected by Rittenhouse, who told Jefferson: "I take an Iron Rod of 3 feet long to be shorter by 1/6 of a line in winter than in Summer." That, Rittenhouse computed, "is 1/2592 part of its whole length." (Rittenhouse to TJ, 21 June 1790, *Papers*, vol. 16, p. 546.)

85. John W. Olmsted: "The Scientific Expedition of Jean Richer to Cayenne (1672–1673)," *Isis* 34(1942): 117–128.

86. Newton recognized that there are no theoretical grounds for supposing that weight should be proportional to mass, as is the case for Newton's law of universal gravity, in which the terrestrial force of gravity acting on a body is directly proportional to the body's mass. In post-Newtonian terms, Newton appreciated that there is no reason why the mass that enters into the laws of motion should be the same as the mass that enters into the law of gravity, that is, no reason why inertial mass should be equivalent to gravitational mass. Accordingly, he devised an experiment which proved such equivalence for a number of different substances and then, by induction, concluded that the result is general, that mass is proportional to weight (at any given place). In relativity, this equivalence of the two types of mass is part of basic theory.

87. TJ to David Rittenhouse, 14 June 1790, *Papers*, vol. 16, p. 510.

88. If the Earth is squashed down at the poles, as Newton supposed, then the pull of the Moon on the nearer bulge will be greater than the pull on the more distant bulge, producing—according to Newton's principles of rational mechanics—a precession of the axis, a phenomenon known quantitatively since the time of Hipparchus, two centuries before the beginning of the Christian Era.

89. There are two separate effects which enter into considerations of a body's weight. One is the gravitational force, the other the result of the Earth's rotation. Both of these depend on the shape of the Earth, and hence the length of a seconds pendulum must be related to the latitude. If the Earth is oblate, squashed down at the poles, the distance from the center will be considerably greater at the equator than at the poles, varying between these two extremes at different latitudes. Similarly, the speed of rotation, determined by the distance from the Earth's axis, will be greater at the equator than at the poles and will also vary with the latitude.

90. For details, see Supplement 7; unfortunately, the presentation of this document in the edition of Jefferson's *Papers* does not make it clear that Jefferson was correcting an error made by Rittenhouse. Furthermore, the quotation from Newton not only is printed in a way that contains a gross error in Latin, but makes no sense; it could hardly serve, as printed by the editors, to correct an error made by Rittenhouse.

91. Carl Becker: *The Declaration of Independence: A Study in the History of Political Ideas* (New York: Alfred A. Knopf, 1922, 1951); Julian P. Boyd: *The Declaration of Independence: The Evolution of the Text as Shown in Facsimiles* (Washington, D.C.: Library of Congress, 1943); Herbert Friedenwald: *The Declaration of Independence: An Interpretation and an Analysis* (New York: Macmillan Co., 1904); John H. Hazelton: *The Declaration of Independence: Its History* (New York: Dodd, Mead & Company, 1906; reprint, New York: Da Capo Press, 1970); Morton White: *The Philosophy of the American Revolution* (New York: Oxford University Press, 1978); Garry Wills: *Inventing America: Jefferson's Declaration of Independence* (Garden City: Doubleday & Company, 1978).

92. John Adams to Timothy Pickering, 6 August 1822, *Works* (see Note 1 to Chapter 4), vol. 2, p. 514.

93. Peterson, p. 80.

94. Ibid., p. 89.

95. Unless an indication is given that a different version is being used, the Declaration is quoted from the engrossed and signed parchment copy as printed in *Papers,* vol. 1, pp. 429–432; cf. the editorial notes on pp. 416–417 and 433. The reader will recall that "Jefferson's words" differ in various drafts; for example, while the "original Rough draught" (*Papers,* vol. 1, p. 423) has "rights inherent & inalienable," the engrossed copy (as in *Papers,* vol. 1, p. 429) has "certain unalienable Rights."

96. Peterson, p. 89.

97. *Papers,* vol. 1, p. 432.

98. Quoted in Peterson, p. 90.

99. For instance, where in the "original Rough draught" (*Papers,* vol. 1, p. 424) Jefferson had written that the king had "suffered the administration of justice totally to cease," the engrossed copy (as in *Papers,* vol. 1, p. 430) said more directly that he had "obstructed the Administration of Justice."

100. Peterson, p. 91.

101. *Papers,* vol. 1, p. 426.

102. Garry Wills concluded that the alterations made by the delegates so changed the message of the Declaration that a distinction should be made between the final version and what he calls "Jefferson's Declaration."

103. I adopt here the language and style of the final text as printed in *Papers,* vol. 1. Carl Becker (p. 185) observed that in the engrossed version "the capitalization and punctuation, following neither previous copies, nor reason, nor the custom of any age known to man, is one of the irremediable evils of life to be accepted with becoming resignation."

104. Jefferson's own drafts did not use capital letters and tend to favor the ampersand over "and." Accordingly, the "original Rough draught" refers to "the laws of nature & of nature's god" rather than to "the Laws of Nature and of Nature's God."

105. Much of Becker's text concerning the Newtonian philosophy consists of listing the number of editions and translations of Newton's own books and of giving bibliographic information concerning books that popularized the Newtonian natural philosophy.

106. Thus, on p. 41, Becker characterizes the difficult mathematical development of universal gravitation in the *Principia* as follows: "by looking through a telescope and doing a sum in mathematics, the force which held the planets could be identified with the force that made an apple fall to the ground." Forsooth!

107. Paul Foriers and Chaim Perelman: "Natural Law and Natural Rights," pp. 13–27 of *Dictionary of the History of Ideas,* vol. 3 (New York: Charles Scribner's Sons, 1973).

108. Ibid.

109. The English terms "law" and "right" are related to but do not exactly coincide with distinctions between "lex" and "ius" (or "jus") in Latin.

110. Lyman H. Butterfield et al. (eds.): *The Earliest Diary of John Adams* (Cambridge: Belknap Press of Harvard University Press, 1966), pp. 54–55.

111. The importance of Burlamaqui was stressed in Benjamin F. Wright: *American Interpretations of Natural Law: A Study in the History of Political Thought* (Cambridge: Harvard University Press, 1931).

112. White, ch. 4. See, further, Oscar Handlin: "Learned Books and Revolution-

ary Action, 1776," *Harvard Library Bulletin* 34(1986): 362–379; also Oscar Handlin and Lilian Handlin: "Words and Acts in the American Revolution," *American Scholars* 58(1989): 545–556.

113. Whoever wishes to explore this topic further will be well advised to study the books by Becker, White, and Wills, supplemented by such more specialized monographs as Wright's *American Interpretations of Natural Law.*

114. Jefferson had in his library both the Latin edition of 1672 and the English translation of 1727. The first is Ricardus Cumberland: *De Legibus Naturae Disquisitio Philosophica, in qua earum forma, summa capita, ordo, promulgatio, & obligatio è rerum natura investigantur, quinetiam elementa philosophiae Hobbianae, cùm moralis tum civilis, considerantur & refutantur* (London: typis E. Flesher, 1672). The English translation has two title pages. One, at the beginning, calls the work *A Treatise of the Laws of Nature* and gives the translator, John Maxwell, and the information on publication (London: printed by R. Phillips, 1727). The second, placed between pages 38 and 39, reproduces the Latin title as *A Philosophical Inquiry into the Laws of Nature, in which their form, chief heads, odrer [sic], promulgation, and obligation are deduced from the nature of things; also the elements of Mr. Hobbes's philosophy, as well moral as civil, are consider'd and refuted.*

115. *Writings*, vol. 1, pp. 470–481.

116. E.g., White, ch. 4. See also Ronald Hamowy: "Jefferson and the Scottish Enlightenment: A Critique of Garry Wills's *Inventing America: Jefferson's Declaration of Independence*," *William and Mary Quarterly*, third series, 36(1979): 503–523.

117. *Notes on the State of Virginia*, p. 31.

118. "An Account of the Book entituled *Commercium Epistolicum Collinii & Aliorum, De Analysi Promota*, published by order of the Royal-Society, in relation to the Dispute between Mr. Leibnitz and Dr. Keill, about the right of invention of the method of fluxions, by some called the differential method," *Philosophical Transactions* 29 (1715): 173–224.

119. The reader will certainly find it odd that Newton wrote and published a long review of this report about himself and his work, but there are additional oddities to this episode. We now know from Newton's manuscripts that he actually drafted the committee's original report. And, as if this was not enough, he ended by ordering a reprint of the report with his own (anonymous) book review as an introduction—and then wrote an anonymous preface to make certain that the reader would get the message. Such are the ways of genius! For details, see A. Rupert Hall: *Philosophers at War* (Cambridge / London / New York: Cambridge University Press, 1980).

120. "An Account," p. 224.

121. John Keill: *Introduction to Natural Philosophy: or, Philosophical Lectures read in the University of Oxford, Anno Dom. 1700*, 4th ed. (London: printed for M. Senex, W. Innys, T. Longman, and T. Shewell, 1745). There were many editions in both the original Latin and in English translation.

122. On the history of the concept of scientific law, see Jane E. Ruby: "The Origins of Scientific 'Law,' " *Journal of the History of Ideas* 47(1986): 341–359.

123. I have used the fourth edition (London, 1731).

124. The first edition was published in 1735. Jefferson owned Benjamin Martin: *The Philosophical Grammar, Being a View of the Present State of Experimented Physiology, or Natural Philosophy*, 6th ed. (London: J. Noon, 1762).

125. See *Papers,* vol. 1, pp. 423–428.

126. In the original draft (see Boyd, *Declaration of Independence;* Becker, p. 142) Jefferson expressed the matter somewhat differently. As White has pointed out, Jefferson had two lines of argument. The first was "that all men are created equal & independent"; the second, "that from that equal creation they derive [in *del.*] rights inherent & inalienable, among which are the preservation of life, & liberty, & the pursuit of happiness." In the final version, the specified set of rights is no longer stated as derived "from that equal creation." John Adams's copy of this draft has "unalienable" for "inalienable."

127. In his *Declaration* (p. 142, n. 1) Becker wrote: "It is not clear that this change was made by Jefferson," and he suggested that the "handwriting of 'self-evident' resembles Franklin's." For a long time scholars accepted Becker's conclusion; I myself, in an earlier publication, followed Becker's interpretation.

Becker was a Cornell professor, one of the most original historians of American thought in the 1920s and 1930s. His most important book was *The Heavenly City of the Eighteenth-Century Philosophers* (1932), which rapidly became a classic, influencing generations of graduate students in American history. Becker, however, was not a handwriting expert.

128. See Boyd's commentary in his *Declaration of Independence* and in vol. 1 of the edition of Jefferson's *Papers.*

129. Possibly, "petitiones" might better be rendered by "position statements." See Edward Rosen: "Copernicus' Axioms," *Centaurus* 20 (1976): 44–49.

130. Copernicus's use of "axiom" could be justified by reference to a passage in Aristotle's logic, his "Posterior Analytics" (I, 2:72*a* 17; I, 10:76*b* 14–15), where he says, "An axiom is a primary proposition which must be possessed by whoever is to gain any knowledge." Also, "axioms are the primary propositions on which a proof depends."

For another interpretation of Copernicus's use of "axioms" in the *Commentariolus,* see Noel Swerdlow's "The Derivation and First Draft of Copernicus's Planetary Theory: A Translation of the *Commentariolus* with Commentary," *Proceedings of the American Philosophical Society* 117 (1973).

131. Newton's first law and others are presented in full also under the heading "*Laws of Motion . . .*" as part of the presentation of "MOTION."

132. White, p. 7.

133. David Hume: *Essays, Moral, Political, and Literary,* ed. Eugene F. Miller (Indianapolis: Liberty Classics, 1987), esp. p. 18.

134. Hume most often writes of "general maxims" and "general truths" rather than axioms.

135. This description of "internal evidence" may be compared with Hamilton's remark in *Federalist* No. 83, where he says that arguing about a certain aspect of the Constitution "would . . . be as vain and fruitless, as to attempt the serious proof of the *existence* of *matter,* or to demonstrate any of those propositions which by their own internal evidence force conviction, when expressed in language adapted to convey their meaning."

136. For an analysis of Hamilton's argument in *Federalist* No. 31, notably the concept of axiom and self-evident truths, see White, pp. 82–92.

137. Hamilton's primary example is "positions so clear as those which manifest the necessity of a general power of taxation in the government of the union." After enumerating these "positions," Hamilton indicates that "it would be natural to conclude that the propriety of a general power of taxation in the national government might safely be permitted to rest on the evidence of these propositions, unassisted by any additional arguments or illustrations." The "experience" of political debate, however, forces Hamilton beyond the realm of the axiomatic.

138. TJ to Benjamin Austin, 9 January 1816, *Writings*, vol. 14, p. 392.

139. See also Charles A. Miller: *Jefferson and Nature: An Interpretation* (Baltimore: Johns Hopkins University Press, 1988).

We may note also that Jefferson would almost certainly have become aware of the importance of "self-evident" truths while an undergraduate, as a result of reading William Duncan's *Elements of Logick* (London, 1748), of which a copy was in his library. His college teacher, William Small, had been a student of Duncan's at Marischal College in Aberdeen. See Wilbur Samuel Howell: "The Declaration of Independence and Eighteenth-Century Logic," *William and Mary Quarterly* 18(1961): 463–484.

3   Benjamin Franklin: A Scientist in the World of Public Affairs

1. The literature concerning almost every aspect of Franklin's career is enormous and continually increasing. The notes include only references to works that have been especially useful in preparing this chapter, although occasionally there are guides to additional discussions of the various subjects mentioned in the chapter.

The primary source for Benjamin Franklin is the edition, currently in progress, of *The Papers of Benjamin Franklin*, of which thirty-one volumes have been published. The first volume (New Haven: Yale University Press, 1959) and volumes 2–14 were edited by Leonard Labaree, volumes 15–26 by William Willcox, volume 27 by Claude A. Lopez. Volumes 28–31 have been edited by Barbara B. Oberg, the current editor. This edition is cited as *Papers*. For the later years of Franklin's life, the chief source is the older edition, *The Writings of Benjamin Franklin*, ed. Albert Henry Smyth (New York: Macmillan, 1905–1907), referred to here as *Writings*. On some issues, Jared Sparks's edition of *The Works of Benjamin Franklin* (revised edition, Philadelphia: Childs & Peterson, 1840; various reprints) is still useful.

Students of Benjamin Franklin and anyone interested in Franklin's life and career will be especially grateful to J. A. Leo Lemay for his superb collection of Franklin's *Writings* (New York: Library of America, 1987). Lemay, the dean of Franklin scholars, has also produced a major new edition (in association with P. M. Zall) of *The Autobiography of Benjamin Franklin: A Genetic Text* (Knoxville: University of Tennessee Press, 1981). See also the useful volume, edited by Lemay and Zall, in the Norton Critical Editions series, *Benjamin Franklin's Autobiography: An Authoritative Text, Backgrounds, Criticism* (New York: W. W. Norton & Company, 1986).

There are many biographies of Franklin, of which the most complete is Carl Van Doren: *Benjamin Franklin* (New York: Viking Press, 1938). Among more recent works are Catherine Drinker Bowen: *The Most Dangerous Man in America: Scenes from the Life of Benjamin Franklin* (Boston / Toronto: Little, Brown and Company, 1974); Thomas

Fleming: *The Man Who Dared the Lightning: A New Look at Benjamin Franklin* (New York: William Morrow and Company, 1971); and Esmond Wright: *Franklin of Philadelphia* (Cambridge: Harvard University Press, 1986). Paul W. Conner: *Poor Richard's Politicks: Benjamin Franklin and His New American Order* (New York: Oxford University Press, 1965) contains many useful insights, as does Alfred Owen Aldridge: *Benjamin Franklin: Philosopher and Man* (Philadelphia / New York: J. B. Lippincott Company, 1965).

Additional references will be found in each of the sections of the chapter.

2. On Franklin's contribution to electrical science, see the discussions below and the works cited in Notes 3 and 11.

3. For example, in the standard work on the science of electricity in the seventeenth and eighteenth centuries, John Heilbron's *Electricity in the Seventeenth and Eighteenth Centuries* (Berkeley / Los Angeles: University of California Press, 1982), one of the divisions of the book is called "The Age of Franklin."

4. Clinton Rossiter: "The Political Theory of Benjamin Franklin," *Pennsylvania Magazine of History and Biography* 76(1952): 259–293, esp. 260.

5. Gerald Stourzh: *Benjamin Franklin and American Foreign Policy* (Chicago: University of Chicago Press, 1954), p. 26. The only reference to Montesquieu that Stourzh was able to find was to "his ideas on criminal law."

6. See Chapter 4.

7. A remarkable job of detective scholarship is embodied in Verner W. Crane's *Franklin's Letters to the Press, 1758–1775* (Chapel Hill: University of North Carolina Press, 1950), a collection of not previously identified letters published in British newspapers under a variety of pseudonyms.

8. For his almanac of 1758, Franklin collected the maxims of Poor Richard, some of them revised, into a speech reported as given by an old man, Father Abraham. Included were some of Franklin's most famous sayings, such as "Three removes is as bad as a fire." This collection of maxims about the conduct of life was published separately a few months later by Benjamin Mecom, Franklin's nephew, as *Father Abraham's Speech*. Later printings were called *The Way to Wealth;* there were many translations.

9. Cf. Van Doren, p. 51.

10. His defense of the unicameral system and the response of John Adams are discussed in Chapter 4.

11. See the introduction to my edition of Benjamin Franklin's book on electricity, *Benjamin Franklin's Experiments: A New Edition of Franklin's Experiments and Observations on Electricity* (Cambridge: Harvard University Press, 1941). Also, my *Franklin and Newton: An Inquiry into Speculative Newtonian Experimental Science and Franklin's Work in Electricity as an Example Thereof* (Philadelphia: American Philosophical Society, 1956; Cambridge: Harvard University Press, 1966).

The change in the understanding and appreciation of Franklin as a scientist has been traced in the opening chapter of my *Benjamin Franklin's Science* (Cambridge / London: Harvard University Press, 1990).

12. These editions are listed and described in my edition of Franklin's book.

13. Humphry Davy: "Agricultural Lectures," part 2, *Collected Works* (London: Smith, Elder, 1840), vol. 8, p. 264.

14. These translations are listed and described in my edition of Franklin's book.

15. Peter W. van der Pas: "The Latin Translation of Benjamin Franklin's Letters on Electricity," *Isis* 69(1978): 82–85.

16. The only rival that comes to mind is Leonhard Euler's celebrated *Letters to a German Princess*, but this was a popular survey of science and not—like Franklin's *Experiments and Observations*—a scientific treatise.

17. Roger Hahn: *The Anatomy of a Scientific Institution: The Paris Academy of Sciences, 1666–1803* (Berkeley / Los Angeles / London: University of California Press, 1971).

18. For details, see my *Franklin and Newton*, p. 121.

19. Joseph Priestley: *The History and Present State of Electricity, with Original Experiments,* 3rd ed. (London: printed for C. Bathurst, . . . , 1775; reprint, New York / London: Johnson Reprint Corporation, 1966, with a valuable intro. by Robert E. Schofield), vol. 1, p. 193.

20. Ibid.

21. For descriptions of these experiments, see my edition of Franklin's *Experiments and Observations on Electricity* and John Heilbron's *Electricity in the Seventeenth and Eighteenth Centuries.*

22. For details see my *Franklin and Newton.*

23. That is, Newton suggested (in the later "Queries" of the *Opticks*) the role of a "subtle" fluid, composed of mutually repelling particles that can permeate bodies and lie in the interstices between the particles of which bodies are made. Newton did not conceive that such a fluid might be operative in electrical phenomena.

24. For details, see *Benjamin Franklin's Experiments.*

25. J. J. Thomson: *Recollections and Reflections* (London: G. Bell and Sons, 1936), pp. 252–253.

26. Robert A. Millikan: "Benjamin Franklin as a Scientist," *Journal of the Franklin Institute* 242 (1941): 407–423, esp. 417. See, further, R. A. Millikan: "Franklin's Discovery of the Electron," *American Journal of Physics* 16 (1948): 319.

27. William Watson: "An Account of a Treatise, presented to the Royal Society, intituled, 'Letters concerning Electricity . . . extracted and translated from the French,' " *Philosophical Transactions* 48 (1753): 201–216, esp. 201–202.

28. *Papers,* vol. 5, p. 131.

29. These include Edmund Hoppe's *Geschichte der Elektrizität* (Leipzig: Johann Ambrosius Barth, 1884), Ferd. Rosenberger's *Die Moderne Entwicklung der Elektrischen Principien* (Leipzig: Verlag von Johann Ambrosius Barth, 1898), Mario Gliozzi's *L'elettrologia fino al Volta* (Naples: Luigi Loffredo, 1937), Park Benjamin's *The Intellectual Rise in Electricity: A History of Electricity from Antiquity to the Days of Benjamin Franklin* (New York: John Wiley & Sons, 1898), and John Heilbron's *Electricity in the Seventeenth and Eighteenth Centuries,* supplemented by any of the editions of Joseph Priestley's *History and Present State of Discoveries relating to Electricity* (3rd ed., London, 1775).

30. BF to Cadwallader Colden, 29 September 1748, *Papers,* vol. 3, p. 318.

31. BF to Cadwallader Colden, 11 October 1750, *Papers,* vol. 4, p. 68.

32. The Mesmer episode is discussed below.

33. Article on Benjamin Franklin in the *Dictionary of American Biography.*

34. *Papers,* vol. 11, pp. 153–173, esp. p. 159.

35. Ibid., vol. 4, p. 233.

36. This essay is reprinted in *Papers*, vol. 4, pp. 90–94; the portion dealing with the polyp occurs on p. 93.

37. This essay is discussed at length below.

38. David George Hale: *The Body Politic: A Political Metaphor in Renaissance English Literature* (The Hague / Paris: Mouton, 1971); I. B. Cohen: *Interactions: Some Contacts between the Natural Sciences and the Social Sciences* (Cambridge / London: MIT Press, 1994), ch. 2.

39. *Papers*, vol. 9, pp. 78–79.

40. Van Doren, p. 220.

41. *Writings*, vol. 10, pp. 57–58; cf. Van Doren, p. 554.

42. Manasseh Cutler: *Life, Journals, and Correspondence*, ed. William Parker Cutler and Julia Perkins Cutler (Cincinnati: Robert Clarke & Co., 1888), vol. 1, pp. 267–269.

43. Verner W. Crane: *Benjamin Franklin and a Rising People* (Boston: Little, Brown and Company, 1954), pp. 161–162.

44. *Journals of the Continental Congress, 1774–1789*, ed. Worthington Chauncey Ford et al., 34 vols. (Washington, D.C.: Government Printing Office, 1904–1937; also reprint, New York: Johnson Reprint Corporation, 1968), vol. 6, p. 1082.

45. (1) "Notes of Proceedings in the Continental Congress," Jefferson, *Writings* (see Note 1 to Chapter 2), vol. 1, p. 324; "Notes of Debates in the Continental Congress," *Journals of the Continental Congress*, vol. 6, p. 1103; (2) "Anecdotes of Benjamin Franklin," Jefferson, *Writings*, vol. 18, p. 167.

46. Jefferson, *Writings* (see Note 1 to Chapter 2), vol. 18, p. 167.

47. Van Doren, p. 216; Wright, p. 81; Crane, *Benjamin Franklin and a Rising People*, pp. 67–69.

48. According to some classifications, demography is not a "hard" science in the sense of physics, chemistry, or zoology, but there is agreement that it is a legitimate science.

49. Cohen, *Interactions*, ch. 2.

50. *Papers*, vol. 4, p. 227.

51. *Papers*, vol. 1, pp. 140–141, 149n, 153n; W. A. Wetzel: *Benjamin Franklin as Economist* (Baltimore: Johns Hopkins Press, 1895; reprint, New York / London: Johnson Reprint Corporation, 1973).

52. Lewis J. Carey: *Franklin's Economic Views* (Garden City: Doubleday, Doran & Company, 1928), pp. 46–47.

53. *Papers*, vol. 4, p. 228: "our People must at least be doubled every 20 Years"; vol. 4, p. 233: "doubling, suppose but once in 25 Years."

54. *Papers*, vol. 4, p. 233.

55. Ibid.

56. Stourzh, p. 60.

57. *Papers*, vol. 4, p. 233.

58. Ibid., pp. 230–233.

59. Charles Darwin: *The Origin of Species* (London: John Murray, 1859; photoreprint, intro. Ernst Mayr, Cambridge: Harvard University Press, 1964), p. 63.

60. Thomas Malthus: *An Essay on the Principle of Population* (London: J. Johnson, 1803), p. iv; a new edition, ed. Patricia James, 2 vols. (Cambridge: Cambridge University Press, 1989), vol. 1, p. 1.

61. Malthus (1803), p. 2; James (ed.), vol. 1, p. 10.

62. Malthus's reference reads: "Franklin's Miscell. p. 9."

63. Malthus (1803), p. iv; James (ed.), vol. 1, p. 1.

64. Malthus (1803), pp. 11, 483–503; James (ed.), vol. 1, p. 18, and vol. 2, pp. 87–103.

65. Thomas Malthus: *An Essay on the Principle of Population* (London: John Murray, 1826), p. 15, n. 1; *The Works of Thomas Robert Malthus,* vol. 2, ed. E. A. Wrigley and David Souden (London: William Pickering, 1986), p. 16, n. 1.

66. See Carey, pp. 57–58. This connection is questioned, e.g., in Alfred Owen Aldridge: "Franklin as Demographer," *Journal of Economic History* 9 (1949–1950): 25–44, esp. 26, n. 4, and 30–32.

67. Malthus (1803), p. 24 with note b, and p. 104 with note e; James (ed.), vol. 1, p. 30 with n. 2, and pp. 88–89 with n. 11. Cf. Carey, pp. 58–59.

68. Thomas Malthus: *An Essay on the Principle of Population* (London: J. Johnson, 1798), p. 21, and cf. p. 20; *The Works of Thomas Robert Malthus,* vol. 1, p. 12.

69. See, e.g., Aldridge, "Franklin as Demographer," pp. 25–26, 28, 32–33.

70. *Papers,* vol. 9, p. 79.

71. Ibid., vol. 9, p. 72.

72. Ibid., vol. 9, p. 73.

73. Ibid., vol. 9, p. 74.

74. Ibid., vol. 4, p. 228.

75. Ibid., vol. 9, pp. 73–74.

76. There are important studies of Malthus's theories of population and their historical background in Joseph J. Spengler: *Population Economics: Selected Essays,* ed. Robert S. Smith et al. (Durham: Duke University Press, 1972). See also Conway Zirkle: "Benjamin Franklin, Thomas Malthus and the United States Census," *Isis* 48 (1957): 58–62.

77. See my *Benjamin Franklin's Science,* pp. 110–117, 137–138.

78. I. B. Cohen: "Prejudice against the Introduction of Lightning Rods," *Journal of the Franklin Institute* 253 (1952): 393–440.

79. See my *Benjamin Franklin's Experiments,* pp. 62, 171–172.

80. *Papers,* vol. 4, pp. 408–409.

81. Ibid., vol. 3, pp. 472–473.

82. *Benjamin Franklin's Experiments,* p. 134.

83. Ibid., p. 135.

84. Ibid., pp. 135–136.

85. Ibid., p. 136.

86. Ibid., p. 137.

87. Ibid., pp. 137–138.

88. There are many important studies of Franklin's life and activities in France during the years of the American Revolution. Among them are Claude-Anne Lopez: *Mon Cher Papa: Franklin and the Ladies of Paris* (New Haven / London: Yale University Press, 1966), French version, *Le sceptre et la foudre: Benjamin Franklin à Paris 1766–1785* (Paris: Mercure de France, 1990). There are many important insights in Alfred Owen Aldridge: *Franklin and His French Contemporaries* (New York: New York Univer-

sity Press, 1957); still of value is an older work, Edward E. Hale and Edward E. Hale, Jr.: *Franklin in France*, 2 vols. (Boston: Robert Brothers, 1888). Of course, every biography of Franklin (such as those by Carl Van Doren and Esmond Wright) deals with this topic.

Of special value in the context of the present discussion are the writings of Jonathan R. Dull, especially *A Diplomatic History of the American Revolution* (New Haven / London: Yale University Press, 1985) and *Franklin the Diplomat: The French Mission* (Philadelphia: American Philosophical Society, 1982—Transactions of the American Philosophical Society, vol. 72, part 1).

89. *Journals of the Continental Congress*, vol. 3, pp. 400–401.

90. *Papers*, vol. 22, p. 290; cf., e.g., vol. 15, p. 178, n. 1.

91. Ibid., vol. 22, p. 628.

92. Elizabeth S. Kite: *Brigadier-General Louis Lebègue Duportail, Commandant of Engineers in the Continental Army 1777–1783* (Baltimore: Johns Hopkins Press, 1933).

93. *Papers*, vol. 22, pp. 627–628; cf., e.g., pp. 289–290, 372. See also Dull, *Franklin the Diplomat*, pp. 12–13 and n. 8.

94. *Papers*, vol. 22, pp. 369–374; cf. Dull, *Franklin the Diplomat*, p. 20.

95. Dull, *Franklin the Diplomat*, p. 20.

96. Ibid., n. 15.

97. I have used the noun "diplomatist" because "diplomat" would be anachronous, while "ambassador" then usually implied nobility. As Jonathan Dull has remarked (*Franklin the Diplomat*, preface, n. 4), "the more accurate title of minister presents possibilities of confusion for modern readers—rather hilarious possibilities in conjunction with Franklin."

98. *Collections of the Massachusetts Historical Society* 4(1878): 446.

99. Francis P. Wharton (ed.): *The Revolutionary Diplomatic Correspondence of the United States* (Washington, D.C.: Government Printing Office, 1889), vol. 3, pp. 332–333.

100. Dull, *Franklin the Diplomat*, p. 2.

101. Paul W. Conner's *Poor Richard's Politicks: Benjamin Franklin and His New American Order*, pp. 13, 173–217, as summarized in Dull, *Franklin the Diplomat*, p. 2.

102. Page Smith: *John Adams* (Garden City, N.Y.: Doubleday & Company, 1962), p. 444.

103. Ibid., pp. 499–501, 504–505.

104. Dull, *Franklin the Diplomat*, p. 26.

105. Ibid.

106. Ibid., p. 27.

107. Ibid.

108. Histories of the Revolution often portray Franklin as a moderate in politics, imbued with a hatred of war, far less "revolutionary" than Sam Adams or Tom Paine. As Dull has noted, however, "most of the revolutionaries who knew Franklin . . . regarded him as one of their own" (*Franklin the Diplomat*, p. 70). Pauline C. Maier has found that the "Loyalists frequently mentioned . . . Franklin as the . . . arch-conspirator"; see her *The Old Revolutionaries: Political Lives in the Age of Samuel Adams* (New York: Alfred A. Knopf, 1980), pp. 8–9.

109. Dull, *Franklin the Diplomat*, pp. 9–10; see, further, Jonathan R. Dull: *The French Navy and American Independence: A Study of Arms and Diplomacy, 1774–1787* (Princeton: Princeton University Press, 1975).

110. Dull, *French Navy*, pp. 89–94; Dull, *Diplomatic History*, pp. 89–103.

111. Conner, p. 156.

112. Simon Schama: *Citizens: A Chronicle of the French Revolution* (New York: Alfred A. Knopf, 1989), p. 48.

113. Ibid., p. 43.

114. *Papers*, vol. 29, p. 613.

115. Adams, *Works* (see Note 1 to Chapter 4), vol. 1, p. 660.

116. *Papers*, vol. 21, p. 582.

117. Schama, p. 44.

118. Carl Bridenbaugh: *Cities in Revolt: Urban Life in America, 1743–1776* (New York: Alfred A. Knopf, 1955), pp. 216–217, esp. p. 217n.

119. Summarized from *L'espion anglois* in Aldridge's *Franklin and His French Contemporaries*, p. 64.

120. Aldridge, *Franklin and His French Contemporaries*, pp. 59–65; Stourzh, pp. 132–146; Dull, *Diplomatic History*, pp. 89–103.

121. For details, see my *Benjamin Franklin's Science*, p. 125.

122. Ibid., pp. 119–121.

123. Ibid., p. 125.

124. For a detailed history of this epigram and its use in France, see Aldridge, *Franklin and His French Contemporaries*, pp. 124–136.

125. John Adams's Diary, *Works* (see Note 1 to Chapter 4), vol. 3, p. 220, entry for 23 June 1779. Adams, with his typical disparagement of Franklin, further declares that although Franklin is "a great philosopher" he "has done very little" as statesman and legislator; cf. pp. 220–221.

126. [Jeanne Louise Henriette Genest,] Madame Campan: *Mémoires sur la vie privée de Marie-Antoinette*, vol. 1 (Paris: Baudouin Frères, Libraires, 1822), pp. 233–234.

127. Ibid., p. 233.

128. Jefferson, *Writings* (see Note 1 to Chapter 2), vol. 18, pp. 167–168.

129. Aldridge, *Franklin and His French Contemporaries*, p. 66.

130. Max Farrand (ed.): *The Records of the Federal Conventions of 1787* (New Haven: Yale University Press, 1911), vol. 3, p. 91.

131. *Writings*, vol. 9, pp. 607–609; Barbara B. Oberg; " 'Plain, Insinuating, Persuasive': Benjamin Franklin's Final Speech to the Constitutional Convention of 1787," pp. 175–192 of J. A. Leo Lemay (ed.): *Reappraising Benjamin Franklin: A Bicentennial Perspective* (Newark: University of Delaware Press, 1993).

132. The Constitution of the United States, Article V.

133. *Papers*, vol. 3, p. 171.

134. *Writings*, vol. 9, p. 489.

135. In addition to sources cited below, see, e.g., Gary B. Nash and Jean R. Soderlund: *Freedom by Degrees: Emancipation in Pennsylvania and Its Aftermath* (New York / Oxford: Oxford University Press, 1991), pp. ix–xiv, and Claude-Anne Lopez and Eugenia W. Herbert: *The Private Franklin: The Man and His Family* (New York: W. W. Norton & Company, 1975), pp. 291–302.

136. Van Doren, pp. 128–129.

137. *Papers,* vol. 3, p. 474.

138. Ibid., vol. 4, pp. 229–230.

139. Ibid., pp. 229–230.

140. Ibid., p. 229, n. 9.

141. In 1756 Franklin had a black slave with him on his journey to Virginia (*Papers,* vol. 6, p. 425), and in his will of 28 April 1757 he merely recorded his intention "that my Negro Man Peter, and his Wife Jemima, be free after my Decease" (*Papers,* vol. 7, p. 203). Later in 1757 his wife recorded the purchase of a "Negrow boy" (*Papers,* vol. 8, p. 425, n. 6) and Franklin and his son took two slaves with them to England (*Papers,* vol. 7, p. 203, n. 9, and pp. 274, 369, 380; vol. 8, pp. 137, 432; vol. 9, pp. 174–175, 327, 338).

142. See John C. Van Horne: "Collective Benevolence and the Common Good in Franklin's Philanthropy," pp. 425–440 of Lemay (ed.), *Reappraising Benjamin Franklin,* esp. pp. 433–435; *Papers,* vol. 7, pp. 98–101, 352–353, 356, 377–379.

143. *Papers,* vol. 8, p. 425; see also vol. 9, p. 38.

144. Ibid., vol. 9, pp. 12–13, 20–21, 174.

145. Van Horne, p. 435.

146. *Papers,* vol. 9, p. 12, n. 1.

147. Ibid., vol. 10, pp. 298–300.

148. Ibid., vol. 10, pp. 395–396, and p. 396, n. 1.

149. Ibid., vol. 17, pp. 37–44; Crane, *Franklin's Letters to the Press,* pp. 186–192.

150. *Papers,* vol. 19, pp. 112–113.

151. Ibid., vol. 17, pp. 137–140, 142, 144–150, 203.

152. Ibid., vol. 19, pp. 112–116.

153. Ibid., vol. 19, p. 113, n. 1.

154. Ibid., vol. 19, pp. 187–188; Crane, *Franklin's Letters to the Press,* pp. 221–223.

155. *Papers,* vol. 21, pp. 151–152; for Condorcet's letter of inquiry see *Papers,* vol. 20, pp. 489–491.

156. Van Doren, p. 774.

157. *Writings,* vol. 10, p. 495.

158. Ibid., p. 86.

159. Ibid., pp. 66–68.

160. Ibid., pp. 127–129; cf. Conner, p. 84.

161. Ibid., p. 128.

162. Ibid., pp. 128–129.

163. Ibid., p. 67.

164. *Papers,* vol. 21, p. 151.

165. *Writings,* vol. 10, p. 67.

166. *Papers,* vol. 21, p. 151.

167. *Writings,* vol. 10, pp. 67–68.

168. Ibid., pp. 87–91.

169. Ibid., pp. 92–93.

170. Ibid., p. 72.

4   Science and Politics: Some Aspects of the Thought and Career of John Adams

1. The primary source materials for any study of Adams are his correspondence, diaries, and published writings. Adams's grandson, Charles Francis Adams, published the first major collection of Adams's writings in a ten-volume set entitled *The Works of John Adams, Second President of the United States, with a Life of the Author* (Boston: Charles C. Little and James Brown, 1850–1856; facsimile reprint, New York: AMS Press 1971), referred to as *Works*.

Beginning in 1954, the Adams Trust began to make available an extensive microfilm edition of papers of the Adams family. A definitive edition of *Papers of John Adams,* begun under the editorship of Lyman Butterfield and published by Harvard University Press, is planned to include correspondence, diaries, and other documents, as well as published writings, of John Adams and other members of the Adams family. This project is still in progress.

Among various other collections, one has proved particularly useful, the *Familiar Letters of John Adams and His Wife Abigail Adams during the Revolution,* ed. Charles Francis Adams (New York: Hurt and Houghton; Cambridge: Riverside Press, 1875).

An older biography, Gilbert Chinard's *Honest John Adams* (Boston: Little, Brown, and Company, 1933), is still worth reading, particularly the discussions of Adams and France, but has been largely superseded by Page Smith's two-volume *John Adams* (Garden City, N.Y.: Doubleday & Company, 1962).

Correa Moylan Walsh's *The Political Science of John Adams* (New York / London: G. P. Putnam's Sons, 1915; facsimile reprint, Freeport, N.Y.: Books for Libraries Press, 1969), although uncritical, remains the most complete study available of the development and context of Adams's political thought. Also useful is George A. Peek's edition of *The Political Writings of John Adams, Representative Selections* (Indianapolis: Bobbs-Merrill Company, 1954). The most recent examination of Adams as a political figure is Joseph J. Ellis's *Passionate Sage: The Character and Legacy of John Adams* (New York: W. W. Norton & Company, 1993).

Of special interest is Zoltán Haraszti's *John Adams and the Prophets of Progress* (Cambridge: Harvard University Press, 1952), an examination of Adams's thought through analysis of the marginal notations in books from his library.

2. Walter Muir Whitehill: "Early Learned Societies in Boston and Vicinity," pp. 151–162 of Alexandra Oleson and Sanborn Brown (eds.): *The Pursuit of Knowledge in the Early American Republic* (Baltimore: Johns Hopkins Press, 1976), esp. p. 151.

3. Recollections of John Adams in 1809, ibid., p. 152.

4. For more information, see I. Bernard Cohen: *Some Early Tools of American Science, an Account of the Early Scientific Instruments and the Mineralogical and Biological Collections in Harvard University* (Cambridge: Harvard University Press, 1950), referred to as *Some Early Tools.*

5. See Chapter 2.

6. JA to Thomas Jefferson, 3 February 1812, in Lester J. Cappon (ed.): *The Adams-Jefferson Letters: The Complete Correspondence between Thomas Jefferson and Abigail and John Adams* (Chapel Hill: University of North Carolina Press, 1959), vol. 2, pp. 294–295.

7. Samuel Eliot Morison: *Harvard College in the Seventeenth Century* (Cambridge: Harvard University Press, 1936), pp. 644–646.

8. Quoted in *Some Early Tools*, p. 33.

9. Smith, pp. 15–18.

10. *Some Early Tools*, p. 2.

11. Clifford K. Shipton: *New England Life in the 18th Century: Representative Biographies from Sibley's Harvard Graduates* (Cambridge: Belknap Press of Harvard University Press, 1963), pp. 349–373. There is, in the Harvard University Archives, an important unpublished study of "John Winthrop and Colonial Science Education at Harvard," written by Michael Carter Mathieu as a senior honors thesis in the Department of the History of Science (March 1989).

12. Winthrop's notes for his course of lectures of the time when Adams was a student are preserved in the Harvard University Archives. In the Archives there is also a later set of notes on these lectures (dating from 1772–1773) made by Timothy Foster. Some extracts from Adams's notes on these lectures are quoted below.

13. *Some Early Tools,* pp. 135–144.

14. Lyman H. Butterfield et al. (eds.): *The Earliest Diary of John Adams* (Cambridge: Belknap Press of Harvard University Press, 1966), referred to as *Diary.*

15. They may be compared with the later set of such notes, made in 1772–1773 by Timothy Foster.

16. According to Winthrop's diary, preserved in the Massachusetts Historical Society Archives; see Adams's *Diary*, p. 64, n. 1.

17. *Diary*, p. 63. Winthrop, like others of that day, used the expression "centrifugal force" for the inertial linear (or tangential) component of a body's orbital motion. See Chapter 2 for a discussion of the relevance to the Declaration of Independence of phrases such as Adams's "Sir Isaac Newton's three laws of nature" and Winthrop's "3 Laws of Motion or Nature."

18. In theory, the equal-arm balance is an example of the equilibrium of two equal forces, the weights in each of the pans. Anyone who has used such balances knows that the more delicate or accurate an instrument it is, the harder it will be to achieve an exact balance. A jeweler weighing a small amount of gold or a precious stone will, therefore, usually not try to achieve a perfect balance but rather determine that a certain weight W is greater than the weight of the object being weighed, while another weight w is less. The desired weight is thus between W and w.

19. Keill offered his students a second example of the equilibrium of three forces. In this case a metal ball is suspended from the ceiling at the end of a cord. A second cord draws the ball away from the vertical at some given angle. There are three forces acting here. One is the downward weight of the ball, acting vertically. Another is the horizontal pull on the ball, causing it to move out or to one side. The third is the tension of the supporting string, now exerted at the above angle to the vertical. Once again, a force diagram of the triangle and simple geometry make it possible to calculate the tension in the supporting string and the force with which the ball is pulled to the side.

It should be noted that a second, and actually equivalent, procedure for computing the forces in these cases is to take components of all the forces acting on a body.

The components are to be taken in two directions, perpendicular to each other. In each direction, the sum of the components must be zero, that is, all the components in each of these directions must balance out, or there would be no equilibrium.

20. *Works,* vol. 4, p. 521.

21. Ibid., vol. 6, p. 323.

22. Ibid., vol. 4, p. 470.

23. Ibid., vol. 6, p. 394.

24. Ibid., vol. 4, p. 354.

25. Ibid., vol. 4, pp. 347, 391, 499; vol. 5, pp. 10, 426; vol. 6, pp. 108, 128, 341.

26. Ibid., vol. 6, p. 323. See the excellent presentation on this subject, but without reference to the science of statics, in Walsh, ch. 5.

27. JA to Benjamin Rush, 1790, quoted in Smith, p. 802.

28. Franklin published some of Winthrop's observations on lightning and the efficacy of lightning rods in the later editions of his book on electricity. See I. B. Cohen: *Benjamin Franklin's Experiments: A new edition of Benjamin Franklin's Experiments and Observations on Electricity* (Cambridge: Harvard University Press, 1941), pp. 150–151.

29. On this topic see I. B. Cohen: *Benjamin Franklin's Science* (Cambridge: Harvard University Press, 1990), ch. 8.

30. *Works,* vol. 2, p. 3.

31. I. B. Cohen: "Prejudice Against the Introduction of Lightning Rods," *Journal of the Franklin Institute* 253(1952): 426, 429.

32. See my introduction to *Benjamin Franklin's Experiments.*

33. Winthrop concluded by averring that he didn't believe there was in "the whole town of *Boston*" so much as "one person, who is so weak, so ignorant, so foolish, or . . . so atheistical" as to have believed that "it is possible, by the help of a few yards of wire, to 'get out of the mighty hand of God.' "

34. Zoltán Haraszti: "Young John Adams on Franklin's Iron Points," *Isis* 41 (1950): 11–14.

35. J. G. Crowther: *Famous American Men of Science* (New York: W. W. Norton & Company, 1937), ch. 1, §3, "Science and the Constitution," p. 140.

36. Richard H. Tawney: "Harrington's Interpretation of His Age," *Proceedings of the British Academy,* 27(1941): 199–223.

37. Samuel H. Beer: *To Make a Nation: The Rediscovery of American Federalism* (Cambridge: Belknap Press of Harvard University Press, 1993). See also my *Interactions: Some Contacts between the Natural Sciences and the Social Sciences* (Cambridge / London: MIT Press, 1994), ch. 2.

38. See *Catalogue of the John Adams Library in the Public Library of the City of Boston,* ed. Lindsay Swift (Boston: published by the Trustees, 1917).

39. James Harrington: *Works: The Oceana and Other Works,* ed. John Toland, with an appendix containing more of Harrington's political writings first added by Thomas Birch in the London edition of 1737 (London: printed for T. Becket, T. Cadell, and T. Evans, 1771; reprint, Aalen [Germany]: Scientia Verlag, 1980), p. xv, referred to here as Toland. For a brief listing of printings and editions of Toland's collection, see Charles Blitzer: *An Immortal Commonwealth: The Political Thought of James Harrington* (New Haven: Yale University Press, 1960), pp. 338–339. For a fuller account see J.

G. A. Pocock (ed.): *The Political Works of James Harrington* (Cambridge / London / New York: Cambridge University Press, 1977), pp. xi–xiv; this edition is referred to as Pocock and is used for quotations from Harrington's text. Of the Toland editions, I have consulted, in addition to the reprint listed above, the original collection by John Toland: *The Oceana of James Harrington and His Other Works* (London: printed [by J. Darby], and are to be sold by the Booksellers of London and Westminster, 1700); *The Oceana and Other Works of James Harrington*, 3rd ed., with Thomas Birch's appendix of political tracts by Harrington (London: printed for A. Millar, 1747); *The Oceana and Other Works of James Harrington* (the London edition of 1771 as noted above); and *The Oceana of James Harrington, Esq., and His Other Works*, with the addition of *Plato Redivivus* (Dublin: printed by R. Reilly for J. Smith and W. Bruce, 1737).

40. Pocock, p. 164; also Toland, p. 37; also James Harrington: *Oceana*, ed. S. B. Liljegren (Heidelberg: Carl Winters Universitätsbuchhandlung, 1924—Skrifter utgivna av Vetenskaps-societeten i Lund, no. 4; reprint, Westport, Conn.: Hyperion Press, 1979), p. 15. I have also consulted James Harrington: *The Common-Wealth of Oceana* (London: printed by J. Streater for Livewell Chapman, 1656); on this and the other "first edition" see Pocock, pp. 6–14. *Oceana* and *A System of Politics* from Pocock's edition of all of Harrington's political works have been reprinted, with a new introduction, as James Harrington: *The Commonwealth of Oceana and A System of Politics*, ed. J. G. A. Pocock (Cambridge: Cambridge University Press, 1992). There is also a useful edition by Charles Blitzer of *The Political Writings of James Harrington: Representative Selections* (New York: Liberal Arts Press, 1955).

41. *Works*, vol. 4, p. 428.

42. *Diary*, p. 62.

43. Julian P. Boyd (ed.): *The Papers of Thomas Jefferson* (Princeton: Princeton University Press, 1956), vol. 13, p. 9.

44. *Diary*, p. 62.

45. Ibid., p. 72; and cf. p. 73.

46. On John Taylor see Henry H. Simms: *Life of John Taylor: The Story of a Brilliant Leader in the Early Virginia State Rights School* (Richmond: William Byrd Press, 1932); C. William Hill, Jr.: *The Political Theory of John Taylor of Caroline* (Rutherford / Madison / Teaneck: Fairleigh Dickinson University Press; London: Associated University Presses, 1977); and Robert E. Shalope: *John Taylor of Caroline, Pastoral Republican* (Columbia: University of South Carolina Press, 1980).

47. John Taylor: *An Inquiry into the Principles and Policy of the Government of the United States,* intro. Roy Franklin Nichols (New Haven: Yale University Press, 1950).

48. *Works*, vols. 4–6.

49. Taylor, p. 36.

50. Ibid.

51. Ibid., p. 37.

52. Ibid.

53. Ibid.

54. Ibid., p. 52.

55. Ibid., p. 83.

56. Ibid., p. 93.

57. Ibid., p. 61.

58. The Adams-Taylor correspondence is published in Adams's *Works*, vol. 6, pp. 443–521.

59. Ibid., p. 466.

60. Ibid., p. 467.

61. Ibid., pp. 467–468.

62. Ibid., p. 468.

63. Walsh, p. 48.

64. See Smith, pp. 690–691.

65. For details, see the works by Walsh, Chinard, and Smith cited above.

66. *Works*, vol. 4, p. 389.

67. On Adams's method of composition by compiling authorities, see Smith, pp. 691–692.

68. The following extracts from Adams's *Defence* are taken from the reprint in *Works*, vol. 4, pp. 389–390.

69. As we have seen in Chapter 1, in Newton's physics a planet (or any other orbiting body) remains in orbit because it has two independent and very different components of motion. One is an inertial component along the tangent to the orbit, a component that tends to make the body move forward uniformly along a straight line. The other is a motion of acceleration, of falling inward toward the Sun, similar to any other motion of falling, such as the free fall of an apple under the accelerating force of the Earth's gravity. These two components of motion are independent of one another, so that the planet or other orbiting body tends to move forward continually while it falls inward. The result is that it keeps falling away from the tangent to the orbital curve. Since this is not a case of equilibrium or a balance of two forces (centripetal and centrifugal), but rather of an unbalanced force, there is a constant acceleration. Professor John Winthrop understood this aspect of Newtonian celestial dynamics thoroughly, even if his former student did not.

70. The same effects of discharge can occur whenever conditions cause a body to have an uneven distribution of electric fluid that may make a region of it charged, even though there is no net charge on the body as a whole. This condition is easy to simulate in the laboratory, as Franklin easily discovered. Let an ovoid metal body be placed on an insulated stand and then bring near to one end a rubbed glass rod (positively charged). Then the region near the rod will become negatively charged while the region far from the rod will become positively charged. That no electric fluid (or charge) has been removed from or added to the body can be demonstrated by removing the rubbed glass rod. The body will now be seen to be electrically neutral, to have no net charge.

71. *Works*, vol. 4, p. 391.

72. JA to Samuel Dexter, 15 September 1801, in Smith, p. 1071.

73. JA to his son Thomas, 15 September 1801, ibid.

74. JA to Francis Adrian Van der Kemp, 20 August 1801, ibid.

75. JA to David Sewell, 12 January 1803, ibid.

76. JA to John Gorham, 1817, quoted in James Kendall: *At Home Among the Atoms: A First Volume of Candid Chemistry* (New York: Century Co., 1929), pp. 7, 75, 78.

77. JA to Abigail Adams, 1780, in *Familiar Letters of John Adams and his Wife Abigail*, p. 381.

5 Science and the Constitution

1. For the meaning of "science" at this time see Supplement 1.

2. Max Farrand (ed.): *The Records of the Federal Convention of 1787* (New Haven: Yale University Press, 1911), vol. 2, p. 321.

3. Ibid., vol. 2, p. 325.

4. Ibid., p. 473.

5. Ibid., p. 505.

6. John Fitch: *The Autobiography*, ed. Frank D. Prager (Philadelphia: American Philosophical Society, 1976), pp. 178–179.

7. I. B. Cohen: "How Practical Was Franklin's Science?" *Pennsylvania Magazine of History and Biography* 69 (1945): 284–293; reprinted in I. B. Cohen: *Benjamin Franklin's Science* (Cambridge: Harvard University Press, 1990). See, further, Supplement 6.

8. Farrand, vol. 2, p. 505.

9. John C. Fitzpatrick (ed.): *The Writings of George Washington from the Original Manuscript Sources 1745–1799* (Washington, D.C.: Government Printing Office, 1931–1944), vol. 30, p. 493.

10. Forrest McDonald: *Novus Ordo Seclorum, the Intellectual Origins of the Constitution* (Lawrence, Kansas: University Press of Kansas, 1985), p. 69.

11. Woodrow Wilson: *Constitutional Government in the United States* (New York: Columbia University Press, 1908); *The Papers of Woodrow Wilson* (Princeton: Princeton University Press, 1966–   ), vol. 18, pp. 69–216, esp. pp. 105–107, 200–201; cf. *Papers of Woodrow Wilson*, vol. 17, pp. 427–428 and 616. See also the news report of an address, given in 1911, in *Papers of Woodrow Wilson*, vol. 23, pp. 572–583, esp. 576–577. Major portions of these documents are printed below in Supplement 12.

12. *Papers of Woodrow Wilson*, vol. 24, pp. 415–416. See further below.

13. See Supplement 2.

14. E.g., letter to Ellen Axson, 1 January 1884 (*Papers of Woodrow Wilson*, vol. 2, pp. 641–642); letter to Houghton Mifflin, 4 April 1884 (*Papers of Woodrow Wilson*, vol. 3, pp. 111–112); cf. the editorial note to the edition of *Congressional Government* (originally published in 1885), in *Papers of Woodrow Wilson*, vol. 4, pp. 6–13. See also Gamaliel Bradford's review in *The Nation* 40 (12 February 1885): 142–143, reprinted in *Papers of Woodrow Wilson*, vol. 4, pp. 236–240, and Wilson's comment on this review in a letter to Ellen Axson, 15 February 1885 (*Papers of Woodrow Wilson*, vol. 4, pp. 254–255). For other comments on Bagehot by Wilson see, e.g., "A Literary Politician," *Mere Literature and Other Essays* (Boston / New York: Houghton, Mifflin and Co., 1896), pp. 69–103 (reprint, Freeport, N.Y.: Books for Libraries Press, 1971); cf. *Papers*, vol. 6, pp. 335–354.

Bagehot's essays were first published in the *Fortnightly Review* between 15 May 1865 and 1 January 1867. The first edition as a separate volume is *The English Constitution* (London: Chapman and Hall, 1867). A scholarly edition based on Bagehot's second edition (1872) may be found in Walter Bagehot: *The Collected Works*, vol. 5, ed. Norman St John-Stevas (London: The Economist, 1974), pp. 161–409.

15. Bagehot apparently based his concept of America on writings such as those of Charles Dickens and of Alexis de Tocqueville; see *Collected Works*, vol. 5, p. 79.

16. *Papers of Woodrow Wilson,* vol. 24, pp. 413–428, esp. 415–418. See Supplement 12.

17. *Papers of Woodrow Wilson,* vol. 24, pp. 415–416.

18. *Papers of Woodrow Wilson,* vol. 24, pp. 416–418.

19. On Norman Kemp Smith (1872–1958) see especially *Papers of Woodrow Wilson,* vol. 16, pp. 398, 452, 454, 465, 507, and vol. 61, pp. 476–477; Lawrence Barmann (ed.): *The Letters of Baron Friedrich von Hügel and Professor Norman Kemp Smith* (New York: Fordham University Press, 1981).

20. *Papers of Woodrow Wilson,* vol. 24, pp. 415–416.

21. For his later use of "Kemp Smith" as surname see Barmann, p. 3.

22. London: Macmillan, 1902; reprint, New York and London: Garland Publishing, 1987.

23. *Papers of Woodrow Wilson,* vol. 16, p. 454.

24. Ibid., vol. 16, p. 507.

25. Ibid., vol. 24, p. 416.

26. The list of writers on American political thought who have endorsed this opinion is impressive, including Morris R. Cohen (philosopher and historian of American thought), Benjamin F. Wright (historian of political theory, later president of Smith College), Edward S. Corwin (America's foremost authority on constitutional law), Ralph H. Gabriel (historian), Karl Deutsch (political scientist), and Richard Hofstader (historian of American political thought). At least one major scholar, Daniel J. Boorstin, voiced doubts concerning the legitimacy of the Newtonian thesis as expounded by Becker.

27. Among others, Robert A. Dahl, Norman Foerster, and Richard Mosier. See James A. Robinson: "Newtonianism and the Constitution," *Midwest Journal of Political Science* 1(1957): 252–266.

28. J. G. Crowther: *Famous American Men of Science* (New York: W. W. Norton & Company, 1937), ch. 1, part 3, "Science and the American Constitution."

29. Crowther, p. 140.

30. Although Crowther (pp. 142–143, 144) devotes a short digressive paragraph to Adams's quotations concerning "the notion of balances" from a variety of authors, including Machiavelli and Harrington, he does not seem to understand the implications of his own conclusion "that 'Newtonian' ideas are older than Newton." His primary concern was rather to show that Boris Hessen's Marxist theory of Newtonian science is valid and that "Newton's mechanics grew out of the social movement named the Renaissance and Reformation," that Newton discovered his "system of mechanics, because the progressive social thought of [his] time, that of the rising trading classes, already interpreted phenomena quantitatively rather than qualitatively."

31. Crowther, p. 149.

32. See Supplement 2.

33. John Patrick Diggins: "Science and the American Experiment," *The Sciences* 27 (no. 6, Nov. / Dec. 1987): 28–31.

34. An admirable survey of this topic had been made by Michael Foley: *Laws, Men and Machines: Modern American Government and the Appeal of Newtonian Mechanics* (London / New York: Routledge, 1990), esp. ch. 1, 2. Foley documents the continuing and widespread acceptance in American scholarly works (and in textbooks of American

history and works on American political thought) of the idea that the Constitution reflects some kind of concept of a Newtonian machine, especially in the notions of balance and separation of powers. Although, in the opening chapter, Foley presents some basic features of the *Principia,* showing that this work has nothing to do with machines or balance of forces, he does not insist on the distinction between "Newtonian mechanics" and the theory of machines or the concept of balance of forces. The "rational mechanics" developed in the *Principia* was a theoretical study of unbalanced forces and the changes in motion they produce and has nothing to do with the theory of machines (as in a clockwork mechanism) or checks and balances. What scientists and philosophers in Newton's day knew as the "mechanical philosophy" was not a theory of machines but a doctrine that all phenomena should be explained by the "principles" of matter and motion.

In an earlier publication, I erroneously said that "the debates on the Constitution of the United States were held up by questions of the meaning and possible applications of Newtonian physical principles"; see p. 170 of I. B. Cohen: "Science in America: The Nineteenth Century," pp. 167–189 of Arthur M. Schlesinger, Jr., and Morton White (eds.): *Paths of American Thought* (Boston: Houghton Mifflin Company, 1963). I should have written that "in the debate between John Adams and Benjamin Franklin on the form of constitutions, the question arose of the meaning and possible application of a Newtonian physical principle."

35. Clinton Rossiter: *Seedtime of the Republic: The Origin of the American Tradition of Political Liberty* (New York: Harcourt, Brace and Company, 1953), pp. 133–134, 440.

36. J. Robert Oppenheimer: *Atom and Void: Essays on Science and Community* (Princeton: Princeton University Press, 1989), p. 12.

37. For details, see Supplement 2.

38. Farrand, vol. 1, p. 153.

39. Ibid., p. 157.

40. Ibid., p. 159.

41. Ibid.

42. James H. Huston (ed.): *Supplement to Max Farrand's The Records of the Federal Conventions of 1787* (New Haven: Yale University Press, 1987), p. 57.

43. Farrand, vol. 2, p. 10.

44. For a reconstruction of Madison's early education, see Ralph Ketcham: *James Madison: A Biography* (New York: Macmillan Company, 1971), ch. 2; also Irving Brant: *James Madison, the Virginia Revolutionist* (Indianapolis / New York: Bobbs-Merrill Company, 1941), ch. 4.

45. Varnum Lansing Collins: *President Witherspoon: A Biography* (Princeton: Princeton University Press, 1925), ch. 4; Brook Hindle: *The Pursuit of Science in Revolutionary America, 1735–1789* (Chapel Hill: University of North Carolina Press, 1956), p. 89.

46. Samuel Blair: *An Account of the College of New Jersey* (Woodbridge, N.J.: printed by James Parker, 1764), quoted in Thomas Jefferson Wertenbaker: *Princeton: 1746–1896* (Princeton: Princeton University Press, 1946), p. 93; on the teaching of science during the first decades of the college's history, see also John DeWitt: "Princeton College Administrations in the Eighteenth Century," *Presbyterian and Reformed Review* (July 1897), esp. pp. 392–409. See also Ketcham, ch. 3.

47. William A. Dod: *History of the College of New Jersey, from its Commencement, A.D., 1746, to 1783* (Princeton: published for J. T. Robinson, 1844), p. 38.

48. Ibid., p. 40. See John MacLean: *History of the College of New Jersey*, vol. 1 (Philadelphia: J. B. Lippincott & Co., 1877), pp. 312–313; Frances L. Broderick: "Pulpit, Physics, and Politics," *William and Mary Quarterly* 6(1949): 50, 52; Hindle, p. 89.

49. Extract from a report on the library published in 1760, quoted in Wertenbaker, p. 107.

50. Ibid., p. 197.

51. Ibid.

52. Ibid.

53. The foregoing two quotations are taken from Howard C. Rice, Jr.: *The Rittenhouse Orrery: Princeton's Eighteenth-Century Planetarium, 1767–1954* (Princeton: Princeton University Library, 1954), pp. 18, 41–42.

54. Wertenbaker, p. 109.

55. James Madison to Thomas Jefferson, 11 February 1784, in William T. Hutchinson and William M. E. Rachal (eds.): *The Papers of James Madison* (Chicago / London: University of Chicago Press, 1962–), vol. 7, p. 419.

56. Brant, p. 276.

57. Ketcham, p. 151.

58. James Madison to Thomas Jefferson, 17 February 1784, *Papers of James Madison*, vol. 7, p. 421.

59. James Madison to Thomas Jefferson, 19 June 1786, ibid., vol. 9, p. 77.

60. Thomas S. Engeman, Edward J. Erler, and Thomas B. Hofeller (eds.): *The Federalist Concordance* (Chicago / London: University of Chicago Press, 1988), p. 484.

61. Jacob E. Cooke (ed.): *The Federalist* (Middletown: Wesleyan University Press, 1961), p. 58.

62. In the more than two thousand pages of the recent two-volume edition of the papers relating to the ratification of the Constitution, there is not a single reference to Newton. See Bernard Bailyn (ed.): *The Debate on the Constitution*, 2 vols. (New York: Library of America, 1993).

63. *Papers of Woodrow Wilson*, vol. 18, p. 105. See Supplement 12.

64. *Papers of Woodrow Wilson*, vol. 18, p. 106.

65. Ibid., p. 106.

66. Ibid., p. 107.

67. Woodrow Wilson: *Congressional Government: A Study in American Politics* (Boston: Houghton, Mifflin, 1885); reprinted in *Papers of Woodrow Wilson*, vol. 4, pp. 13–179.

68. *Papers of Woodrow Wilson*, vol. 4, pp. 16–19.

69. Ibid., vol. 5, pp. 54–92.

70. Ibid., vol. 5, pp. 81–82.

71. Ibid., vol. 6, pp. 221–239; cf. p. 221, n. 1.

72. Ibid., p. 233.

73. Woodrow Wilson: *The State: Elements of Historical and Political Politics* (Boston: D. C. Heath, 1889), pp. 597–598; these pages are reprinted in *Papers of Woodrow Wilson*, vol. 1, pp. 256–257, although the whole book is not included in the *Papers*.

74. *The Nation* 49(26 December 1889): 523–524; reprinted in *Papers of Woodrow Wilson*, vol. 6, pp. 458–462.

75. *Papers of Woodrow Wilson*, vol. 6, p. 461.

76. Ibid., p. 335, n. 1 to "Wilson's Critique."

77. Ibid., p. 335.

78. See Clement Eaton: "Professor James Woodrow and the Freedom of Teaching in the South," *Journal of Southern History* 28(1962): 3–17; cf. *Papers of Woodrow Wilson*, vol. 1, p. 42, n. 1, and vol. 3, p. 218, n. 1, p. 543, n. 1; also William Diamond: *The Economic Thought of Woodrow Wilson* (Baltimore: Johns Hopkins Press, 1943), p. 43.

79. See J. David Hoeveler, Jr.: *James McCosh and the Scottish Intellectual Tradition from Glasgow to Princeton* (Princeton: Princeton University Press, 1981).

80. *Papers of Woodrow Wilson*, vol. 3, pp. 216–217.

81. Ibid., vol. 18, p. 106.

82. Bagehot, *Collected Works*, vol. 5, p. 364.

83. Ibid., vol. 7, pp. 65–66.

84. Ibid., p. 30.

85. Ibid., p. 42.

86. *Papers of Woodrow Wilson*, vol. 18, p. 106.

Supplements

1. *Papers* (see Note 1 to Chapter 2), vol. 23, p. 370.

2. Irma S. Lustig and Frederick A. Pottle (eds.): *Boswell: The English Experiment, 1785–1789* (New York / Toronto / London: McGraw Hill Book Company, 1986), p. 236.

3. John P. Fitzpatrick (ed.): *The Writings of George Washington from the Original Manuscript Sources*, vol. 35 (Washington, D.C.: Government Printing Office, 1940), p. 222.

4. *Works* (see Note 1 to Chapter 4), vol. 9, pp. 450, 512, 523.

5. In a passage which first appeared in the final "Query" of the Latin version of the *Opticks,* translated by Samuel Clarke (London, 1706), later published as Query 31 of the second English edition (London, 1717 / 1718), Newton pointed out that "Comets move in very excentrick Orbs in all manner of Positions," but that "all the Planets move one and the same way [that is, in one and the same direction] in Orbs concentrick." He did take note, however, that in the observed uniformity and regularity of the motions of the planets, there are some "inconsiderable Irregularities" which "may have risen from the mutual Actions of Comets and Planets upon one another, and which will be apt to increase, till this System wants a Reformation."

6. Thomas Jefferson to Elbridge Gerry, 26 January 1799, *Writings* (see Note 1 to Chapter 2), vol. 10, p. 78.

7. For details, see my *Benjamin Franklin's Science* (Cambridge / London: Harvard University Press, 1990), ch. 3, "How Practical Was Franklin's Science?"

8. Ibid.

9. Benjamin Franklin to Peter Collinson, 17 August 1747; for the context, see my

Benjamin Franklin's Experiments: A New Edition of Franklin's Experiments and Observations on Electricity (Cambridge: Harvard University Press, 1941), p. 63.

10. It was for this reason that George Gaylord Simpson held that Jefferson's influence on paleontology was "retrogressive" (*Proceedings of the American Philosophical Society* 86[1942]: 155).

11. Howard C. Rice, Jr.: "Jefferson's Gift of Fossils to the Museum of Natural History in Paris," *Proceedings of the American Philosophical Society* 95(1951): 597–627.

12. José Maria López Piñero and Thomas Glick: *El megaterio de Bru y el presidente Jefferson* (Valencia: Instituto de Estudios Documentales e Istóricos sobre la Ciencia, 1993).

13. Silvio A. Bedini: *Thomas Jefferson, Statesman of Science* (New York: Macmillan Publishing Company, 1990), pp. 270–271.

14. Julian Boyd: "The Megalonyx, the Megatherium, and Thomas Jefferson's Lapse of Memory," *Proceedings of the American Philosophical Society* 102 (1958): 420–435.

15. *Papers* (see Note 1 to Chapter 2), vol. 14, pp. xxv–xxxiv, 498–505.

16. Georges Cuvier: "Notice sur le squelette d'ine grande espèce du quadrapède inconnu jusqu'á présent, trouvé á Paraguay, déposée au cabinet d'histoire naturel de Madrid," *Magasin enclyclopédique* 1(1796): 303–310; "Sur le megalonyx, animal de la famille des paresseux, mais de la taille de boeuf, dont les ossmens ont été découverte en Virginie," *Annales du Muséum d'Histoire Naturelle* 5(1804): 538–576.

17. Bedini, p. 485.

18. Memorandum dated 19 April 1788, *Papers* (see Note 1 to Chapter 2), vol. 13, p. 27. See also Edwin Morris Betts (ed.): *Thomas Jefferson's Farm Book, with Commentary and Relevant Extracts from other Writings* (Philadelphia: American Philosophical Society, 1953), p. 49.

19. *Betts*, pp. 47–64.

20. All of the correspondence relating to Jefferson's study of the plow and the design of moldboards (including the correspondence with Patterson) is printed in Betts's *Farm Book*.

21. Ibid., p. 51.

22. Ibid., p. 52.

23. See, further, Edwin Morris Betts (ed.): *Thomas Jefferson's Garden Book 1766–1824* (Philadelphia: American Philosophical Society, 1944), Appendix VI, "Jefferson's Description of His Mouldboard of Least Resistance in a Letter to Sir John Sinclair," 23 March 1798. Jefferson's own account of his plow may be found in *Transactions of the American Philosophical Society* 4(1799): 313–320.

24. This curve, technically known as a "brachistochrone," is a form of cycloid. Although it may be thought that the answer to the challenge problem is simply a straight line, this is incorrect. At issue is not the least distance, but rather the least time. Essentially, a curved path is required so that the bead can fall rapidly and quickly gain a high speed with which to traverse the rest of its path. On a straight line, the speed would increase gradually and the total time would be longer.

25. I. B. Cohen: "Isaac Newton, the Calculus of Variations, and the Design of Ships: An Example of Pure Mathematics in Newton's *Principia,* Allegedly Developed for the Sake of Practical Applications," pp. 169–187 of Robert S. Cohen, J. J. Stachel,

and M. W. Wartofsky (eds.): *For Dirk Struik: Scientific, Historical and Political Essays in Honor of Dirk J. Struik.* (Dordrecht / Boston: D. Reidel Publishing Company, 1974—Boston Studies in the Philosophy of Science, vol. 15). See, further, D. T. Whiteside (ed.): *The Mathematical Papers of Isaac Newton,* vol. 6 (Cambridge: Cambridge University Press, 1974), pp. 456, 463, 470–480.

26. See, further, my "Thomas Jefferson Corrects David Rittenhouse's Gloss on Newton's *Principia,*" *William and Mary Quarterly* (in press).

27. Thomas Jefferson to James Monroe, 17 June 1785, *Papers* (see Note 1 to Chapter 2), vol. 8, p. 229.

28. Charles Thomson to Thomas Jefferson, 2 November 1785, ibid., vol. 9, p. 38.

29. John Adams to Thomas Jefferson, 22 May 1785, in Lester J. Cappon (ed.): *The Adams-Jefferson Letters: The Complete Correspondence between Thomas Jefferson and Abigail and John Adams* (Chapel Hill: University of North Carolina Press, 1959), vol. 1, p. 21.

30. Merrill D. Peterson: *Thomas Jefferson and the New Nation: A Biography* (London / Oxford / New York: Oxford University Press, 1970), p. 260.

31. Thomas Jefferson to Chastellux, 1 June 1785, *Papers* (see Note 1 to Chapter 2), vol. 8, p. 184.

32. Bishop James Madison was president of the College of William and Mary, and was cousin to the James Madison who became the fourth president of the United States.

33. Thomas Jefferson to James Madison, 11 May 1785, and James Madison to Thomas Jefferson, 15 November 1785, *Papers* (see Note 1 to Chapter 2), vol. 8, pp. 147–148, and vol. 9, p. 9.

34. Thomas Jefferson to Chastellux, 1 June 1785, ibid., vol. 8, p. 184.

35. Ibid., vol. 1, pp. 130, 318; vol. 6, p. 608. A discussion of Jefferson's view of "slavery as a moral wrong and an outrage to humanity" is given in the extended note (pp. 286–288) of Peden's edition of *Notes on the State of Virginia* (see Note 37 to Chapter 2).

36. Peterson, p. 259.

37. Ibid., p. 261.

38. E.g., by John Chester Miller: *A Wolf by the Ears: Thomas Jefferson and Slavery* (New York: Free Press, 1977). For a perceptive analysis of this topic in its historical context, see Edmund Sears Morgan: *American Slavery, American Freedom: The Ordeal of Colonial Virginia* (New York: W. W. Norton & Company, 1975). A particularly insightful analysis of this problem, together with an examination of the literature, may be found in Charles L. Griswold, Jr.: "Rights of Wrongs: Jefferson, Slavery, and Philosophical Quandaries," pp. 144–151 of Michael J. Lacey and Knud Haakonssen (eds.): *A Culture of Rights: The Bill of Rights in Philosophy, Politics, and Law—1791 and 1991* (Cambridge / New York: Cambridge University Press, 1991).

39. On Banneker, see Silvio Bedini: *The Life of Benjamin Banneker* (New York: Charles Scribner's Sons, 1972); also Bedini's *Thomas Jefferson, Statesman of Science,* p. 232.

40. Thomas Jefferson to Benjamin Banneker, 19 August 1791, published in Bedini's *Banneker,* p. 150; see also Bedini's *Thomas Jefferson, Statesman of Science,* p. 222.

41. Thomas Jefferson to Condorcet, 30 August 1791, *Papers* (see Note 1 to Chapter 2), vol. 22, pp. 98–99.

42. For details, see Bedini's *Banneker,* pp. 180–283, 287–289.

43. Thomas Jefferson to Henri Grégoire, 25 February 1809, *Writings* (see Note 1 to Chapter 2), vol. 12, p. 255.

44. Max Hall: *Benjamin Franklin & Polly Baker: The History of a Literary Deception* (Chapel Hill: University of North Carolina Press, 1960), p. 167.

45. For the many reprintings, see Hall, pp. 168–176, 177–184.

46. Hall, ch. 6.

47. This report of the Polly Baker episode from Jefferson's anecdotes about Benjamin Franklin is published in Jefferson's *Writings* (see Note 1 to Chapter 2), vol. 18, pp. 171–172. The story, as told by Jefferson, contains one inaccuracy. The hoax was never published by Franklin in his own Philadelphia newspaper, where he did publish other hoaxes of different kinds.

48. On the general subject of "nature's god," see Alfred Owen Aldridge's admirable *Benjamin Franklin and Nature's God* (Durham: Duke University Press, 1967).

49. *Works* (see Note 1 to Chapter 4), vol. 3, p. 378.

50. The printed text has "in a vacuo," but Adams, who prided himself on his Latin, would never have made such a gaff in print. He certainly would have used either "in vacuo" or "in a vacuum" in any published work.

51. This is the source of our college or university usage of "faculty of arts" or "arts curriculum" and the degrees of "bachelor of arts" or "master of arts" and the phrase "graduate in arts."

In more recent times the "liberal arts" have come to embrace new subjects in the academic curriculum: philosophy, languages, economics, history, political science. Furthermore, the ancient seven "liberal arts" and these new "arts" subjects have been joined in our colleges and universities since the eighteenth century by the modern experimental (or observational) sciences, so as to produce a "faculty of arts and sciences" and even such a learned body as the American Academy of Arts and Sciences.

52. *Oxford English Dictionary,* s.v., "science."

53. These examples, and others, are cited in the *Oxford English Dictionary.*

54. Harvey Mansfield: "The Silence of a Mechanism," *Government and Opposition,* 28(1993): 126–129. Mansfield refers specifically to works by James Ceasar, Robert Eden, and Charles Kesler.

Acknowledgments

. . . . .

I should like to acknowledge with gratitude the institutions that have given me permission to reproduce illustrations from manuscripts and drawings, prints, and books in their possession. In assembling these illustrations, I have drawn heavily on the resources of the Harvard Collection of Historical Scientific Instruments. I especially wish to acknowledge the help and kindness of William J. H. Andrewes (the David P. Wheatland Curator of the Collection of Historical Scientific Instruments). In particular, the color photograph of the B. Martin orrery that appears on the jacket has been made available by the Harvard Collection of Historical Scientific Instruments. Over many years I was able to discuss the subject of scientific illustrations with David P. Wheatland, founder of the collection.

Several illustrations come from the Burndy Library of the Dibner Institute for the History of Science and Technology (Cambridge, Massachusetts), a rich source of illustrations on almost every historical subject in science and technology. I am fortunate in having been able to discuss the choice of illustrations with the late Bern Dibner, founder of the library.

Lynn Farrington of the Special Collections, Van Pelt-Dietrich Library, University of Pennsylvania, has been especially helpful in locating and helping to identify the photographs of the Rittenhouse orrery. Claude Lopez of the Franklin Collection, Yale University, and Aimee Felker of the Isaac Newton Collection, Babson College Library, have also been of the greatest help. Michael Wentworth of the Boston Athenaeum has very kindly shared with me his knowledge of eighteenth-century death masks and the history of the death mask of Isaac Newton associated with Thomas Jefferson.

Many of the illustrations from eighteenth-century journals and books have been taken from copies in the Harvard College Library (H.C.L.). Further information concerning many of these illustrations and on the visual history of science in the age of the Founding Fathers may be found in my *Album of Science: From Leonardo to Lavoisier, 1450–1800* (New York: Charles Scribner's Sons, 1980).

Fig. 1, Collection of Historical Scientific Instruments, Harvard University; Figs. 2, 3, 10, the Abbé Nollet's *Leçons de Physique Expérimentale*, Paris, 1748 (H.C.L.), the text does not specify that the microscopists are mother and daughter; Fig. 4, Robert John Thornton's *New Illustrations of the Sexual System of Linnaeus*, London, 1797–1807, Gray Herbarium, Harvard University; Figs. 5 and 6, Trembley's *Mémoires*, Paris, 1749 (H.C.L.); Fig. 7, the Abbé Nollet's *Essai sur l'Electricité des Corps*, Paris, 1746, Burndy Library; Fig. 8, William Watson's *Expériences et Observations pour Servir à l'Explication de la Nature et de Propriétés de l'Electricité*, Paris, 1748, Burndy Library; Fig. 9, *A New Universal History of Arts and Sciences, showing their Origin, Progress, Theory, Use and Practice*, London, 1759, vol. 1 (H.C.L.); Fig. 11, *Universal Magazine*, 1748 (H.C.L.); Fig. 12, *Universal Magazine*, 1749 (H.C.L.).

Figs. 12–15, Collection of Historical Scientific Instruments, Harvard University; Figs. 16–17, Special Collections, Van Pelt-Dietrich Library, University of Pennsylvania; Fig. 18, Department of Rare Books and Special Collections, Princeton University Libraries; Fig. 19, Isaac Newton Collection, Babson College Library; Figs. 20 and 21, Burndy Library; Fig. 22, "Bowdoin MS," American Academy of Arts and Sciences, on deposit in the Boston Athenaeum; Figs. 23 and 26, Burndy Library; Fig. 24, Jacques de Romas's *Mémoire sur les Moyens de se Garantir de la Foudre dans les Maisons*, Bordeaux, 1776, Burndy Library; Benjamin Franklin's cartoon of a "joynt snake" (Fig. 25) is reproduced through the courtesy of the Franklin Collection, Yale University; Fig. 27, Jean-Baptiste Chappe d'Auteroche, "Expérience sur l'Electricité Naurelle," separate printing of a plate from Chappe d'Auteroche's *Voyage en Sibérie, fait par Ordre du Roi en 1761*, Paris, 1768, I. Bernard Cohen Collection.

Fig. 28, Dominicus Beck's *Kurzer Entwurf der Lehre von der Elektrizität*, Salzburg, 1787, David P. Wheatland and Collection of Historical Scientific Instruments, Harvard University; Fig. 29, Massachusetts Historical Society; Fig. 30, Collection of Historical Scientific Instruments, Harvard University; Fig. 31, Franklin Collection, Yale University; Fig. 32, *Le Magnétisme Dévoilé*, an eighteenth-century engraving, Cabinet des Estampes, Bibliothèque Nationale, Paris; Fig. 33, John Keill's *Introductio ad Veram Physicam*, Cambridge, 1741, Burndy Library.

Figs. 34 and 37, W. J. 'sGravesande's *Mathematical Elements of Natural Philosophy*, London, 1731, I. Bernard Cohen Collection; Fig. 35, George Adams's *Lectures on Natural and Experimental Philosophy*, London, 1794, I.

Bernard Cohen Collection; Fig. 36, William Jones's *An Essay on the First Principles of Natural Philosophy,* Oxford, 1772, Isaac Newton Collection, Babson College Library.

Figs. 38–40, Madison Papers, Library of Congress; Fig. 41, *Monthly Magazine and British Register,* September 1796 (H.C.L.); Fig. 42, Jefferson Papers, Library of Congress.

Index

· · · · ·